D1257255

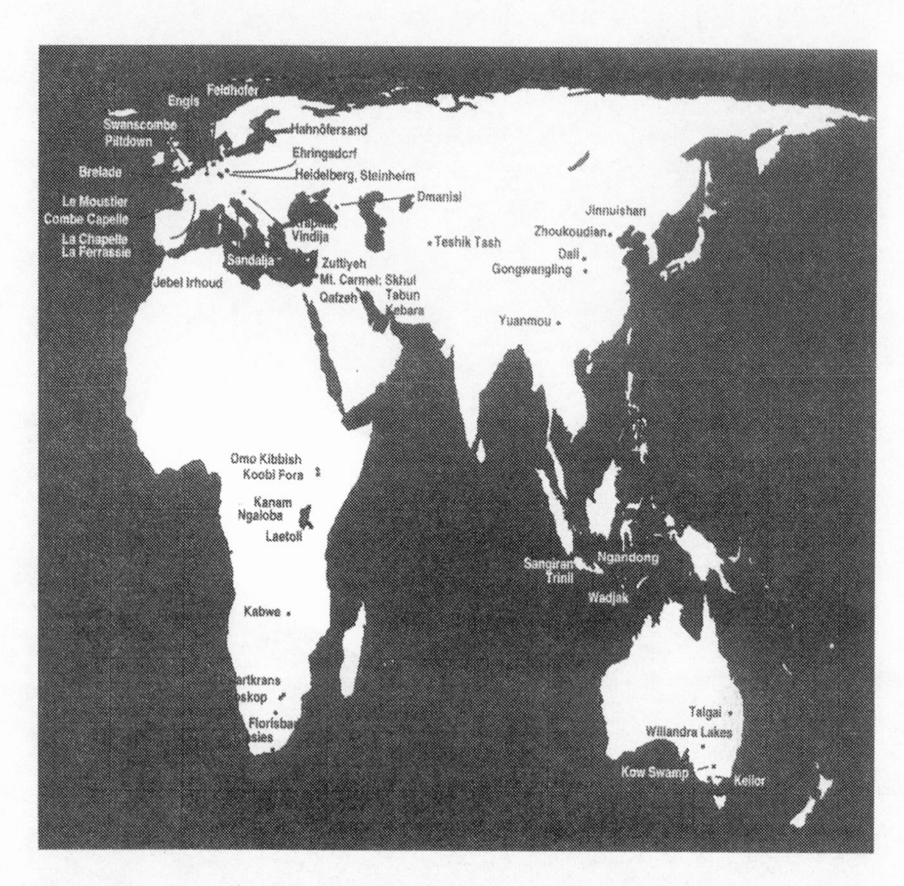

Map of the Old World, indicating the location of fossil human sites mentioned in the text. This is a Peter's Projection, showing the earth's surface in proportion to the actual distances across it—this is the way the planet would be mapped by someone counting his steps from place to place. It is the way it looked to humans as they colonized their world. (Map prepared by Tracey Crummett.)

RACE AND HUMAN EVOLUTION

Milford Wolpoff and Rachel Caspari

SIMON & SCHUSTER

SIMON & SCHUSTER
Rockefeller Center
1230 Avenue of the Americas
New York, N Y 10020

Designed by Irving Perkins Associates

Manufactured in the United States of America

3 5 7 9 10 8 6 4 2

Library of Congress Cataloging-in-Publication Data
Wolpoff, Milford H.
Race and human evolution / Milford Wolpoff and Rachel Caspari.
p. cm.
Includes bibliographical references and index.
1. Human evolution—Philosophy. 2. Fossil hominids. 3. Racism—History.
4. Racism in anthropology—History. I. Caspari, Rachel, date. II. Title.
GN281.W6418 1996
573.2—dc20
96-33466
CIP

ISBN 978-1-4165-7796-6 ISBN 1-4165-7796-3

To Franz Weidenreich. He understood.

CONTENTS

INTRODUCTION

Once da rockets are up,
who cares vhere zey come down.
Dats not my department
sez Wernher von Braun.

TOM LEHRER[1]

WHEN I WAS A LITTLE GIRL I spent hours talking to my father about the mysteries of our existence. I remember curling up beside him as he attempted midafternoon snoozes, asking him again and again about the beginning of the universe. Every day we played the same game. Working backward, prompted by my endless questions of ". . . and then, where did *that* come from . . . ," my father would skim through the history of life on earth, to the history of the solar system, to the creation of the universe. His narration ended with the Big Bang, which for me, at the time, was a disappointing little fizzle.

Invariably, I would ask the final question, ". . . and then, where did all the matter come from?"

To this my father would smile and shrug, indicating that the game was over. "I guess that's where God comes in," he would say, picking up his *New Yorker*, and turning ever so slightly away.

I found this ending very unsatisfactory; after all, my father was a physicist and was supposed to know these things.

I grew up to be a paleoanthropologist. I study the fossil evidence for human evolution, to which we can pose smaller, perhaps more answerable questions. There are only a handful of professionals in our field, but there are many little children—and even some adults—all over the world who ponder the same questions we do. This is partially because fossil hunting, finding the remains of our ancestors, is glamorous and well publicized. When I was a child, when paleoanthropology was even smaller than it is today, there already were many popular books about human fossils and their finders. Raymond Dart's *Adventures with the Missing Link* and Ralph von Koenigswald's *Meeting Prehistoric Man* were on my parents' bookshelves as far back as I can remember, and by the mid-1960s, Louis Leakey was already a household name in our family and in many others. Fossil research was exciting; it took place in exotic places and it stimulated the imagination. But, for the most part, people ponder paleoanthropological questions because the issue is *origins*, a topic that captures virtually everyone's interest and attention. Thus paleoanthropology is a public discipline. More than most other sciences,[2] the study of human evolution deals with issues that people think about and care about. It's *our* story—our beginnings and our history. All cultures possess cosmologies; people need to feel rooted in their world. Many people "*know*" the answers to questions of our beginnings, and everybody has an opinion. This makes the question of human origins and evolution an emotional issue, often rooted deeply within us along with our perceptions of ourselves as human beings and our place in the social and natural world around us. Scientists who study human evolution must answer to a great many authorities. The flip-side to this is that there are many who hold themselves to *be* authorities on human evolution. The public has a vested interest in our work which sometimes affirms or refutes deeply held beliefs.

Aside from but frequently associated with these deeper, often emotional interests, human evolution attracts public attention for a potentially sinister reason. Intrinsic to work on origins is the recognition of differences. Part and parcel with human evolution is the study of biological histories and relationships of human populations, that is to say the study of race. To understand the genesis of humanity is to grasp the foundation, history and dynamics of the races of humankind. In this link between human evolution and race lies a danger. History shows again and again that anthropological work has been used to justify an array of repressive social practices. This justification is not always based on the work of well-meaning scientists taken out of context by repressive political agencies; all too frequently the investigators themselves have promoted a conscious social agenda, and it would be a serious mistake to think of them as helpless victims, entrapped in political systems beyond their control. Of course, anthropological work has also been used to support progressive social causes by both the scientists themselves and political agencies using the results of their work. But one generation's progressive social cause can become the repressive policy of the next, as the history of the eugenics movement so clearly shows.[3]

Until recently, I paid little attention to the social and political implications of our theories, or the converse, the impact of society on science. Naively, I focused on fossils, without true regard for the impact of politics and society, past and present, on our field. This is particularly ironic because I work on Neandertals and their successors, perhaps the most contentious and controversial members of the human fossil record. For well over 100 years scientists have quibbled, with great intensity, over whether these ancient Europeans were ancestral to modern humans. Emotional political stakes have surrounded various Neandertal interpretations throughout history. Indeed, Neandertals, mirroring us, reflect our views on race and human relationships. Yet all this was lost on me, until the Eve theory.

In recent years these complex social and scientific relationships have become less easy to ignore. My husband, Milford, is one of the architects of Multiregional evolution, a theory of human evolution and the origins of modernity which has political implications. These implications only became clear to us with the "Eve debates," the Multiregional theory, which posits an ancient origin for the human species and its regional differentiation, entered the domain of public science and contradicted the Eve theory, which considers the species to be very young, appearing only with the advent of so-called modern humans.

The controversy we have become entangled in is about when and how we became *human*, the issue of *modern* human origins.[4] The debate, as it has developed over the past decade, pitches two dramatically different interpretations of the events that led to modern humanity against each other. Was the advent of humanity a single incident, like hominid origins when we parted from our last common ancestor with chimpanzees, or was it the culmination of many small changes that happened at different times and in different places? The Multiregional evolution hypothesis argues modernity was approached over a long time period as successful new features and behaviors appeared in different places and spread across the human range as people migrated or exchanged genes. There was no singular event of modernization, no Rubicon to be crossed by widespread populations. Instead, their continued contacts through migrations and mate exchanges created a network of genetic interchanges that linked even the most far-flung peoples.

The other interpretation, dubbed the Out-of-Africa theory, implies humans became modern because modernity appeared in one place—one modern population expanded because of its advantages and replaced all the others. Its most extreme version, the Eve theory, is based on a high-tech analysis of the DNA that exists outside of the cell's nucleus. According to the Eve theory, the first modern population was an African one and it replaced native peoples around the world rather than mixing with them, because the first moderns were a new species (by definition members of different species cannot interbreed with each other and have fertile offspring). There can be no reconciliation between the Eve theory and Multiregionalism because the two explanations are so incompatible that both of them cannot be correct.

The current controversy is largely a reflection of different scientific philosophies, linked to ideas about race through their treatment of variation. But there is more to it. Even if they have no conscious[5] social agenda, scientists are bound by the same preconceptions as everyone else—their social, religious, and educational backgrounds influence their choices of theories and, perhaps more important, their philosophies of science. Karl Popper[6] noted more than once that it doesn't matter where hypotheses come from, only whether they explain the evidence they are based on, whether they are subject to disproof, and whether they can hold up to enthusiastic attempts to disprove them. This philosophy forms the basis of deductive science. But hypotheses *do* come from somewhere, often the underlying assumptions of society. Moreover, not only the differences in sources of ideas, but also different premises scientists have held about evolution, human nature, God, and how science *should* be done, have always underscored the controversies about human evolution.

The modern human origins debate encapsulates these problems and more and has become the most public ever in this public discipline. Its publicity has proven to be both a help and a hindrance. Much of Eve's popularity rests on social factors: the appeal of science, technology and "newness," and political correctness. These variables have influenced theories of human origins for centuries. The sociopolitical and scientific conflicts at the heart of the issue have been played out over and over again in the course of the history of science.[7] This is but the latest bout in what is actually an age-old dispute about human origins and the origin of human diversity, an argument that has continued throughout the history of anthropology, even preceding those over evolutionary theory, and a debate that will continue long after the specific Eve issues are settled.

What brings the elements of today's controversies together is found in the first question anthropology asked—do the races have a single origin as the Bible pronounces, or are there multiple origins, different and separate developments, and fundamental and essential racial differences? Arguments over single versus multiple origins of the human races centered in Europe, because of its history of colonization, and in America, because slavery forced scientists to face the question of whether the enslaved Africans were human. They are older than anthropology itself. Multiple origins ascribed racial variation almost completely *to* history, and in doing so became the first formal link between race and human evolution. Single origins ascribed racial differences to adaptation. The issue irrevocably linked race and human evolution.

The modern human origins debates reflect an internal tension within paleoanthropology that comes from bringing together two fundamental aspects of biological science. These are *variation* and *history*, combined in the complex relationship that has developed between *race*, as a means of organizing human variation, and *human evolution*, as a means of understanding how it developed into its present form. The changes in theories about them reflect the intellectual history of paleoanthropology. Yet, these don't necessarily reflect intellectual continuity, and few of the past notions are explicitly used in current theorizing. The influence of past ideas, even if they are now known to be wrong, lies shallowly below the surface of our science because the understanding of evolution, like evolution itself, did not somehow progress steadily from the wrong ideas to the right ones.

Putting together race and human evolution, in the context of how multiple origins theories developed and, later, in the genesis of our own ideas, covers part of the history of paleoanthropology. At the same time it also covers the story of later human evolution, the fate of the Neandertals, and the rise and definition of modern humanity. The story of human evolution doesn't exist without a theoretical framework, and the story of the framework is inextricably tied to that which it explains. You cannot divorce the two; it makes sense to tell them together, and that is the essence of what we have written. From racist, pre-Darwinian beginnings, through its adaptation to evolutionary principles, to its modern development, we have chronicled the idea of multiple centers of human evolution. We reveal the implications this idea has for understanding races and their histories, and its part in a long-standing controversy—fueled by the fires of conflict in religion, social values, and the battle to apply evolutionary theory to that most unusual species, *Homo sapiens*.

The clash of theories forced a certain introspection upon us, and prompted late-night discussions about society, politics, and philosophies of science. We looked at Multiregionalism historically, and sought to understand how our own philosophies of science were related to social and scientific history and training. We reached the conclusion that scientists who deal with issues of race and evolution must recognize the complexity of the interrelationship between them, and that these scientists have an obligation to make their thinking as clear and accessible as possible. This is necessary because biological determinism remains such an active and dangerous topic in the world of public science.[8] When commenting on the heat sometimes generated in our field by even inconsequential debates, our good friend the Israeli paleoanthropologist Yoel Rak likes to quip, "After all, we are not the engineers of Chernobyl." But the fatal attraction of race and human evolution has created more than one disaster that dwarfs Chernobyl in terms of human suffering, and if the responsibility does not lie in our profession, then where? To address issues that could become social concerns and to put these ideas into perspective historically, socially and scientifically, Milford and I decided to write this book. It is an intellectual history of the fundamental problem in the study of human origins, an age-old, yet strangely ongoing controversy over multiple vs. single beginnings. Linked to this problem are various explanations for modernity—ideas about the rise of humanity and, ultimately, what it means to be human.

—Rachel Caspari

1

MULTIREGIONAL EVOLUTION AND EVE

SCIENCE AND POLITICS

SOMETIME AROUND 2 MILLION YEARS AGO something remarkable happened. Actually, at the time it was not so remarkable, because the new species that was born was one of many of its kind. The dry tropical environment of what is now eastern Africa was home to many species of hominids that we collectively call australopithecines, small bipedal apes that eked diverse existences out of anything they could on the wooded terrain. They had evolved some 3 or 4 million years earlier, radiating at various times and producing species that were specially adapted to the harsh demands of this hostile environment. For most of the course of human evolution, at least two thirds of it, these australopithecines represented what it meant to be hominid. Among them, not special in any obvious way, was the species that became our direct ancestor. The most unique and important thing that australopithecines had in common was not their intelligence but their locomotion—they were completely bipedal. They had so many anatomical adaptations for standing erect and striding on two legs that they had to develop tremendously robust and heavily muscled shoulders and forearms in order to climb trees, because the price paid for bipedalism was the loss of a foot that could grasp. To greater or lesser extents these species evolved huge faces and jaws, some with large gorilla-like crests on the tops of their skulls, for attachment of their massive chewing muscles. They had molar teeth as much as twice as large as ours, with thick hardened-enamel caps to grind the gritty roots and tubers, leaves, and grass products like seeds available to them. They were well-adapted vegetarians, but chemical analysis of their teeth reveals them also to have been meat eaters, probably scavenging the remains of carnivore kills during the dry season, when edible vegetation retreated underground or behind unbreakable husks or shells. In fact they ate any-

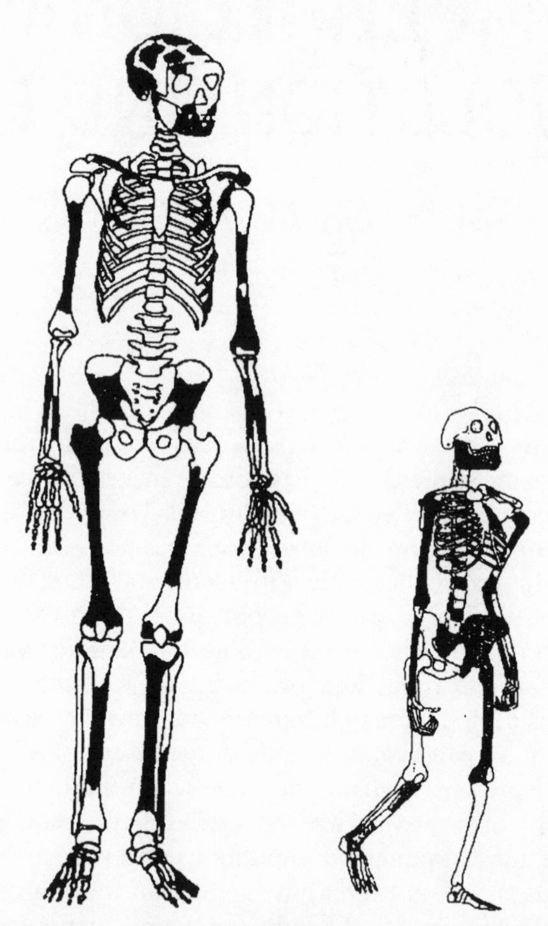

The first members of our own species, early Homo sapiens, *are really quite distinct from their australopithecine predecessors and contemporaries. Perhaps the most fundamental dissimilarity, dramatic size difference, is shown here in this correctly scaled comparison of the reconstructed skeletons of two women: "Lucy"* (right), *a 3-million-year-old australopithecine,*[1] *and ER 1808, a woman of our species about half that age.*[2] *Australopithecine contemporaries of ER 1808 were as small as Lucy.*

thing—roots and tubers, a variety of plants and nuts, even other animals' garbage, scavenging up the leftovers of hyena kills for their meat and the fat-rich marrow in their bones. They were occasionally lucky enough to steal freshly killed carcasses when they could chase the predators off, throwing rocks and brandishing sticks, and sometimes, no doubt, they were able to run down some game of their own. If we could see them, their heads and faces would seem enormous, but their brains were small and ape-sized (the impression of big heads results from the flesh-covered crests), and many were midget-sized, only 3 to at the most 5½ feet tall.

Isn't it strange to imagine the scenario of many species of humans inhabiting a limited habitat and a small geographic range, like the gazelles and bovids that they shared the landscape with? It's a different evolutionary pattern from the one that applies to us today, where a single human species is spread widely and thoroughly around the entire planet, yet it is a pattern that characterized most of our evolutionary history. Two million years ago, one of these hominid species founded our lineage and with it a whole new way of dealing with the world. Members of the new species rapidly became much larger, brain size increased dramatically, they left varieties of stone tools behind in large number, and shortly after the speciation they were able to break away from the restrictions of their dry African habitat and began to colonize other parts of the globe. But did the old bushlike evolutionary pattern of multiple speciations continue to characterize the new lineage? Did different adaptations continue to require different types of hominids? Or did the modern evolutionary pattern of a single widespread, highly adaptable species emerge at that time? Did the modern pattern of evolutionary changes in dispersed populations, and all that it implies, originate when these humans first left Africa? Or did it arise only recently as a characteristic of fully modern humans?

In 1984 Milford, Alan Thorne, and Wu Xinzhi included their first detailed discussion of Multiregional evolution, a theory that addresses these questions, in the *Origins of Modern Humans* edited by Fred Smith and Frank Spencer.[3] In it, they laid out their hypothesis about evolutionary processes in subdivided species with far-flung populations and the application of these processes to an understanding of the past. We humans, of course, are currently a widespread species with regional variations often called races, and Multiregionalists believe that this polytypism (the existence of observable average differences between populations) can also explain the patterns of diversity in the past, since the initial human migrations from Africa. But although it is fundamentally a model about processes, Multiregional evolution became specifically known as a theory of modern human *origins*, perhaps initially because it was included in the Smith and Spencer volume where attention focused on its applicability to origins questions. Later, however, there is no doubt that the widespread characterization of Multiregional evolution as a human origins theory was a result of its role in the Eve debate where, because of its recognition of the longevity of the human species, it represents the polar position to the popular theory of recent African origins.

When they were asked to contribute to the Smith and Spencer volume, it was luck—bad luck at that—that Milford and Alan already had a partially finished paper. They had first met at the 8th Pan-African Congress of Prehistory and Quaternary Studies, held in Nairobi in September of 1977. Milford was a little embarrassed to meet anybody at the time—PanAm had lost his suitcase and it proved hard to find replacement clothes that fit him in Nairobi. At the conference, with Milford quietly sitting far in the back, in ill-fitting clothes, Alan read his first paper on "Center and Edge,"[4] his idea that the initial colonizations of the Old World (now thought to have occurred one and a half million years ago, or even earlier) led to different homogeneous human populations at the peripheries of the human geographic range. There at the edges, evolutionary forces have different effects than they do at the center of a species' range which, for humans, is Africa. Alan discussed Indonesia, New Guinea, and Australia as the peripheral region he knew best. It was an exciting paper that was to become an integral part of the Multiregional hypothesis as it developed, but at the time it was most important in establishing the parameters of a long-lasting friendship and effective working relationship. Milford and Alan argued about "center and edge" at length and with vigor (often, it must be admitted, hidden away in pubs), setting the standard for the tenor of their subsequent theorizing. Their joint ideas evolved and continue to evolve the same way. After much heated discussion they mutually derive the implications for the final idea, honed down into a workable hypothesis. A Discovery Channel camera once caught the "heated" part on tape being made for the *Paleoworld* series, causing the cameraman to quip "and these two *agree?*"

After the Nairobi conference Milford's suitcase reappeared, and he and Alan stayed on in Nairobi. They set upon a systematic examination of the hominid remains in the National Museums of Kenya, where Richard Leakey had kindly allowed them to examine all the specimens he and his colleagues had discovered. Most of those remains were of *Australopithecus* (they reconstructed several specimens together, including the only cranium from the Koobi Fora site on Lake

Turkana, of the early australopithecine species Lucy represented, *Australopithecus afarensis*[5]). But there were also various examples of the lineage that first appeared 2 million years ago, our own, and this gave them the opportunity to study, and argue, about the center part of the "center and edge"[6] and the beginnings of humanity.

The following summer Milford was in London studying fossils in the Natural History Museum, and Alan was there with his family, finishing up a calendar-year sabbatical that was spent in Britain. Alan had brought a cast of an important specimen, part of the Kow Swamp sample from Australia,[7] for presentation to the museum. The specimen was only some 10,000 years old, more recent than the Pleistocene[8] fossils Milford usually worked on, but its size and robusticity caught Milford's interest. Milford invited the Thornes for an extended visit at Christmas, on their way back to Australia. They stayed for three weeks, and during that time lines were drawn and conflicting positions taken that had to be reconciled before Alan and Milford could begin to think and work together productively on their ideas on Multiregional evolution.

Primarily, their conflicts were a reflection of differences in training, particularly in where that training took place. Alan was trained in Australia and sensitive to the overbearing Eurocentrism of paleoanthropology. Historically, as Europeans wrote about it, the European prehistoric sequence, biological or archaeological, was always taken to be central, and evidence from other regions was more often than not "shoehorned" into it. It was common for Europeans to write about a "Neandertal Phase" for human evolution, or for Late Pleistocene fossils from Indonesia to be called "tropical Neandertals," even though the Neandertals (the so-called cavemen of the last Ice Age) were people who lived only in Europe and the western part of Asia. Alan's training was with materials a world apart, involving problems, such as the origins of the Australians, that were unrelated to any aspects of European prehistory. His tendency was to focus on regional parameters of human evolution, as his contrast of the center and the edge exemplifies. If any of his interests could be considered worldwide, these would be interests in worldwide *diversity*. For Alan, similarities in anatomy meant closeness of relationship.

In contrast, Milford began his university work in physics, and changing majors, subsequently was trained in the intellectual tradition of American anthropology that emphasized the four-field approach (the melding of biological anthropology with archaeology, linguistics, and sociocultural anthropology). His tendency was to focus on biocultural evolution, deemphasize differences, and seek generalities and

Alan Thorne and friend.

their explanation. For Milford regionalism of any sort was unimportant; his interest was in the worldwide stages of human evolution and the general factors that caused them to change. It seemed to him that the changes happening everywhere, the dramatic expansions in brain size affecting populations from Jerusalem to Jokjakarta to Johannesburg, the reductions in the teeth and the jawbones that supported them, were the important signals sent by the fossil record. With his background in physics, he was often considering biomechanical hypotheses that related similarities in form to similarities in function. Therefore, for Milford, similarities in anatomy meant similarities in adaptation. His dissertation adviser at Illinois, Gene Giles, was a geneticist from Harvard University, a student of the famous physical anthropologist William W. Howells. Gene did pioneering research in New Guinea on the importance of genetic drift (the random genetic changes that can occur when population size is very small). This provided Milford with an understanding of this evolutionary mechanism that served him well later in his career. But in the 1970s, the message he read from drift was the broader problem of sampling error in general—that the small sample sizes of human fossil remains meant it was unlikely that regional differences, even if they were present, could ever be found. Small samples would tend to differ by accident, without truly reflecting biological differences between the populations they were from. So for Milford, the questions most likely to be answerable were those that addressed the largest samples, and these were questions of worldwide evolutionary stages (worldwide samples are big) and their causes.

These greatly different theoretical orientations and experiences, and the two paleoanthropologists who embodied them, clashed for weeks in Milford's Ann Arbor laboratory, and even when it appeared that the two were going to survive the encounter, it often seemed as though the plaster casts they were all but throwing at each other were not!

Round 3. The following spring of 1979 Milford was traveling again, first to London, on the way to Zagreb (his favorite Neandertal stopover—the world's largest collection is at the Croatian Natural History Museum), Beijing (as the first, lone paleoanthropologist from America to study the Chinese fossils), Indonesia, and finally Canberra to visit Alan. The London visit was short, but it was important to spend some time talking to Chris Stringer, paleoanthropologist at the Natural History Museum, as their personal relationship had recently taken a turn for the worse. They were at odds over the Neandertal issue: Milford considered them to be at least one of the ancestors of Europeans while Chris considered them unrelated and extinct. Milford had written a paper that discussed the issue and sent a prepublication copy to Chris for comment. Chris wrote back a scathing letter, signing it, "Desperately yours, Chris Stringer." Puzzled over the heat of the comments, Milford gave the paper to Mary Russell, then one of his graduate students and an especially good writer, for one more reading. Mary found the problem: Milford thought he had written about "Stringer's disparate analyses," but actually wrote "Stringer's desperate analyses." A telegram went to the press and the change was made just days before publication, but Chris never seemed sure it was *really* a mistake. Milford wanted to reassure him it was.

Waiting in Chris's office, Milford noticed an issue of volume 12 of the British anthropological journal *Man* on his desk. A paper by R.

Hiorns and G. Harrison[9] particularly caught his attention; it discussed the combined effects of natural selection and migration on the human evolutionary process (Darwinian or natural selection is the process in which genes can change in their frequency over time because of differences in reproduction and survival—some individuals have more surviving offspring than others). Hiorns and Harrison wrote on a rather different way of looking at clines,[10] and their importance in human evolution, than Milford had thought about before. Milford, like most biological anthropologists of his generation, was steeped in the postwar credo that human races were arbitrary divisions of humanity, and human variation could best be described as clinal. He learned human variations almost always followed independent, gradually varying gradients, because the causes of the variations were themselves gradients. For instance, skin color varies gradually from lighter in the north to darker in the south of Northern Hemisphere continents because it responds to solar radiation. Dark skin is adaptive when there is more intense radiation, so natural selection creates a gradient of skin color that conforms to solar radiation, only interrupted by discontinuities created by dietary or cultural factors, or migrations so recent that the populations did not have enough time to respond to natural selection. The Hiorns and Harrison paper discussed a very different way that gradients could form, as a *dynamic* response to constant but opposing forces rather than a *static* response to a varying force. They pointed out that when populations were spread over a broad geographic range, genetic exchanges (genes that go from one place to another because populations move, or because they exchange mates on a regular basis) between them could have the opposite effect of selection. Genes weeded out of populations at one end of the range could be reintroduced because of genetic exchanges bringing the same genes from the other end of the range. Hiorns and Harrison showed gradients will form even when their causes are discontinuous, because whenever there is genetic exchange and selection opposing each other there will be a balance—one cannot overwhelm the other. The balance between them creates a gradient, whose steepness (how rapidly a feature changes from one population to another) depends on the relative magnitudes of the genic exchanges and selection. They argued this process could be important in human evolution. It explained genetic variation broadly across a region, even a continent, and how its clinal variation could be maintained for long time spans.

After London came Zagreb, Athens, and the long flight to Beijing where Milford met Wu Xinzhi and other Chinese paleoanthropologists. The Chinese fossils were important for Milford's masticatory studies, and the issue of whether there were Chinese australopithecines (the evidence was based on a few isolated teeth, and the answer was no). In a triumph of the power of expectation over objective observations, there was plenty of evidence of regional continuity in the Chinese fossils, features that uniquely related them to living North Asians, but none of it registered because Milford was not there to ask questions about regional continuity. When the questions are not asked, data are generally not forthcoming with answers, because contrary to the common phrase, data do *not* speak for themselves.

Milford had a diversified research strategy, in that he collected information about the masticatory apparatus (the reason the trip was funded), and also about other problems he found interesting such as the Chinese australopithecine issue, and finally he collected a broadly comparative data set. He learned the importance of a comparative set from his australopithecine studies and in his interactions with the famous South African paleoanthropologist John Robinson. Milford's first passion was with australopithecines, and he spent the last year of his graduate studies at the University of Wisconsin where Robinson worked. Robinson was the first paleoanthropologist Milford had studied with and his research strategy was quite different. Robinson focused on the whole organism and believed there was much more to be learned from a few complete individuals than from piles of small fragments (the australopithecine sample provided good examples of both). His understanding of australopithecine varieties and their relationships was based on analyzing and comparing the best-preserved individuals of each. Milford was more focused on the population and its attributes, which made him a reductionist (this also reflected his genetics training and the statistics forced on him as an undergraduate physics major), interested in the variation of features in samples that included few whole specimens and many small piles. He believed individuals could be best understood in this population context. But Robinson had a good point—the small piles could not be adequately studied because they preserved portions of the anatomy that were rarely reported. Since there was little literature and virtually no previous analyses, these anatomical details could not be compared. Part of Milford's long-term research strategy addressed this problem, and wherever he went he collected systematic information about anatomical details that could be used to compare large samples of small fragments. In his notes, photographs, and measurements of the Beijing specimens was unrecognized evidence for regional continuity.

Many of the Indonesian fossils were in Jokjakarta, but the high point of Milford's Indonesian visit was at the laboratory of a fine geologist and scholar in Bandung, Darwin Kadar. Kadar then possessed what is still the most complete adult skull known for the early human species, "*Homo erectus*" (now called "early *Homo sapiens*" by many). The African examples are much better known and get better press, but this Indonesian skull, number 17 of the Sangiran remains, was more complete than any of the African ones. Discovered in 1969, it had been only briefly described in print, and its pictures and drawings looked strange and were difficult to interpret. Yet it was the only Indonesian fossil with anything approaching a complete face (little is missing from it) and it clearly wanted further study. When Milford first saw it, the face was sloppily put together and attached to the braincase (the cranial vault or skullcap) at only one spot, the top of the nose, by a piece of latex that let it swing back and forth like a pendulum. Darwin allowed Milford to reconstruct the face and permanently attach it to the vault, before the bony contacts were ground to dust by people trying to fit them together time after time. With Duco glue and Satay skewers at hand (these are an integral part of Milford's traveling kit), Milford began a reconstruction that, when completed, left him quite surprised. First working on the face, he took it apart and reconstructed the facial pieces on a desk in front of him, propping the individual fragments up with clay. But when the face was finished, fitting it onto the braincase was a problem. It is usual to use a sandbox to fit large pieces together, as the sand can be arranged to support pieces in any position. But there was no sandbox in the laboratory, nor sand to make one. In desperation, Milford held the vault on his lap, facing him, and fit the face onto it, holding it together until the glue dried. It was not difficult to use bone contacts to get the pieces to fit together correctly, very much like working a jigsaw puzzle when only some of the pieces can be found but a few key fits outline the picture. But reconstructing bone is not objective, especially when the joints are few, far between, and involve eroded surfaces. The subconscious mind has the potential to guide a reconstruction, to make it appear as the scientist expects, not always as the specimen actually was. But circumstances conspired to reduce the influence of Milford's subconscious, because he had to hold the drying reconstruction on his lap and couldn't see the most critical thing about it, the angle at which the face met the vault. He had to let the bone contacts dictate that angle, without knowing what it was.

With the glue finally dry, Milford was able to turn the skull resting on his lap to its side. The side view, which he now saw for the first time, was shocking: it looked much like the Kow Swamp skull Alan had shown him in London the year before. It shouldn't look like a

Holocene Australian (showing the influence of region), it should be more like the Beijing fossils from the famous Chinese "Peking Man" site, Zhoukoudian (showing the influence of "stage"), Milford thought. Alan *must* be wrong on this issue of regionalism; he surely can't be so right. After all, brain size was increasing and masticatory structures reducing in this part of the world just as surely as everywhere else. To be absolutely certain of the reconstruction, Milford decided to redo it, so the specimen was measured and photographed and then taken apart and Milford worked on other materials for a week. The last thing he did in Bandung was once again to rest Sangiran 17 in his lap facing him and try reconstructing its face and positioning it on the vault. When the work was finished the second time, the side view was the same, the measurements almost identical, and with a special roll of film in his pocket, Milford flew on to Canberra.

Alan met him at the Canberra airport the next morning, and Milford told him he had a roll of film in his pocket that it might be interesting to develop. Without further discussion they visited the film service at the Australian National University, and a few hours later Alan shared the most powerful and convincing evidence of regional continuity in Australasia possible. Alan always had suspected it because of the anatomy he recognized in Aboriginal Indigenous Australians, but Sangiran 17 gave the first real evidence from the other end, the beginning of the local evolutionary process. It was now for them to articulate *why* it was evidence, and explain how there could be *any* special link be-

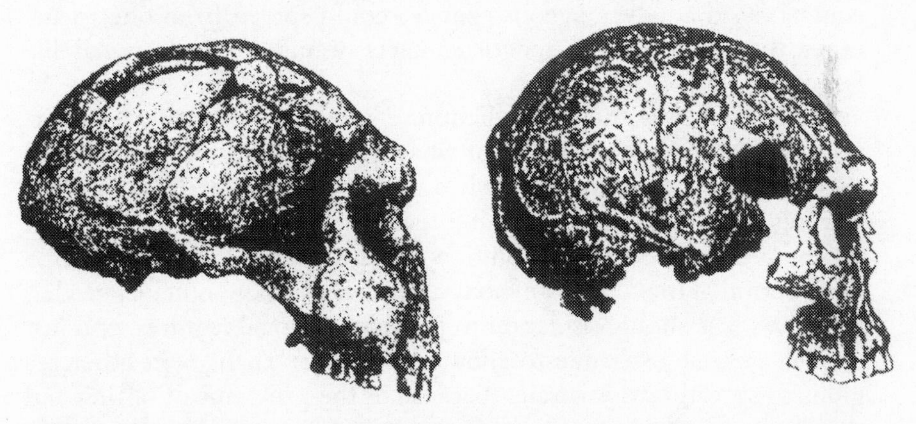

Milford's Sangiran 17 reconstruction (the left cranium) *and Kow Swamp 1.*[11] *These specimens, from the same region of the world, lived some three quarters of a million years apart. This view from the side shows the angle of the face to the vault which Milford could not see as he worked on the reconstruction.*

tween a man who died 750,000 years ago in Indonesia and recent and living Australians.

The problem, and the reason there was all that fighting in Ann Arbor that last Christmas, lay in a contradiction between regional continuity and worldwide change that neither Alan nor Milford could resolve. Unique elements of continuity, anywhere, would seem to imply isolation. Otherwise, how could genetic combinations sustain distinction and not merge back into the gene pool? Wouldn't they be lost unless they could be separated away so the merger was impossible? This would seem to imply a sort of polygenism, a pre-Darwinian theory of human origins that extended in various forms through much of the 20th century, positing separate origins and independent evolution of the races. Polygenism is a theory that is contradicted by Darwinian evolution, and neither Alan nor Milford considered it. Yet a unique Australasian pattern of continuity existed, in the face of the same evolutionary tendencies that could be found everywhere. If Sangiran 17 was thought of as a base, the recent and living Australians differed from it in much larger brains, smaller teeth, and a general level of gracility—*exactly the same kinds of changes that happened during the same time span everywhere else in the world.* The polygenism theory explained this by parallelism, the independent development of the same changes in different, isolated lines. But according to the tenets of Darwinism, such independent developments are improbable to the extreme. A much more reasonable explanation would be that different populations evolved the same way because they were in genetic and perhaps social contact, so that advantageous changes could spread from one to another. But wouldn't such genetic contacts swamp out the regional differences?

Here, then, was the contradiction. There were opposing explanations for the different aspects of what Alan and Milford now agreed they could see: isolation seemed to explain the unique links in some features across a long time span; interconnection seemed to explain a common direction to the evolution of other features across the same time period. How could anybody expect the lucky coincidence that there was just enough isolation to preserve regional features, and just enough genetic exchange to allow evolutionary changes in other regions to spread into Australia (because of the problems of rafting and the direction of ocean currents, population movement was probably one-way)?

But wait! Problems of reconciling change and stability created by a balance of forces made Milford think of the long-lasting clines described in the Hiorns and Harrison paper. These authors had detailed how opposing evolutionary forces *can* create stability. But this was a stability based on interconnections, not isolation. By applying this insight to the contradictions Milford and Alan faced, the Multiregional evolution explanation was born. Center and edge laid out how peripheral distinctions and local homogeneity first came to be, and now there was a solution for why it was that some of those distinctions persisted for long periods of time, even as changes in other features enveloped all of humankind. In the heat of discovery, with fresh evidence and their first attempt at its explanation at hand, Alan and Milford collaborated on a paper,[12] comparing Milford's Sangiran 17 reconstruction with Alan's Kow Swamp specimens, and explaining the pattern of evolution revealed.

Having begun this line of investigation it was hard to stop. Alan and Milford realized they had much to work out if their Multiregional explanation was to be accepted. It was important to develop its details, accurately model the evolutionary forces involved, and focus on ways to attempt its refutation. Alan came to Michigan for a long visit in 1981, and they spent a good deal of time articulating the details of how Multiregional evolution could work and developing an understanding of its genetic constraints. This was a very productive period of reading and exchanging ideas and developing a new and much more detailed perception of the major pattern to human evolution and how it works. The differences in training and experience that put them so at odds during Alan's last Ann Arbor visit now became a great advantage in their collaborative efforts. More and more evidence seemed to fit the Multiregional pattern, a good deal of which could be found in the systematic notes and observations Milford had collected. But it became clear that the most critical data needed at the moment required a survey of the distribution of regional features in the skeletal remains of recent populations. This data could be systematically compared with the fossil data already at hand so an accurate picture of the data base could be developed. To this end, Milford and Alan submitted a grant proposal to NSF, the National Science Foundation.

To show NSF why their project was worthwhile, Alan and Milford reviewed the literature and put the Multiregional hypothesis in a historic context, discussed the genetic and evolutionary literature that addressed various aspects of how it worked, and then detailed the fossil data it could explain. Although the proposal was ranked highly by the NSF's review panel, it was not funded. However, when they were invited to contribute to the Smith and Spencer volume, much of the work had already been done and Milford and Alan were able to take some two thirds of the text of their paper from the grant proposal.

Preparing a paper for the volume created another fortunate circumstance, as it brought Wu Xinzhi into the authorship and, more important, into the development of Multiregional evolution. Wu, a senior scholar at the Institute for Vertebrate Paleontology and Paleoanthropology (IVPP) in Beijing, had been a good host and friend to Milford and Alan during their various trips to Beijing to study the fossil human remains, although never at the same time—the three had not yet been in the same room when they began to work together, although that was soon to change. Wu is a gentle man, whose experience ranges from writing a book on gibbon anatomy to heading several expeditions to search for Yeti hairs on trees near villages where they had been reported (analysis of their hair could provide important information about which primates Yetis are most closely related to—if, of course, they actually exist, which Wu doubted). Most important was Wu's authorship of critical papers about the human fossil record in China, based on his long and detailed studies.

Wu Xinzhi.

Wu's interest was in the pattern of human evolution in China, whose details were somewhat different from those in Australasia. Like many Chinese paleoanthropologists he was predisposed to favor Multiregional evolution because he was already well versed in the main observations it was developed to explain—the regional continuity in the Chinese fossil record—and because of the influence of Franz Weidenreich, the paleoanthropologist who a half-century before oversaw the excavation at the "Peking Man" site of Zhoukoudian and who wrote the descriptions of the human remains that were found. Weidenreich[13] believed the Chinese fossil record included evidence of certain unique features that linked prehistoric and modern populations. He contended these ancient Chinese had made a significant genetic contribution to modern ones, although he realized that they were ancestral to other populations as well and that modern Chinese had other ancestors. Continuity, in Weidenreich's mind, involved lines of descent and not racial identity; he regarded races as short-lived and ephemeral, and believed the people who lived at Zhoukoudian a half-million years ago had Mongoloid features, but were not of the Mongoloid race. Weidenreich's intellectual influence on subsequent generations of Chinese paleoanthropologists was immense, and so Wu was very interested in becoming involved with the Multiregional hypothesis as he heard about it from Alan and Milford. He wanted in, and the invitation to prepare a paper provided part of the opportunity.

Luckily, Milford received a grant to visit China through the NSF's Scholarly Exchange Program with the People's Republic. The grant was to study human fossils and visit fossil sites, and as part of the exchange it brought Wu (Milford's IVPP host) to New York and Ann Arbor for comparative research. The long stay in Ann Arbor in the spring of 1983 gave Wu the unanticipated chance to think about and write on Multiregional evolution and add his precepts and unrivaled experience with the Chinese fossils to the paper.

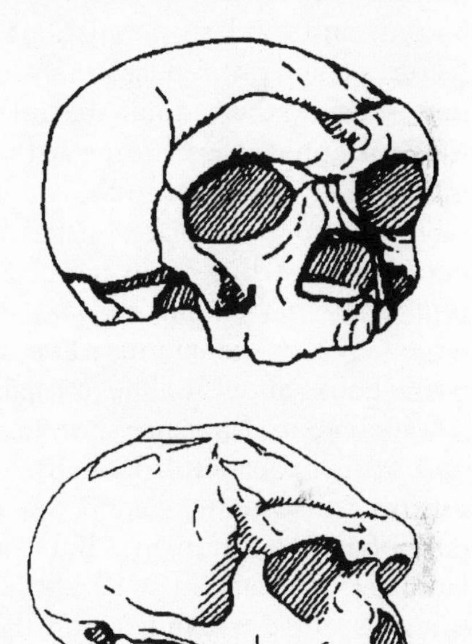

The Chinese fossil remains were quite different from those to the south, even though they were evolving in much the same way. Compared[14] *with Sangiran 17* (bottom), *specimens such as Dali* (top) *had similarly long and low skulls with thick brow ridges but differed in other features such as facial flatness and the angle between the face and cranial vault.*

Thus a long paper detailing Multiregional evolution was in print in 1984. The three authors thought a shorter and more focused statement might be appropriate, perhaps for a broadly read journal like *Science*, but there was no rush. There were no funds to initiate the intensive data-gathering they knew was necessary, and Alan, already distracted by the heated reburial issue, as political pressure increased to rebury the Australian Pleistocene human fossil record, became involved in filmmaking and produced a valuable documentary series on the peoples of the Pacific Rim. Wu was made head of the IVPP, and had to set aside his research and writing agenda. Milford's department at the University of Michigan was in turmoil. Besides, they were sure there was no rush. Then the Eve theory exploded on the scene.

First published in 1987,[15] the Eve theory posited a very recent origin of modern humans, as a new species, because of molecular data that came from analysis of mitochondrial DNA (mtDNA), obtained through increasingly sophisticated techniques of DNA mapping. The Eve theory contradicted Milford's and Alan's work directly. If modern humans were a new species, there were no connections between most ancient and modern populations. Regional continuity in anatomical features could not reflect genetic continuity; if modern humans were a new species, there could be no genetic links between ancient and modern populations except at the site of its origin. The Eve theory immediately appealed to the public, partially because of the appeal of "science and technology," partially for sociopolitical reasons. Soon it became the darling of the popular press. There were articles on Eve in *Newsweek*, in *Time*, in *The New York Times* and virtually every other large paper in the country; Eve was on the evening news and people wrote books about it. The journal of the American Association for the Advancement of Science, *Science*, turned to Chris Stringer and a second British paleoanthropologist, Peter Andrews, both of the Natural History Museum in London, to write an evenhanded review of the origins of modern humans. But the review[16] was anything but evenhanded;[17] how could it be? Science by its very attempts to explain things is not evenhanded, and these British scientists already had an explanation to defend—Eve. The debate was joined. The Eve theory had accomplished what we could never do as well on our own. It was a focused, well-thought-out attempt to refute our hypothesis by the scientists who believed they had a better explanation.

Next, we summarize the competing theories: they could not be more disparate. They posit different patterns for recent human evolution, they primarily rest on different data sets, and their proponents hold different scientific philosophies.

WHAT IS MULTIREGIONAL EVOLUTION?

The fundamental question that has been asked historically is how people all over the world could evolve in the same way, all becoming modern humans, and yet maintain some regional differentiation for long periods of time. This is an observation, and the fundamental problem is how to resolve the contradiction that seems to lie at its heart. The genic exchanges between populations that would seem critical for one would seem equally destructive of the other. Confronted with compelling similarities between Australasian specimens separated by three quarters of a million years, and somewhat different similarities across perhaps an even longer time span in China, patterns of details not shared by fossils from other regions, Milford, Alan, and Wu developed a model to explain these sets of seemingly contradictory observations that have puzzled paleoanthropologists for close to a century—evidence of long-standing regional differences between human groups in the face of evidence of important global similarities in the direction of evolution.

Over the years, many students of the human fossil record have grappled with the contradictions of differentiation and common evolutionary direction, but they have been unable to satisfactorily reconcile them; solutions have ranged from denying any regional features actually have antiquity, or even exist, to invoking supernatural powers to account for parallel development. Franz Weidenreich, working in China, developed a theory he called Polycentric evolution that is the intellectual precursor of the Multiregional model; in fact, the Multiregional model can be considered the marriage of modern ideas of population biology and the Polycentric theory. Weidenreich described the relationships of fossils to each other and modern humans, recognizing regional evolution through time with genetic interchanges between regions. Although fully aware of it, he could only partially explain the paradox inherent in his model: that between regionalism and common evolutionary trends.

Our key insight for resolving the paradox of local continuity in the midst of worldwide change came, not from Weidenreich's work, but from the Hiorns and Harrison paper Milford read at Chris Stringer's desk, the paper percolating through his mind all the time as observations and ideas about Multiregional evolution sorted themselves out. Genic exchanges were not the opposite of differentiation, they were its cause. They were not the problem but its solution! The far-reaching gradations described in the Hiorns and Harrison paper were not disrupted by genic exchanges, *they depended on them*, and it was along these gradients that populations toward the extremes could differentiate and remain distinct.

Multiregional evolution provides resolution of the contradictions between genetic exchanges and population differentiations in a broad-based theory that links gene flow and population movements, and natural selection, and their effects on populations both at the center and at the peripheries of the geographic range of the human species. In a nutshell, the theory is that *the recent pattern of human evolution has been strongly influenced by the internal dynamics of a single, far-flung human species, internally divided into races. Human populations developed a network of interconnections, so behavioral and genetic information was interchanged by mate exchanges and population movements. Gradients along these interconnections encouraged local adaptations. These and other sources of population variation that depended on population histories developed, and stable adaptive complexes of interrelated features evolved in different regions. But, at the same time, evolutionary changes across the species occurred as advantageous features appeared and dispersed because of the success they imparted. These changes took on different forms in different places because of the differing histories of populations reflected in their gene pools, and the consequences of population placements in terms of habitat and their relations to other populations. Some evolutionary changes happened everywhere, because of these processes and because of common aspects of selection created by the extra information exchanges allowed by the evolving cultural and communications systems. Consequently, throughout the past 2 million years humans have been a single widespread polytypic species, with multiple, constantly evolving, interlinked populations, continually dividing and merging. Because of these internal divisions and the processes that maintain them, this species has been able to encompass and maintain adaptive variations across its range without requiring the isolation of gene pools. This pattern emerged once the Old World was colonized, and there is no evidence of speciations along the human line since then that would suggest there were different evolutionary processes, such as complete replacement, at work.*

Over the past decade Multiregional evolution has itself evolved into a broad and malleable frame. It is a *general* explanation for the pattern and process of human evolution within which virtually any hypothesis about dynamics between specific populations can be entertained, from the mixture, even replacement, of some populations to the virtual isolation of others.[18] To be valid, the model must be able to incorporate a wide range of population dynamics, from expansion to extinction, leaving paleoanthropologists room to derive more detailed understandings of specific evolutionary patterns for particular times and places. Various groups of people behave in different ways that affect their demographic structure (that is, the specific attributes of their population, such as its size, mortality rates, sex ratios, age profiles). If you are trying to predict patterns of evolutionary change, this demographic information is absolutely essential, since the major evolutionary forces of natural selection and genetic drift operate differently on populations with diverse demographic structures. As with all social animals, every human population has a different evolutionary story, with its own historical, biological, and social constraints that affect its evolution. The human evolutionary pattern is even more dynamic than that of other species, because cultural and linguistic factors are added to the list of constraints, even as they expand the different ways in which populations can exchange and share information. Culturally prescribed marriage systems, trading networks, religious practices, likes and dislikes, all affect reproduction, death, and breeding group size and therefore the evolution of these populations. Consequently, *detailed* understanding of the course and processes of human evolution is unusual, and can be obtained only for small temporal and geographic windows, where many ecological, demographic, and cultural variables are known. Multiregional evolution can be thought of as the structure in which these windows sit. It is compatible with all the windows we've looked through so far, a structure that allows them all to exist together. In other words, it is a model that fits the skeletal and genetic data we have today, and we also think it works in the past, where the information is much less precise, and there is much less of it.

Using the Multiregional model for interpreting the past assumes that the modern pattern of human evolution is the best model for interpreting the human condition ever since the first colonizations of the world outside of Africa began. If this assumption is valid, the present can be used as a model for explaining the past; this is the principle of "uniformitarianism" that the geologist Charles Lyell so successfully applied during the last century to interpreting the geological and paleontological record, work that was critical to Darwin's emerging theory of evolution. We consider this the most logical approach to understanding the recent pattern of human evolution, and treat it as the null hypothesis (the hypothesis of no difference, or no change, is the hypothesis to disprove, or try to disprove, with ongoing research and discoveries) for interpreting the past. It is the simplest hypothesis, one that models the evolutionary patterns of our behaviorally complex, geographically widespread predecessors after the living species most like them: our own.

We are quite aware that people have not always been the same. The evolutionary dynamics of modern humans are far from fully understood, and there are many factors in modern human populations that cannot be applied to the past. People have changed dramatically in recent times—their cultures have become incredibly complex, their demographics have altered remarkably, populations have expanded dramatically—and there is no way that evolutionary processes at work today can be expected to be identical with those of the past, just as the evolutionary processes at work today vary from population to population. But stepping away from the details, there is a frame of conditions these processes work within, and it is here that we draw the basis for applying the uniformitarian principle. It seems to us that the bases for approaching the past this way are twofold: we recognize no biological species formation in humanity once *Homo sapiens* appeared some 2 million years ago, and the fundamental shift from a solely African scavenging/gathering species to a colonizing species[19] taking place at *Homo sapiens* origins or early in their evolution set up the conditions of polytypism across a broad geographic range that allowed Multiregionalism to work.

Thus Multiregional evolution is a gradualist model, with the primary tenet that humans *are* a single polytypic species and *have been* for a very long time into the past. It interprets the fossil record to show that human beings—that is, our species *Homo sapiens* and its main attribute *humanity*—happened only once, and once on the scene they evolved without a series of speciations and replacements. No speciation events seem to separate us from our immediate ancestors, and cladogenesis, the splitting of one species into two, last characterized our lineage at the origin of *Homo sapiens* some 2 million years ago, when members of what we once called "*Homo erectus*" first appeared in East Africa. For 2 million years, from the end of the Pliocene until now, ancient and modern *Homo sapiens* populations are members of the same species. This doesn't mean they didn't change—*au contraire*—but we think these changes neither led to nor required a speciation. The broad-based evolutionary processes proposed in Multiregional evolution are formulated to explain patterns of variation *within* a polytypic species: the same evolutionary processes shown to be important in other polytypic species have shaped our patterns of diversity in the past and do so in the present.

The ability to account for all the data it is supposed to explain is only one hurdle for a hypothesis. It also must, at least in principle, be refutable. Multiregional evolution would be wrong, a disproved and invalid hypothesis, if the evolutionary changes it accounts for and the contradiction between genic exchanges and local continuity of features it resolves were explained instead by a series of successive speciations and replacements. Multiregional evolution would also be wrong if the pattern of human evolution it describes never existed—that is, if the interpretation of long-standing polytypism in the human fossil record is incorrect, since the explanations would then be elucidating a pattern that did not exist. Evidence of multiple speciations, indicating a different *pattern* of human evolution and, in particular, a recent speciation for modern humans, could provide this refutation, and the Eve theory claimed to rest on just such evidence.

So Eve came as a wake-up call for Multiregionalism. Although not particularly aimed that way at first, it was soon correctly perceived by all as the first serious attempt at its disproof, and for "Popperian" scientists, refutation is the key way that science proceeds. Milford is a deductionist, strongly influenced by the philosopher of science Karl Popper and most concerned with refuting hypotheses. The role of deduction comes after a hypothesis is framed; what matters most is whether it is explanatory, is testable, and requires the least number of assumptions. Multiregional evolution is our null hypothesis, the simplest and most explanatory hypothesis that covers the pattern of Pleistocene human evolution. But it was just recently developed, at least in its modern form, and until the Eve theory there were no significant attempts to disprove it.

After the publication of the 1984 paper, Alan, Wu, and Milford didn't think too much about Multiregional evolution in a theoretical way. Prior to the marketing of Eve, they had each proceeded to treat Multiregional evolution as a working hypothesis. Many others accepted the hypothesis as well, and research was initiated within a Multiregional paradigm, which in itself was not the focus of the investigations. As Multiregionalists studied human evolution, they were consciously aware of geographic variation and its confounding effects in understanding human evolution as a whole; Multiregional scholars were, and are, careful to view temporal trends as potentially regional phenomena and cautious not to generalize too quickly between one region and another, often avoiding a kind of ethnocentrism applied to the fossil record.

Before Eve, the few attempts to show species change in the recent human fossil record were focused on the seemingly unending Neandertal issue and were unconvincing to most scientists. Nothing effectively challenged the explanatory value of Multiregionalism as an explanation for worldwide change. As Alan made movies all over the world, the patterns of variation in the people he visited fit the Multiregional model. He collected and bred snakes, and the patterns of their variation fit the Multiregional model. Wu struggled with the long-awaited completion of his monograph on the Dali skull. Milford returned to focused, problem-oriented research. Always interested in patterns and causes of variation, he wrote papers on allometry, on sexual dimorphism, and on biomechanics, seeking explanations for trends that extended across broad periods of human evolution. But they all returned to issues of Multiregional evolution after the 1987 publication announcing what was soon widely called the Eve theory,[20] and several publications following in the next year cited evidence that could refute the Multiregional model and our entire understanding of the human fossil record.

The Eve theory played *the* lead role in what quickly became a confrontation between paleontological and molecular genetic interpretations of the past. The development of Multiregionalism owes a great deal to her. When the Eve publications emerged, the Multiregional camp quickly responded to them, pointing out several problems that prevented them from refuting the Multiregional hypothesis. The Eve debates made us very introspective about our proposals and their implications. We were forced to think about Multiregionalism's development and testability and the things we think make it a good hypothesis.

EVE OF THE MITOCHONDRIA

Molecular genetics offers a tool to study the past, in particular to reconstruct evolutionary relationships. The data are quite different from paleontological data, where there are different-sized windows into the past. In the realm of molecular genetics there is no time depth; the data are the *results* of the evolutionary process that survived until now. Because the studies focus on the DNA molecular structure itself, the phylogenetic trees (depictions of the evolutionary relationships between groups, similar to family trees) that can be developed are, at least now, necessarily all based on the relationships of living populations; there is no direct evidence from the past. This makes molecular genetics an independent way to assess the relative closeness of different extant groups. From their genetic similarities, the length of time that has passed since different groups shared a last common ancestor can be estimated, if one makes the assumption that the greater the number of gene differences, the greater the genetic distance measured, and therefore the longer the time separated. Basically, this is like reconstructing a tree from the branch tips.

The assumption that genetic difference measures genealogical divergence is similar to the assumptions often made in studies of morphology: the greater the difference between two species, the longer since they have diverged. Here species rather than populations must be considered, because populations can exchange genes and two populations that have done so after they diverged will appear much more similar than two populations that diverged and did not exchange genes. Species, on the other hand, cannot exchange genes.

But do genetic or anatomical differences measure time since divergence? Can closeness of relationship be estimated by the number of shared genes or traits? On the face of things this is counterintuitive. It does not fit our experience in life, where our relatives and the closeness of their relation to us are defined by common ancestry (same parents? same grandparents?), not similarity. We have all seen people in crowded cities that look more like us than our own sisters or brothers. To be able to conduct our research from the contention that similarity measures closeness of relationship, we must be able to assume that differences accumulate at a constant rate, so species that have been separated twice as long are twice as different.

As far as genetic evolution is concerned, the basis for making such an assumption is that mutations, physical changes in the DNA, are caused by random forces in the environment, like radiation. Mutations alter the sequence of base pairs that are the basis of inherited information. The DNA molecule has two strands that are linked together by bonds between pairs of bases, just as the rungs on a ladder hold the sides together. But the bases differ from rung to rung, *and the sequence of these differences is the genetic code*, the source of inherited information. There are only four bases to carry this information, just like the limited number of letters in an alphabet. Similar to English, different sequences of bases describe the actual information (as letters form words). Unlike English, though, the genetic words are always only three letters long, but that is more than enough to carry a limitless number of messages in the nuclear genes because the messages are strings of words, rarely individual words themselves, and these strings can be quite long. When mutations occur, the messages can change. Yet more often they do not, because only a small part of the DNA carries critical information, and mutations can accumulate unaffected by natural selection.

Each mutation is like the ticking of a clock: it advances the hands one notch. In fact, the idea that genetic differences measure, or "clock," the time since species diverged is called the "molecular clock." Crucial to the concept of molecular clocks is neutral gene theory. Developed in the mid-1970s by the Japanese geneticist Motoo Kimura,[21] the central tenet of this theory is that most genes are redundant or nonfunctional, and most changes in the code for these genes are random because they are caused by mutations that do not influence the organism in either a helpful or harmful way: neutral mutations. A majority of the changes in the genome, according to this theory, are the consequence of neutral random mutations that occasionally become established in populations by accident. Accordingly, the argument goes, molecular clocks can work since selection (the effects of advantageous or disadvantageous changes) affects very few genes, and if thousands of genes can be sampled, the influence of selection on genetic variation is negligible.

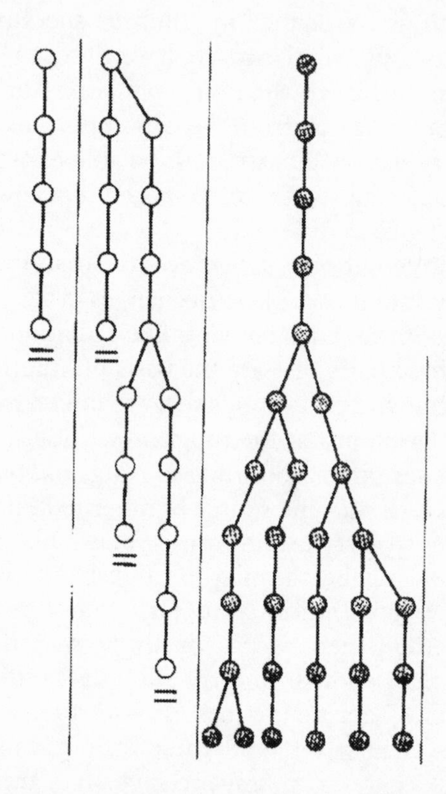

Most new mutations become extinct,[22] because they only happen once and sooner or later the original change, or its daughters if it is lucky enough to multiply, encounters a generation when it is not passed on—perhaps because of independent assortment on the nuclear chromosomes if it is not assorted into a fertilized egg's genome, perhaps if there are all sons if it is on the mitochondrial chromosome (mitochondria are passed on in mother's cytoplasm). These extinctions are represented by the fates of the open circles. But extinction is not inevitable. Rarely, as the dark circles depict, a new mutation is passed on, and multiplies often enough so that no accidents of inheritance are likely to extinguish it. The chances of this are very small, but if many mutations are considered there will be numerous incidents of establishing new mutations this way.

The best molecular clock, then, would be in an area of the genetic code where there was no Darwinian selection acting. Molecular biologists at the University of California, Berkeley, focused on what they believed was such an area, part of the code inherited outside of the nucleus, in organelles called mitochondria existing in the cytoplasm. Their study of lines of descent of mitochondrial DNA led to the theory that all modern humans have a common, recent origin: the Eve theory. This theory comes from the dissertation research of Rebecca Cann, a Berkeley Ph.D. student of Alan Wilson, a well-known, slightly maverick, molecular biologist. This theory was based on relating human

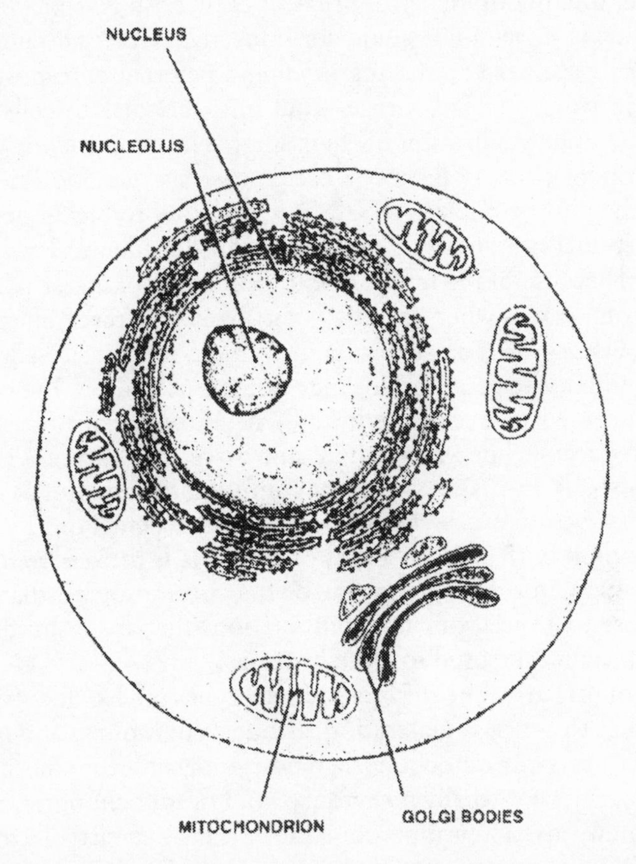

A cell such as this one,[23] perhaps under one ten-thousandth of an inch in diameter if it is like most nucleated cells, has more than one system of inheritance. One of these, the one most of us are familiar with, is found in the DNA of the paired chromosomes in the nucleus: one of each pair is inherited from each parent. The other is found in the unpaired single chromosome of the mitochondrion, one of the organelles of the cell's cytoplasm. This DNA is passed on from only the mother because at fertilization we receive only our mother's cytoplasm, with all of its contents.

population histories to the histories of mitochondrial lineages—that is, the evolutionary histories of the mitochondria themselves. Phylogenies and a molecular clock based on this small DNA segment are the genetic "facts" underlying the Eve theory.

For perhaps a billion years, beginning when what some scientists think was a viruslike ancestor of modern mitochondria entered a cell and began functioning in it, this organelle has enjoyed a symbiotic relationship with its host. In return for a home, the mitochondrion produces ATP, a source of energy for the cell. The mitochondria of today are descended from an independent organism and have their own genetic code, totally different from that of their host. As in a virus, that genetic code is stored on a single chromosome. Viruses do not reproduce sexually so there are no maternal and paternal chromosomes to combine in pairs. Instead, viruses, and mitochondria in cells, reproduce by the exact replication of their single chromosome, or cloning.

The cloning process makes it easier to trace genetic lineages in mtDNA than in the nuclear DNA. In sexually reproducing organisms every individual is genetically unique (except for identical twins, who are clones) because of the independent assortment of maternal and paternal chromosomes when the egg is fertilized. If a researcher tries to trace the history of a particular gene variant, she is faced with two problems. First, the gene variant may not get passed on. Every time a zygote is formed it receives only one chromosome out of each pair from each parent. The variant may not be on the chromosome that went to the baby-in-making. Further confusing the attempt to recreate genetic history, the expression of most genes depends on the genetic information on both chromosomes of a pair. It is the *combination*, not just the genetic information on one of the chromosomes, that underlies the observed variation, and combinations that are reshuffled each generation cannot be used to trace histories.

Mitochondria are quite different. Their genes are like family names, passed intact to the next generation, although by women and not men as each of us gets our mitochondria (and cytoplasm) from our mother's egg (the sperm has virtually no cytoplasm). For mitochondria, the only source of new variation is mutation. This means that virtually all of the genetic differences between mitochondria are due to random events, as most mutations are, that then were passed on to the next mitochondrial generation. Each mutation, like a tree branch, begins a new mitochondrial lineage because it can be uniquely identified.

It is much easier to study mitochondrial DNA than the DNA in a cell's nucleus because there is much less of it. The amount of genetic information in mtDNA is very small. Only 37 genes or some 16,500 base pairs long, it is a mere 0.0005 percent of the over 3 billion pairs in the nuclear DNA. And there is another important difference. The molecule evolves at a much faster rate than nuclear DNA because it lacks the self-repair facility of the nuclear genes. Mutations can accumulate faster when they are not weeded out.

These advantages—maternal inheritance, rapid evolution, small information content—make it possible to use the mtDNA of today to reconstruct mitochondrial evolution. All of the base pairs on an mtDNA molecule can be identified in their proper order, and matrilineages (female lines of descent) based on variations from one individual to another can be reconstructed and traced into the recent past, just as family names sometimes are.[24] It is most useful over the range of thousands of years to several million years, which is the span of the evolution of *Homo*. And because mitochondrial lines only differ by splitting (forming new branches), never by merging together, all mitochondrial DNAs must come from a single source, just as all branches of a tree must be traced to the stem of a single trunk. That source, for mitochondria, has been dubbed the mitochondrial Eve (because inheritance is maternal), and the Eve theory of modern human origins is a consequence of claims made about where and when she lived, based on the conclusion that the mitochondrial Eve was an Eve for all humanity.[25]

The number of genetic differences between mitochondria is in effect an indication of how closely related the mitochondria are. This is because it is assumed that the more mutations there were, the longer the period of separation they must mark. The assumption comes from the fact that mutation rates seem to be relatively constant. Thus the Eve theorists added temporal considerations to the picture: if there are only a few genetic differences between two mtDNA lineages, the shared common ancestor is recent; if there are many genetic differences, the common ancestor is more ancient. In her dissertation and subsequent research Cann studied the mtDNA of a small sample of living people from diverse populations (147 at first—it was hard to get mtDNA that could be analyzed in those days). From her analysis she calibrated a "mitochondrial clock" for human evolution. The "clock" indicated a very recent Mitochondrial Eve, the common ancestor of all the mitochondria she studied.

How recent? Remember that the lineage of our species *Homo sapiens* originated some 2 million years ago, and by a million years ago the tropical and subtropical regions of Eurasia were inhabited by human populations. Eve was much later, perhaps 200,000 years old and possibly even more recent according to the original thinking of the Eve theorists. Here is the problem that Cann and others faced. If all mitochondrial lines can be traced to a woman who lived that recently, what happened to the mitochondrial lines of the other women who were alive when she was? Two hundred thousand years is not enough time for them to disappear by accident (not all women have female children), so, it was reasoned, those other women have no existing descendants because they were replaced by Eve's descendants. Replaced? What exactly does that mean? For the Eve interpretation to be correct it must mean replaced without mixing, because if there was mixture some of those other female mitochondrial lines would have been passed on and gotten into surviving gene pools. And that is the genesis of the Eve theory: an origin for today's mtDNA so recent it must have been spread around by population replacement without mixture.

Replacement without mixture does not merely mean that invading men *did not mate* with the women in the populations they were replacing. It means they *could not successfully fertilize* the women in the populations they were replacing. This is because, as human history shows over and over again, when people can mate they do. When fertility is lacking it means there are two different species, and this is how the Eve theory became a theory about modern human origins. Modern humans, according to it, are a new species. This is why Cann's and Wilson's research seemed to indicate that it was not just a mitochondrial Eve they had identified in 1987, but a human Eve as well. They had linked the evolutionary histories of the organelle and its host.

Cann's and Wilson's work also suggests that the common ancestor of us all was of African origin. This was in accord with Multiregional evolution and all of the fossil evidence and was not at all problematic for Milford and Alan. What was totally incompatible with their notion of Multiregionalism as an explanation for recent human evolution was the time scale the Eve theorists envisioned. Evidence supporting a recent speciation and worldwide replacement would refute Multiregional evolution, and Milford wanted to examine it carefully. As broad as Multiregional evolution was, it could not accommodate a recent speciation in human prehistory. If the Eve theory were correct, regional continuity, the observations Multiregionalism was developed to explain, could not indicate genetic continuity. Moreover, it would indicate that a "branching pattern" characterized recent human evolution, not a networklike one. He was initially extremely concerned, although this certainly wouldn't be the first time he changed his mind in the face of new evidence. Milford had learned from experience the advantage that came from admitting when your theory is wrong, and moving on to something else. But first he needed to be convinced that this refutory evidence existed, and it would have to be strong evidence to throw out virtually the entire human fossil record. He looked over the published work carefully, learned about mitochondrial genetics of many animals from a variety of sources, and continually pondered the relationships between mitochondrial histories and human populational histories, not always without confusion. It seemed, though, that the closer he looked at the Eve theory, the messier it got. It became increasingly clear to him that Cann's and Wilson's research started out with many logical flaws, some extremely obvious, some less so, and he was sure there wasn't enough there to refute the more explanatory and less assumption-ridden Multiregional hypothesis.

Cann's and Wilson's position had been brewing for some time. However, even those paleoanthropologists who were aware of it paid relatively little attention. Long before the 1987 paper was a 1982 conference in Australia, organized because a book, written by two journalists,[26] cited Wilson as speculating that his research pointed to Australia as the place of origin of modern humans. The idea of a single, recent origin for moderns was first discussed internationally there (Wilson denied the attribution), and both Alan and Milford were present and presented papers summarizing the evidence for regional continuity. However defining and important the 1987 paper was, it really was not until it entered the realm of public science that the Eve idea gained momentum in all circles, becoming for a short time the answer to the question of modern human origins. In 1987, engineered by Wilson's well-oiled publicity office, the press picked up on Cann's work, and the Berkeley geneticists were heralded for bringing understanding and truth to the pattern of human evolution. Hurtled into the spotlight, the Eve theory overnight became the most popular theory of human evolution, for both scientists and lay people alike. And Milford and Alan learned much about the role of public science from a superb teacher.

THE POLITICIZATION OF EVE

Eve was new. Eve was modern. Eve was glamorous and sexy. Eve was a simple theory that made science reporting easy and fun. Eve gave answers and represented 20th-century technology providing answers—telling us about our origins. Eve implied the brotherhood of all humankind and was politically correct. Eve was perfect in every way, actually too good to be true. How is it that a theory so flawed could be embraced by so many?[27] Why was she so uncritically accepted? The answer incorporates politics,[28] and Eve gained political favor two ways: first by the appeal of new science, new technology, and new ideas replacing old-fashioned ones. It was a demonstration that public tax dollars were not really being misspent, that the results tell us something new about ourselves and something we can all understand. Second, it underscored the genetic unity of the human species, something we all need to be reminded of in the face of so many factious elements in our world. Both factors contributed to Eve's appeal to the public, and both entered the scientific discourse because science, in the end, is a human activity.

Unlike many scientific debates, where different sides may write quiet (or not so quiet) articles back and forth in professional journals for decades, the Eve debate quickly became politicized for a variety of reasons discussed in this book. As the science of human origins has always been, this debate is public. It is sometimes pitched as a battle between the paleontologists (using archaic science) and the geneticists (modern scientists, exploiting the advantages of new techniques and modern technology), although this is far from true.[29] Much hay is made over personal differences between the investigators in the different disciplines and even more over the differences in technology used. In a sense the fuss over Eve is an appealing topic because it illustrates the advantages of the modern age. Images are fostered of bright young geneticists using modern techniques the doddering old fossil hunters, ill prepared to understand, let alone participate in real science, cannot hope to. This is actually actively promoted by a few of the Eve researchers in statements such as those of Wilson and Cann in *Scientific American*, no less, where they contrast paleoanthropologists with "biologists trained in modern evolutionary theory" who "reject the notion that fossils provide the most direct evidence of how human evolution actually proceeded."[30] Cann then held paleoanthropology in especially low regard, once saying "it is too much to hope the trickle of bones from fossil beds would provide a clear picture of human evolution any time soon."[31] The paleoanthropologists themselves are portrayed as poor scientists engaged in circular reasoning. For instance Wilson and Cann quipped in *Scientific American*, "fossils cannot, in principle, be interpreted objectively . . . [paleoanthropologists'] reasoning tends to circularity."[32]

In fact, there is nothing particularly difficult to understand about mitochondrial genetics, or the Eve position. We find the fuss over technology something less than relevant, since quality of science is not measured by the level of technology employed, but by the design of questions asked and testing methodology (not technology). Many people seem to believe something seen with the naked eye is less valid or scientific than something seen with a microscope, and the more powerful the microscope the more valuable the observation. But microscopes "showed" scientist after scientist that humans had 48 chromosomes, when they actually have only 46.[33]

Scientists working today do have great advantages over their historical counterparts, and some of the advantage comes from technology and its applications in research. Advances in instrumentation are extremely beneficial, but by themselves do not make "good" science. In fact, there has been a real tendency for the technologies themselves to drive the direction of scientific research, as they become techniques in search of questions to answer. Real advantages we enjoy come from our recognition of the understandings arrived at by our predecessors and their incorporation into our consciousness, our world view. As Newton said of himself, "If I see so far, it is because I stand on the shoulders of giants." Multiregional evolution was only derived now, in spite of age-old grappling with many of the same problems, not because of the advance of technology, but because our world view has changed. The triumph of Darwinian thinking and an appreciation of population dynamics are actually very recent. Whatever insights we may have into the evolution of humans as a polytypic species are due to the influence on our thinking, both conscious and unconscious, of the prior work of others.

Misconceptions about the power of technology are generated by the press, not the geneticists (with one or two notable exceptions). For the most part, there are very good feelings between geneticists and paleontologists, two groups of scientists who study different data bases, but who sometimes ask the same questions of them. Both kinds of data can give us information about evolutionary history and relationships. One kind of data, whether from genes or bones, is not "better" than the other, and if data from different, independent sources seem to bring totally conflicting evidence to bear on a single question, it is not time to choose between them, but rather to see what is wrong with our hypotheses and methods of analysis.

There are other, far more serious ways than the technology issue in which the public aspects of the debate have influenced it, the positions taken by its participants, and its perceived outcome. These evolved over the question of political correctness.[34] It is possible, as the late Glynn Isaac reportedly said, that Multiregional evolution holds the high ground on the political correctness issue because by positing an ancient divergence between races it implies that the small racial differences humans show must have evolved slowly and therefore are insignificant. But the high ground is widely perceived to be held by the Eve theory, not Multiregional evolution, and in any event Multiregional evolution does not mean that the modern *races* are particularly ancient: groups of features, not groups of populations, are ancient according to this model.

Even as the debate was first joined, Eve theorists claimed the high moral ground for themselves. In 1987 S.J. Gould wrote: "We are close enough to our African origins to hope for the preservation of unity in both action and artifacts."[35] In 1988 he proclaimed: "Human unity is no idle political slogan . . . all modern humans form an entity united by physical bonds of descent from a recent African root."[36] Of course, if the Eve theory means the *unity* of humankind, what could Multiregional evolution mean? And why should either side be more politically correct? A paper[37] read by Fatimah Jackson at the 1994 meetings of the American Anthropological Association tars all of the modern human origins theories with typology and racism in one form or another.[38] As she sees the debate, it begins with the presumption that there are typologically distinct races. She believes that this assumption is a reflection of Eurocentrism: the races must be distinct for Europeans to be distinct from the others. Writing with L. Lieberman, she goes on to conclude of all the theories, "Each . . . relies to varying degrees on static, typological definitions of human biological variation at some point in its analysis, and this reliance limits the explanatory power and utility of each model for understanding the origins and maintenance of human diversity."[39]

Although there is much truth in what she says, Dr. Jackson misses a fundamental point. Far from having the same view on race and human variation, different views on these topics underscore the various theories of modern human origins. The different origins theories hold very different positions on how to model human variation, or race, and therefore on evolutionary pattern. Or perhaps it is the other way around: proponents of different origins theories have different ideas about evolutionary pattern, and this influences their views on race and modern human variation. Whatever the case, the two issues (race and modern human origins) are inextricably related. Multiregional evolution is clearly tied to race: it was developed to explain regional continuity which, given our own views of race, *should not exist*. But by accepting the existence of regional continuity, we recognize morphology that has been interpreted to both elevate and rank the importance of human differences, we believe incorrectly, with horrendous consequences. In order to understand the nature of the mutual influences of race and human evolution we need to examine how these mutual influences developed, as we do in the next few chapters. There, we can also find clues to the origins of modern predispositions toward one theory or another.

2

A FIRST LESSON IN THE POLITICS OF PALEOANTHROPOLOGY

LATE ONE SPRING MORNING, we were shivering over coffee in the Croatian Natural History Museum. In spite of the impressions of sunny southern climes and proximity to warm sandy Adriatic beaches promoted by travel offices, spring in Zagreb is like spring in the rest of Europe: rainy, cold and without central heating. But even in the cold and damp the Croatian Natural History Museum is one of our favorite places to be, not only because it is the home of the finest and most extensive Neandertal collection in the world, but also because it is home to some of the world's finest people. Chief among these is Jakov Radovčić, curator of the famous Krapina Neandertal fossils. Nowadays, when we think of the former Yugoslavia, it is hard to get past the images of war and devastation that cloud our thoughts and attempt to rob us of our memories of the happy times we've spent there. But our thoughts of Jakov transcend all that—images of his smiling figure, lean, yet expansive and welcoming, dominate the ugliness, and in spite of its shell scars, Croatia remains a place we'll always love and return to. The museum and its valuable fossils only narrowly missed destruction in 1992. During the single attempt to bomb the Croatian government, whose offices are located a stone's throw from the museum, one bomb hit the house across the street from the museum, home of Dragica, the museum's motherly and garrulous longtime caretaker. She survived, and we are told moved into the museum and then her sister's home until the repairs on her house were realized the next year. A second bomb hit a pavilion in the valley behind the museum, where the single unfortunate casualty of the bombing was enjoying his last lunch.

It seems fitting to begin a historical essay with Jakov because part of Jakov belongs to the past. His presence reminds us of the influence of the past on the present, the extent to which we owe our scientific thinking to different intellectual figures and climates in the last century and the funny parallels that exist between past and present. He loves to talk about history, but more than that, he embodies the past. As a fish paleontologist turned human paleontologist by circumstance, his professional life parallels that of the great Croatian fish paleontologist Karl Gorjanović-Kramberger, who excavated the Krapina Neandertals at the turn of the century. Using methods that were superior to many used since, Gorjanović left an incredible legacy to science: the

Jakov Radovčić, on the occasion of a visit to Milford's laboratory.

fragmentary remains of over 70 Neandertal individuals whose bones are remarkably well preserved and provenience well detailed. His fastidious descriptions of the site and the remains, timely published only two and five years after their excavation, document his shifting interest to human paleontology. Jakov, having also abandoned fish to study human prehistory, now curates the Krapina remains from Gorjanović's old office at the museum. This room remains a memorial to the last century; Gorjanović's books and notebooks line the walls, his 19th-century instruments are on display, his desk remains where he left it, with only the low steady hum of an IBM reminding you this is a 20th-century office. At the museum our perspectives of time change in subtle ways; our proximity to the past and all the other scholars who have worked here over the last century is palpable. They studied the same remains and their questions have framed our own.

The fossils we study here are one hundred thirty thousand years old, very young in terms of our evolutionary lineage. They remind us how short a century is. Our grandmothers were born a century ago, and my mother's mother's mother saw the Civil War as a child. Had they moved in the right circles, our grandparents could have known Gorjanović, our great-grandparents, Darwin. We would have been lucky if they had; we could have heard firsthand stories of these figures who were so important in our fledgling discipline, even as we were able to hear stories from the next generation from my uncle Ernst, a geneticist who was in Germany during the coming of age of the German genetics community. For a long time, really until after Ernst's death, we treated these only as family stories; we had no context to place them in, no way to learn from them. Previous generations didn't lose their connection with the past as readily as we do. For 19th-century scholars, workers of earlier centuries were a conscious part of their lives and discipline; they were always aware of whose shoulders they stood on. For us, it is different. We feel that our precursors belong to the dark ages, separated from us by myriad things and part of a different culture in which many seemingly familiar words and phrases have different meanings.[1] Now, technology and many of its concomitant social changes, not just the passage of time, distance us from our predecessors. The incredible technological changes over the last century separate us from our past more effectively than the years.

But if you strip away the differences in technology, our science has not changed much in these hundred years. Situations and controversies—issues about how evolution works, relationships between short-term (micro) and long-term (macro) evolutionary processes, how species are defined and classified, varying notions of "progress" in evolution and in anthropology, arguments about the definition and the origin of human races—crop up again and again and are never permanently laid to rest. They may be altered slightly, appearing in new guises in new generations. They may go in and out of fashion, but in the end, they never seem to die. They are never truly resolved. It is surprising to us, for example, that virtually the same words are used by some modern "splitters" of the human fossil record—biologists who create classifications based on minimal differences and therefore recognize many species—that were used by "splitters" in the past. Some of these words were used over a century ago to try to justify slavery; others to show that races had independent origins. There were many reasons to argue against the unity of the human species and maintain that humans are not a special case, that the diversity and species proliferation in the rest of the animal kingdom should be a guide for understanding diversity within the genus *Homo*. But, of course, the issues themselves were different and were reflections of different social and theoretical contexts; yet, the underlying scientific principles had much in common, as we will see. Chronic issues like these demonstrate how the controversies of today in many ways repeat past battles. We cannot truly understand the present, the creation of and reaction to modern theories of the pattern of human evolution, without understanding its link with the past. History influences ideas and the interpretation of ideas, and scholars often will unconsciously read present situations through lenses distorted by historical nightmares.

THE NATURAL HISTORY

We had been in Zagreb for two months and would be staying another two as we completed various projects that depended heavily on the analysis of the Krapina Neandertals. This particular morning, Jakov appeared in the workroom unusually early. Generally his mornings were committed to the administrative affairs of the museum, and we were lucky to see him for a late coffee. But this morning we put the fossils aside, had coffee, and joked about invented scenarios of mishaps associated with a huge anthropological congress convening in Zagreb later that summer. Changing the subject, Jakov leaned across the table and handed us a copy of *Natural History*, the popular magazine published by the American Museum of Natural History, that had just crossed his desk.

"Thought you might want to take a look at this issue's 'This View of Life,'" he smiled. "Neandertals are in it."

We understood what the smile meant: another popular article proclaiming that "new evidence" showed Neandertals could have nothing to do with the ancestry of later humans. In this case, the "new evidence" was both the "Eve" interpretation of mtDNA variation put forth the year before by the Berkeley group[2] and the publication of some new and very surprising dates for human fossil remains from Israel.

It is true that people see what they want to see; it is certainly not unusual for scientists, especially those writing across disciplines, to focus uncritically on studies that support their pet theories. The pet theory of Stephen J. Gould, who writes "This View of Life," is "punctuated equilibrium." In this theory, developed in the early 1970s,[3] Gould and several others addressed some contradictory evidence about how evolution works. The contradiction first became obvious for this century's evolutionists as information being developed by two different disciplines became available, as so often is the case. Biologists studying microevolution, the yearly to century-spanning changes that can be observed in nature, were coming up with rates of evolutionary change that were a gallop compared with snail-pace rates of macroevolution calculated across the eons by paleontologists studying the fossil record. It seemed as though 10 million years was more than enough time for a horse to evolve into a human, at microevolutionary rates. Of course, a horse *did not* evolve into a human, but to understand why such major changes were not prevalent, these conflicting observations needed to be reconciled.

Gould and others[4] argued that the popular concept of Darwinian gradualism, slow and gradual evolutionary change, does not best explain major evolutionary developments.[5] Species, they contend, are for most of their lives characterized by morphological stasis when there is no change, interrupted by short periods of rapid change when the conditions are right for one species to split into two—the speciation events that have created the rich biological diversity we enjoy. Apart from these periods of major change, species only react to local oscillations in their habitats, according to this theory. Long periods of equilibrium are punctuated by rapid change at the time of speciation. When the average rate of change in macroevolution is calculated, they contend, it seems to creep, but the actual rate of change, in those few times when there *is* change, is quite rapid. Even more than other explanations of the evolutionary process, punctuated equilibrium is wedded to the idea that many more incipient speciations occur than are successful; most cases of speciation fail. But the few times a new species develops striking advantages, it replaces the old, and there is a short period of rapid anatomical, behavioral, and genetic change. Species are constantly going extinct and others forming, so evolutionary patterns more closely resemble bushes than trees.

We agree a punctuated evolutionary pattern best describes the evolutionary histories of many phyletic groups, including, we think, the earlier and much longer part of human prehistory when humans were only another African primate species. But we believe punctuated equilibrium does not reflect what happened to humans in the later part of human evolution as they became successful colonizers and when there *was no* macroevolutionary change. As we read the fossil record, there is no evidence of speciation events in the recent past; in fact, there is strong evidence against them. But the Eve interpretation promised to support a punctuated model for later human evolution that was denied by interpretations of the fossil evidence such as ours.

We knew if Gould were writing about Neandertals and the "Eve" debate, we probably weren't going to agree with it, in spite of generally liking Gould's writing. But even with these expectations, we were unprepared for the new level of politics in the Eve controversy we recognized in that issue of *Natural History*.[6] Many papers have been written that we disagree with, and people frequently publish popular articles we think are wrong, but here was a paper subtitled "Human unity is no idle political slogan or tenet of mushy romanticism." This was the first time we saw our position implicitly placed on the politically incorrect side of the stands. Quite frankly, we were amazed.

The impetus for this particular essay came from some new dates based on recently discovered technology that had been determined for human occupation sites in Israel.[7] These dates were surprising and therefore controversial because they seemed to show that some early "modern" people lived there about 100,000 years ago, evidently earlier than the Neandertal folk of the region. If the dates were correct and the fossil remains validly diagnosed as "modern" and "Neandertal," it would mean the two human groups in Israel must be branches (they could be in a linear progression somewhere else, for instance, with Neandertals evolving into moderns, but the Israeli evidence implies branching was part of that process). Gould accepted all this and went a step beyond. To fit his punctuational model this cannot be any old kind of branching, for instance, different races or populations. It must be a species branch because there was significant change, as evidenced by the "modern" population, and punctuated equilibrium restricts major changes to the speciation process. Gould's further step was an assumption he made, that "Neanderthals and modern humans lived in the

Levant—and maintained their integrity without interbreeding."[8] If this was true and there was no interbreeding, it would be the proof he needed to show they were different species. The assertion of no interbreeding was several steps away from the data, and an unsupported contention few would agree with. Yet this was not the part that so raised our hackles. It was what came next.

Gould's essays invariably have a not-so-hidden agenda; it is one of the things that makes him such an interesting writer. The agenda here was not about Neandertals, or even punctuated equilibrium, but about the moral implications of the Eve theory, for he was quick to link his presentation of fresh evidence showing modern humans were a new species when they first appeared in the Levant and the mtDNA-based Eve theory that posits modern humans arose as a new species just to the south, in Africa. Gould was addressing the issue of human unity, and the proof of it paleoanthropology (at least his rendition of it) so clearly proclaimed. An important part of his agenda lay in the "human unity" line used in the subtitle and what followed: "All modern humans form an entity united by physical bonds of descent from a recent African root; we are not merely the current state of a tendency as the multiregional model suggests. Our unities are genealogical; we are an object of history."[9]

Gould not unexpectedly likes the Eve theory ("I advocate this . . . view with such delight for it sits so well with my own");[10] it is the only model of later human evolution that employs punctuated equilibrium. But in this issue of *Natural History* he went beyond supporting a pet theory. By appealing to the implication that it demonstrated we are all brothers under the skin, the unspoken but implicit charge is that the opposing view (ours) somehow shows we are *not* "all brothers under the skin." In the *Natural History* article Gould, for the first time, placed the debate in the arena of political correctness, and political *in*-correctness was clearly attributed to our side of it. To the best of our recollection, it was just about this time we first heard Jakov describe paleoanthropology as opera.

RACIAL POLITICS

It didn't take long for further ramifications of this sort of political characterization in widely read magazines and the popular press to become increasingly clear. After we returned home in the fall of 1988, I answered a phone call at our house from an admirer of Milford's who wanted to discuss the Eve issue with him. This had been happening fairly frequently, although most often callers phoned the office and usually had some background in the natural sciences or were journalists who were prepared to listen. In this case, the caller was a lawyer from upstate New York who once lived in Michigan and still returns to vacation. Upon learning that Milford was out of town, he discussed the Upper Peninsula that he loved, establishing rapport as a fellow Michiganer, and then in a conspiratorial tone asked me to tell Milford that he "really appreciated what he (Milford) was doing." It took me some minutes to realize that what he was talking about and what he appreciated had nothing to do with the two models of recent human evolution being debated. The man was an unabashed racist who erroneously believed Multiregional evolution postulated the independent origins of different human races. He viewed Milford as a hero, single-handedly fighting the bastions of political correctness to espouse "the truth"—that the human races had separate and independent histories. He assumed that we share his extremely racist social agenda and continued to talk about his views of inferior races and misplaced public spending. I hung up, badly shaken. What the caller was discussing was anathema to our sociopolitical views. It was not after reading Gould, but only then we realized how thoroughly we had inadvertently grasped the tar-baby of racial politics.

Because of his assumptions about independent origins and parallel evolution the caller had a rather different take on the political implications of Multiregional evolution than Gould did. The connection he drew was a more ominous one. But where did the idea of independent origins and evolution come from? Not from our writings. The insight of Multiregionalism is that elements which seem to be opposing and contradictory—the *separation of populations* needed to explain regional distinctions, and the *links between populations* needed to explain the worldwide spread of advantageous features—do not actually contradict each other and disrupt the Multiregional pattern. Instead, their very opposition is the basis for that pattern, creating the equilibrium conditions that make Multiregional evolution work. But because they are aspects of evolution that interact with each other, neither alone could explain the pattern we find in human prehistory, and certainly not the idea of parallel evolutionary lines. Yet we could see that Multiregional evolution was becoming associated with just that idea, and the topic became something we realized we needed to explore.

The mischaracterizations of Multiregional evolution we were encountering were more than simply annoying; they could be downright dangerous. Yet as unpleasant as they were, the lawyer's misunderstandings about the tenets of Multiregional evolution were not his fault; he only knew what he read. And he could have read in an increasing number of places, from the pages of *Scientific American* to the local paper, that Multiregional evolution is the theory of the parallel evolution of the human races. Of course, nothing could be farther from the truth, yet the model was repeatedly misunderstood as being one of separate origins for different segments of humanity, a concept that carries its own ugly history of scientific racism that is an embarrassment to anthropology. In our search we came to realize the idea of a separate, parallel evolution of races is deeply rooted in the debates about independent origins of the races and their humanity that began centuries before we or, for that matter, Gorjanović were born.

3

POLYGENISM, RACISM, AND THE RISE OF ANTHROPOLOGY

WE FOUND OURSELVES face-to-face with polygenism, one of those theories we learned about in history of anthropology courses, but ignored as another example of wrongheaded thinking from long ago with little bearing on the present. It is true that polygenism is an old idea about the races that predates Darwin's theory of evolution, and even anthropology itself. But we were surprised to find how adaptable this particular wrongheaded idea was, persisting even while changing dramatically as the process of evolution, and the human fossil record, came to be understood. In fact, polygenism has remained an important element in theories about the origin of races, and it is alive and well in popular science today. Polygenism interprets the human races as separate, *really* separate. It is the precept that races have different origins, different characteristics, and different histories: even that they are different species.

As ideas about the causes and significance of human variation developed, polygenism became the first link between race and human evolution, and it continues to be a link between race, *racism*, and human evolution. Mischaracterizations of Multiregionalism, we soon came to realize, describe it as a polygenic model, and the inheritance of polygenic thinking influences scientists and lay people alike: polygenism is built into the often typological views of race that still can be found in anthropology, and it is closely wedded to the very emergence of anthropology, as a discipline. Polygenism is pivotal in the relationship between race and human evolution, as it developed and as it continues to exist. How this relationship matured gave us great insight into the problems we were facing.

Half a millennium ago, Columbus returned from his voyages to the New World, bearing to his patrons in the Spanish court material evidence that their investment was sound. He brought tales of beauty and bounty, and gifts of fruit, vegetables, spices, precious metals, and human beings of the Arawak nation. The people he and other explorers encountered and introduced to Europe and the stories that developed about the appearance and customs of others presented quandaries to the Europeans of the age of expansion that remain unresolved to the present day, through all the ensuing various faces of the colonialism that emerged from that age. Who is "the other"? How is "the other" related to us? Is he our brother? Our father? Our feebleminded distant ancestor? Or cousin? Does he represent Adam before the fall, or is he a product of a separate creation?

Motivated by the prospect of riches beyond their wildest dreams, and empowered by technological advances—the invention of the lateen sail which made ships faster and more efficient in the use of wind power, improvements in magnetic compasses and chronometers that vastly increased the precision of navigation, and major advances in metallurgy—15th- and 16th-century European nations and mercantile associations began an exploration process that would ultimately touch every corner of the globe. The purpose of the explorations was to secure gold, but the consequence was contact and its concomitant slavery, murder, and colonialism. The Europeans found foreign lands already occupied. To justify their right to the riches, many assumed the nonhumanity of "the other." For some, this was an obvious fact; for others, with pangs of conscience, it needed to be verified. This need contributed to the birth of anthropology three centuries later, and it is therefore not surprising that the question of origins, whether single or multiple, represents the oldest controversy in the field. But the origins question predates the discipline, since the problem of variation between the races was debated by various members of European and American society long before anthropology was established as a science.

With increasing knowledge of native peoples encountered by the colonizing European nations, it became commonly recognized that races existed and that their differences were enough to account for the perceived inferiority of native minds and cultures. These differences were first and most commonly explained as the result of history and environment—not of independent creations—because the dominant religious conservatives maintained that all races were descended from Adam and Eve in the Garden of Eden. In 1512 Pope Julius decreed that the just-discovered Aboriginal Indigenous Americans were descendants of Adam and Eve, which made polygenists heretics.

The complete separateness of racial lineages was well accepted, even if they had a common origin. Different races were believed to have long, independent histories as descendants of different sons of Noah. These long histories explained human differences by providing ample time for nonwhite races to degenerate from the original state in different (generally tropical) environments, and similarly accounted for other racial differences perceived. At first these independent histories seemed to isolate races enough to accommodate ideas of European separateness and superiority, without resorting to belief in separate origins. But did they?

Polygenism gave a compelling answer to what the increasingly obvious human variation meant. It provided a fundamentally different way of justifying racial inequality because it assumed that different races *acquired their humanity separately*—if in fact the nonwhite races were human at all. The rise of polygenism was almost certainly linked to continued discoveries of the extent of human diversity, that of American Indians on both continents, of the Koi and San of South Africa, and of Melanesians—a much greater diversity than had been expected.[1] One after another these discoveries disturbed the long-held European idea that the human races were made of three groups: the whites of Europe, yellow peoples of Asia, and African blacks. As discoveries continued, explaining the source of this unexpected diversity became as pressing a problem as the need to distinguish it from the Christian world. For some, long histories following a single creation did not seem sufficient.

As early as the mid-17th century, polygenism appeared in European writings in the work of French writer Isaac de la Peyrère, who wrote of separate pre-Adamite creations in a book that was burned in Paris, quite possibly because of its claim that of all of humankind, only the Jews descended from the creation represented by Adam and Eve.[2] Polygenism was first clearly articulated in the latter 18th century by the Scottish philosopher Henry Home, Lord Kames. He proposed the idea in 1774 in *Sketches of the History of Man*. Upon noting what he considered to be vast amounts of human variation, he suggested God had "created many pairs of the human race, differing from each other both externally and internally; that he fitted these pairs for different climates and placed each pair in its proper climate; and that the peculiarities of the original pairs were preserved entire in their descendents."[3]

Although he suggested this and argued for its feasibility, he admitted its heterodoxy and concluded racial variation occurred later, "as divine punishment for the presumption of the Tower of Babel."[4] Nevertheless, this work began the monogenist/polygenist debates that characterized early anthropology and lasted for most of the 19th century, with vestiges still apparent today. Lord Kames may not have been able to afford being branded unpious, but he nevertheless clearly and sympathetically spelled out the ideas of polygenism. He was, in effect, extending the baton for those less orthodox to run with.

By the early 19th century polygenist ideas had diffused throughout Europe and America, but although its initial treatment was carefully worded, polygenic theory was not highly regarded. Even after it became well established, especially in America and France, polygenism remained unpopular with large segments of the public and the scientific community. It was not so much that popular sentiment favored the equality of human races, but rather the view of polygenism as heresy, that led to its disfavor. This view was exploited by many of the staunchest opponents of polygenism, who themselves were egalitarians, to gain support within largely racist societies in their drive to discredit polygenism. Appeals to religious piety rallied popular opinion against polygenism. Accepting polygenism required a very flexible view of Genesis, incorporating the idea of multiple Edens for the different races, a flexibility that was hardly the hallmark of the age. In addition, many people, in spite of their assumptions of European superiority, retained a sympathetic view toward the "savage" that predated the onset of "racial thinking."[5]

MONOGENISM IN THE GERMANIC WORLD:
A HIERARCHY OF RACES

In the scientific world, the typological concept of race was coalescing. In Central Europe the study of human races was largely in the hands of Johann F. Blumenbach (1752–1840) of the University of Göttingen, who has often been described as the father of physical anthropology because of his interest in skull shapes and his wide range of knowledge about human variation. He set the stage for the continued interest in human crania and their various forms, collecting many specimens in his museum. Perhaps more important, Blumenbach is considered the father of the study of race and, as such, provided much fuel for the fires of polygenism. The author of *De Generis Humani Varietate Nativa*, published in 1775, Blumenbach was the first to develop a modern racial classification. He was a typologist, an essentialist, yet paradoxically he believed all humans fell within one variable type and was himself a confirmed egalitarian and monogenist.

Essentialism, a Western world view, influenced virtually all biology before Darwin. It is based on the Platonic idea that the natural world is made up of fixed and distinct ideal types, or essences, and that variation, if not significant enough to reflect a different type, is the result of imperfections or deviations away from the ideal. Many of Plato's ideas, while important philosophically, became a disaster for the natural sciences, and none more so than the concept of essentialism, which effectively crippled biology for two thousand years. Based on his preoccupation with geometry and his observation that a shape like a triangle will always be a triangle, discontinuous and distinct from all other geometric forms, Plato arrived at the concept of fixed *eide*. The *eide*, or essences, were unchanging forms that made up the physical world and the world of ideas. All variation was considered to represent imperfections of the underlying essence, and the discontinuity between all types of things was stressed. The essences were constant and unchanging, and the only source of change could be the independent, discontinuous origin of new essences. Many of Plato's views fit particularly well with Christian world views and thus were promoted for the millennia following his death. What made them impediments to an understanding of biology was not only the concept of essences, but also the idea of a creator, as opposed to spontaneous generation, and the concept of a soul. These deeply held assumptions produced a stumbling block to an understanding of evolution that still exists in both lay and scientific understandings of the natural world.[6]

There is a natural link between essentialism, typological thinking, and polygenism. Separate origins imply differences in essence, or type, and polygenists sought evidence of this in the features used in racial classification. Conversely, many typologists, by seeking differences, supported polygenic interpretations of variation. Anatomists discovered detailed similarities between people of different races, but nevertheless essentialist thinking focused on the search for underlying distinctions in type that polygenists could attribute to independent origins. Yet in spite of the link between essentialism and polygenism, Blumenbach was a major scientific voice *against* the polygenist view.

He was an outspoken monogenist, who found human racial variation minor. If the races were different types, as the polygenists insisted, there should be no gradations between them. In Blumenbach's classification there were "minor" racial categories providing links between "major" races, and this, Blumenbach believed, was evidence for monogenism.

He emphasized that morphological and behavioral distinctions between races were not fixed and static, and he believed racial differences were the results of different climates. For his times, Blumenbach was a racial egalitarian, pointing out that with education, there was no racial difference in the potential for human achievement. He cited examples of fine African scholars educated in Germany and the Netherlands, which supported his belief that racial differences in intellectual and other behavioral capacities are inconsequential,[7] and actively collected the literary works of non-Europeans as evidence of equality.[8] He and his associates were most unusual in this regard; most monogenists, even those with strong egalitarian political ideals, still assumed innate European intellectual superiority. Blumenbach was somewhat differently Eurocentric, as he saw Europeans as the ideal type from which the other races deviated.

Here is how his explanation went. Blumenbach believed the human races had a single origin and that some populations came to differ from the original type as they moved into different environments. New climates, he thought, caused changes that after a few generations became inherited. It was widely believed by most natural scientists then that when animals or people were consistently exposed to a different environment or behaved differently, and responded by anatomical or physiological changes, those changes would be inherited if they continued for a few generations.[9] Europeans living in the tropics would become black; giraffes stretching their necks for food near the tops of taller trees would become longer-necked. Blumenbach's essentialist approach led to the assumption that within all human variation, there had to be an ideal, original template that all variations extended from. He thought Caucasians were that ideal, original race—he described them as the most beautiful of the human races—and wrote that others varied from the Caucasoid archetype as they changed from the ideal form. Blumenbach envisaged five races: the original Caucasoids; two races that varied most from them, Asians and Africans (whom he called Ethiopians); and two transitional races between the ideal type and the extremes, Aboriginal Indigenous Americans for the Asians, Malays for the Africans. These other races came to vary from the Caucasoids as they moved into different climates.

Yet ironically, this view of racial variation was perhaps *the* fundamental change that came to underlie modern racism[10] when others went to apply rank and value to the different racial groups. The key to understanding Blumenbach's thinking is the importance of the Malays in his scheme. A major race? They didn't appear in Blumenbach's original formulation or in the four races of Carl Linnaeus that preceded it. Linnaeus, the father of all modern classification, and Blumenbach in his early writings, defined the races by geography. Later, though, Blumenbach turned them into taxonomically ranked entities, "major" races, and linking them were less broad "minor" races. These extended in two directions, from the ideal to the races farthest removed from it. The Malays were critical to his scheme because without them there was no link between the Europeans and the Africans. Thus he saw these minor races as gradations, proof that races were not discontinuous types and that all humans were members of the same species. But others saw a way to rank their variation.

The difference between natural history and anatomy backgrounds was significant for the 18th- and early 19th-century scientists. In a natural history framework there is an emphasis on classification, a tendency to treat varia-

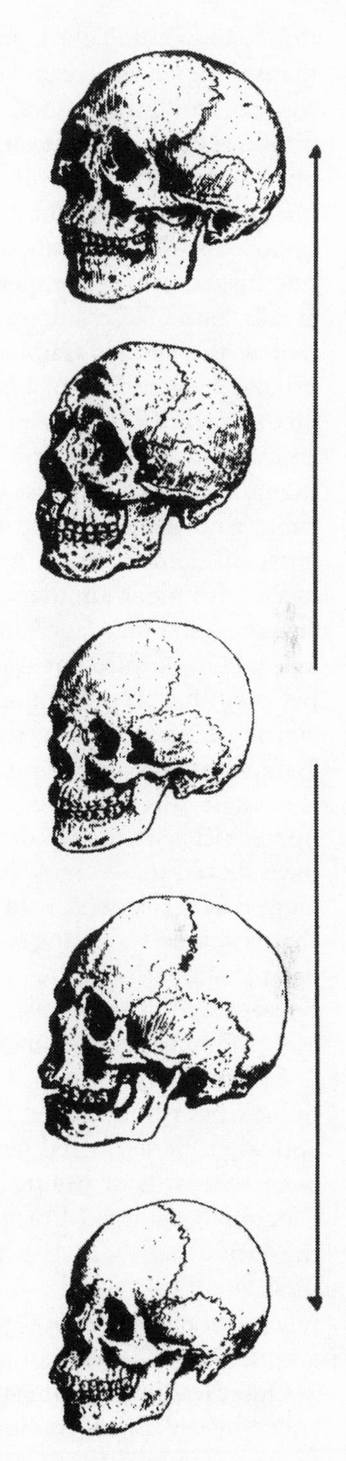

Blumenbach's arrangement of the races. The central cranium is a European, a Caucasian from Georgia. Directions of variation from it are toward the Mongolian (above) *from Tungus via the Native American Carib and the African* (below) *from Ethiopia via the Tahitian Malay.*[11]

tion narrowly and allow small differences to indicate new types. If humans were simply treated as another part of nature, this outlook could support polygenic views, and the polygenists advocated these ideas. The intellectual traditions of Blumenbach and other Central European anatomists were very liberal: for them race was considered a minor human variant, not a fixed part of creation. These anatomists all opposed slavery and advocated racial equality. Among this group of scientists is Pieter Camper (1722–1789) of Leiden, who is best known as the "father of craniometry." He developed, among other measurements, the famous Camper facial angle. This angle measures the degree of prognathism of a face, the extent to which the lower part of the face protrudes. Camper argued that different races have different amounts of prognathism, notably that Africans are more prognathic than other races. The use of this criterion was denounced by Blumenbach, who argued that it did not distinguish races or skulls from "the most different nations, who are separated as they say by the whole heaven from one another . . . and on the other hand many skulls of one and the same race, agreeing entirely with a common disposition, have a facial line as different as possible."[12] However, because other workers believed that more prognathic races were more apelike, Camper's facial angle was used to denote ranking in terms of "primitiveness," and Camper attained a bad reputation among many anthropologists. Since this angle was later used as the basis for many racist and sometimes polygenist scientific studies, it is often assumed Camper himself must have shared these views, but this is not true. Camper was not racist if viewed in the context of his times. In fact, his discovery of a facial bone that is shared by all ape species but is absent in humans helped establish the similarity of the races through their contrast with apes. He was one of the first anatomists to attempt to categorize the distinctness of apes and humans systematically.

Camper developed his facial angle during that effort, but it was part of his work on aesthetics. Camper was an artist as well as an anatomist, and when he perfected his facial angle he was trying to define European standards of beauty that he believed universal.[13] It is true that Camper considered European physical qualities more beautiful than those of other races, but he never tied physical appearance to behavioral qualities. He, like Blumenbach, held egalitarian views about intellectual and behavioral capacities of different races, and outside his artistic realm, he was not a racial thinker.

Other leading anatomists of the Germanic countries were also influenced by the liberal tradition personified by Blumenbach, and this influence extended to at least the century's end. Most used their science to undermine arguments that racial differences were large, intrinsic, and unchangeable, and they championed abolitionist causes. Germany also had its share of polygenists, many of whom were most active after 1859. However, Blumenbach's influence on anthropology, as it developed in the 19th century, was quite strong, and his monogenic views held. But his influence was strong in another way. The mere classification of races by listing attributes that distinguished races along gradations, however minor those differences were to some, inadvertently provided fodder for polygenism. This is because *ranking* races was an immediate consequence of Blumenbach's classification according to similarities, and the implications of ranking the races eventually were realized throughout Central European science, a development that was a world apart from the liberalism Blumenbach expounded. Racist thinking in Germany finally prevailed in scientific ideology, with Ernst Haeckel's famous polygenic interpretations of Darwin emerging before the end of the century.

POLYGENISM IN THE WEST:
THE EMERGENCE OF ANTHROPOLOGY

The egalitarian humanism developed in the French Enlightenment was reflected in scientific and social views of human nature and society, in disciplines we might now consider under the broad umbrella of anthropology. But then citizen Robespierre, with bouquet in one hand and torch in the other, signaled the beginning of changes, the likes of which the Western world had never seen. The French Revolution was, in some respects, the culmination of the popular understanding of the Enlightenment in France. This was reflected in the works of scholars and politicians who affected thinking throughout Europe. Many of the French Enlightenment's principles shaped the thinking of the founding fathers of the fledgling United States even as they, in turn, helped shape the Enlightenment. Much of the egalitarian optimism voiced in the American constitution and Declaration of Independence can be traced to France. Thomas Jefferson and Benjamin Franklin were strongly influenced by the Enlightenment. However, a conservatism followed the French Revolution that was likewise reflected in intellectual spheres, markedly contrasting with the period preceding it.

As the 19th century began, the face of science was changing along with the rest of the world; the grand schemes and unifying theories of the Enlightenment were falling into disfavor, and harder, colder attitudes were taking hold in Europe that provided a strong social and theoretical foundation for polygenism. Although polygenism was initially unpopular, the turn of the 19th century was a time ripe for its acceptance; the concept of race was becoming an important part of the way Europeans viewed the world. Natural history traditions developed to deal with an increasing awareness of diversity in ever more typological ways, and the relationship between science and religion was changing. There are key words for any age, words whose constant repetition but uncertain meaning signal that they reflect the unconscious assumptions of the time. If the key words of the 18th-century Enlightenment are nature, natural law, first cause, reason, sentiment, humanity, perfectibility, for the 19th they must be matter, fact, progress, and evolution.[14] The adoption of the concept of polygenism and the development of anthropology as a discipline mirrored these changes. Closely linked together, they were tied to a European shift toward racial thinking. Reflecting this focus, polygenism gave maximum emphasis to deep and intrinsic causes of human variation and had the added advantage of *not* being the position of the pious.

Also fundamental to the acceptance and success of polygenism was the need to recognize and define "the other" as something separate and remote. The "other" needed to be fit into the prevailing views of man's place in the universe. Achieving definition only in contrast with "us," "the other" forced Europeans to reexamine themselves in light of what the new people represented. In turn, concepts of self-identity themselves determined just what the new people did represent. This determination was affected by political and social constraints. The most important of these for the rise of polygenism is related to the need to justify exploitation of indigenous peoples and their lands, and the practice of slavery. If indigenous peoples were not human, the Europeans had discovered unoccupied lands they then could justly claim without compromising the moral tenets of a religion that was a powerful influence in most people's lives.

However, the determination of the identity of "the other" was complicated, if not actually disputed, by various aspects of the social and intellectual climates of various periods. For example, at first the optimism and egalitarianism of the Enlightenment were reflected in the "noble savage" ideal and the unquestioned assumption of monogenism. This view saw members of primitive cultures as having equal capacities and as pristine members of the human race—what a European would be like without the buffers of civilization. Humanity was considered fundamentally good, and therefore the romantic image of the pristine savage was generated. The savage in this view had an enviable childlike innocence but also could grow up if blessed (or cursed) with access to civilization. Thus, during the most liberal periods of the Enlightenment the determination of the identity of "the other" was that of all humanity before the fall. It was also true that justification for exploitation could be facilitated within this view; the savage was a child, innocent but also irresponsible and incapable of owning, managing, or understanding the value of the resources on the land he occupied. Additionally—and this view prevails today—European civilization was considered a blessing; contact with it could not help but better the savage. Exploitation was seen as providing benefits for the conquered.

When the egalitarianism of the Enlightenment gave way to much more conservative principles and world views at the turn of the 19th century, views of "the other" were strongly affected. The noble savage idea was replaced by an increased repugnance for primitive peoples, even as the image of "the other" turned from the romanticized red-skinned Americans and tawny Pacific Islanders to the black Africans and Australians.[15] These attitudes corresponded to the beginnings of true racial thinking that was to dominate social policy ever after, and they set the stage for the acceptance of polygenism. Public and scientific views on the nature of the relationship between civilized men and savages were extremely complicated, influenced by politics, religion, philosophy and, of course, the omnipresent ulterior motive of profit. Polygenism had become an attractive way to solve the major conflict between greed and conscience that had been developing for centuries.

Polygenism found some of its largest support in American antiabolitionist circles, and became the central tenet of the "American school" of anthropology, because of its importance in the debates over slavery and its promotion by key American scientists. And it was increasingly accepted internationally. However, the popularity of polygenism was not influenced just by sociopolitical factors. Part of the world being classified was human, and among the naturalists doing the classifying there was a widely prevalent "splitting" mentality that, while not necessarily essentialist, was often a product of typological thinking. "Splitters" are the natural scientists who focus on difference, using taxonomic categories to accommodate, if not explain, the variation existing within a group. They often see minor physical variants as reflections of species. A splitter might regard closely related doglike animals—coyotes, wolves, and dogs—as different species because they can be easily distinguished. "Lumpers," the antithesis of "splitters," focus on similarities, recognizing much more variation within taxonomic groups, and at the extreme might subsume many legitimate taxa within a single species. A lumper would place the three—coyotes, wolves, and dogs—in the same species in spite of their differences, because of shared anatomical features and because they can breed together and have fertile offspring. Since typological approaches seek differences to support the standards of different types, they can foster and benefit from splitting methodologies. This tendency to split was even further accentuated by the descriptive and classificatory emphasis of the 18th- and 19th-century natural historians, many of whom were employed to describe and classify the new world that the colonizing Europeans were encountering. The young Charles Darwin was a beneficiary of the Royal Society's policy of placing a natural historian on every British exploratory vessel to identify and describe new fauna and flora, especially potentially profitable and useful varieties.[16] As they emphasized the high level of difference between human races, polygenists used and were no doubt influenced by this classificatory approach to the natural world.

Anthropology as a discipline developed out of this, and not without its own ironies. Anthropology's emergence around the turn of the 19th century was related to a change in European attitudes toward the relation of race and civilization. Enlightenment scholars saw civilization as a *human* attribute, not just a quality of certain races. But at the turn of the 19th century human diversity began to be seen primarily in racial terms; that is, diversity was seen as representing intrinsic, deep-seated and essentially unchangeable differences between groups. It followed that civilization and, by extension, humanness were attributes of only certain races. In a sense this was the beginning of true "racial thinking," and the development of this kind of outlook went hand in hand with the emerging doctrine of polygenism. Racial thinking and polygenism are historically inseparable; as Europeans saw diversity in racial, biological terms, attempts to classify or to analyze relationships brought the notion of polygenism to the fore as one of two competing possibilities. Early anthropology was really the study of race, sometimes simply in a typological, anatomical sense and sometimes as it related to cultural or behavioral stereotypes.

When it emerged, around the turn of the 19th century, anthropology was itself made up of many disciplines: natural history, medicine, and various humanities and social sciences. It is difficult to identify the specific beginnings of anthropology because the discipline itself is so hard to define, but it clearly developed to deal with European understandings of civilization, race, and human nature. We think of its emergence as a consequence of the political and social "human" problems superimposed on a scientific climate of increased focus on natural history. Its appearance straddled this period of changing attitudes, and by the time people were calling themselves "anthropologists," racism and ethnocentrism had become strongly developed components of early anthropology, as members of the new discipline sought verification of and explanations for physical and cultural variation. The primary biological question asked dealt with the relationships between Europeans and other races, and the differences between races were focused on and magnified. The practitioners were often natural historians, who described human variation in the same typological way they dealt with the rest of the natural world. As anthropology became a science, biological (as opposed to ethnological) aspects of the discipline predominated, even as racism became more and more of a social issue. Abolitionist groups were at work on both sides of the Atlantic.

A contradiction thereby became incorporated in anthropology, reflected in the disparate sources it emerged from. It is one which then, and now, continues to weaken the fabric that holds so many different elements together in the discipline. From the beginning, anthropology was defined by the attempt to meld the biological and social aspects of humanity in a single discipline and search for a unified paradigm that would bind them together. This way anthropology, and not the separate emerging biological and social sciences, held claim to be the proper study of mankind. But polygeny did not provide the glue that could effectively bind these, and the union of biological and social science that was (and is) anthropology's strength is also its weakness. The quest for understanding the true basis of humanity has seesawed back and forth between biology and culture. The oscillations reflected a much more widespread nature *versus* nurture debate that was developing over whether human differences are historic, the stigmata of racial history, or developmental and thereby changeable. Each age gives a different answer to this question, answers often phrased in the key words of the time: from natural law to natural (Darwinian) selection, from human nature to universal culture. Nature versus nurture is also the Mendelism of "racial hygiene" versus the Lysenkism of the "new Soviet man." The biological and social sides of anthropology never really merged comfortably, and to this day there never has been a truly universal anthropological paradigm.

Lacking true intellectual bonds, the creaking strains from this poor fit can be heard today, in the bristling reactions of ethnologists to the increasing number of sociobiologists in anthropology,[17] those who seek to find and define a Darwinian (and thereby biological) basis for human behavior. By promoting Darwinian evolution as the long-sought universal theory, too much of the basis of modern ethnology would be contradicted, or lost. In the North American system, anthropology is dominated by a four-field approach, a conglomeration of biological anthropology, cultural or social anthropology, archeology, and linguistics that was brought together by the European immigrants who defined anthropology in America. While American anthropologists specialize within one of these four subdisciplines, our American training includes all of them, and we are particularly sensitive to the ways culture affects all aspects of the field, including biological aspects. In much of Europe today, where ironically anthropology first emerged as a holistic discipline, the union was incomplete or even nonexistent. The anthropological subdisciplines are largely separate, and students of one rarely have formal training in the others. Reflecting the 19th-century focus on race, in these European countries anthropology often still refers only to physical anthropology, a discipline that is virtually unrelated to ethnology, a different discipline which plays no part in the training of anthropologists, who are by and large anatomists.

The varying nature of the discipline today shows the consequences of failing to resolve the contradictions sewn into the fabric of emerging anthropology, as the 18th century gave way to the 19th. As the biological explanation of variation prevailed, differences were ranked on scales of increasing perfection. Racial thinking permeated early anthropology, and it soon, in reality, became the study of race and little more. But anthropologists did more than study races; they evaluated and judged the importance of their differences, and finally their merit and their humanity. This is how Blumenbach, finally, gave way to Haeckel.

LAMARCK, CUVIER, AND THE LAST OF THE ENLIGHTENMENT

At the end of the Enlightenment, there was one last attempt to link behavior and biology, in the work of Jean-Baptiste de Lamarck (1744–1829). Lamarck's ideas were evolutionary, and could serve to bind behavioral and biological sciences, through the physical and mental changes Lamarck thought took place in tandem as humanity originated. But they were not taken seriously. Much of 18th-century science was philosophy, but by the 1790s the relation of theory and observation in science was being carefully scrutinized, and the grandiose system building that was at first the pride of 18th-century science came to be increasingly seen as an obstacle to scientific progress. Lamarck's view of evolution, part of his attempt to form a synthesis of biology with physical and chemical phenomena, was the last grand scheme of the period.

Lamarck was fundamentally a naturalist,[18] one of the first to coin the term "biology" to describe his scientific studies. His grand scheme was the first to describe the animal kingdom as the result of an evolutionary process, and he was the first to devote an entire book to a theory of change.[19] He rejected the biblical notion of fixity of species, instead arguing for an organic mutability. Lamarck envisaged an inner drive to nature, in which simpler forms of organisms are successively transformed into more complex ones. He tried, and failed, to exhibit the hierarchy of species as a linear progression, and finally convinced that this arrangement was logically impossible, he was the first to develop a phylogenetic tree to express how they are related to each other.

The tree notion of branching relationships, which is what a phylogeny illustrates, is important because it means that species are related in a hierarchy of descent, and determining their relationships involves assessing the nature of the genealogical ties among fossil groups, and between fossil and living ones. A phylogeny is a genealogy, but a genealogy of related species rather than the genealogy of related people that we are each familiar with. For we humans, working out our phylogeny means determining who our ancestors are and how they are related to us—drawing our family tree. Lamarck's phylogenetic tree was the first to have extending and bifurcating branches from a common trunk rooted in the ancient past. He addressed how this hierarchical arrangement developed in *Philosophie zoologique.*

Like Darwin later on, Lamarck was a uniformitarian, believing that the environmental forces of the present account for changes in the past and disclaiming any role for catastrophic change.[20] But he saw that the unfolding fossil record showed differences between ancient and living organisms. This was a problem because all the species of life were supposed to be there at the time of creation, and various scholars arrived at different conclusions. Embracing the idea that organisms could be modified, that species were mutable and had changed,[21] Lamarck also presaged Darwin in being a transformationist. But unlike Darwin, Lamarck did not consider adaptation the major cause of the mutability of organisms.

Lamarck's theory of biological evolution was grounded in 18th-century natural history. He never used words like "evolution" or "transformation," referring instead to "*la marche de la nature.*" Change, for him, was the natural course of events, the way things would happen were it not for constraining circumstances. This idea was an old one, with a history back to at least Aristotle's time. Aristotle, in contrast to

Plato, had a real affinity for biology, unusual in a world dominated by both an emphasis on the physical sciences and a belief in outside sources for biological development.[22] Aristotle was an empiricist and devoted a lot of time to observations of the life histories of many organisms. Most important, however, he was very interested in diversity and is considered the founder of the comparative method. He tried to explain the phenomena he observed by asking "why" questions as opposed to "how" questions.[23] It really was not until very recently, when the biological sciences "emancipated" themselves from the physical sciences, that Aristotle became appreciated.[24] Like most natural philosophers, Aristotle was interested in natural laws. In a natural philosopher tradition (of which Lamarck was one of the last representatives), he extrapolated from the biological to the physical worlds, applying an assumption of functionalism (or adaptationism) to the universe. He tried to explain physics and cosmology in terms of biology and was then derided by generations of physical scientists. Aristotle postulated the existence of *eidos*, or "forms," defined by internal natural tendencies or forces that could only be understood through examination of the end result. In a sense these are analogous to the genetic program, and it has been argued that if Aristotle's views are reinterpreted with that analogy in mind, his teleological statements (as applied to organisms) are in accord with evolutionary theory.[25] Aristotle's vitalism described all biological structures or activities as the end results of processes that gave them what we would now call adaptive significance.

This "power of life," as Lamarck called his interpretation of vitalism, could be, and was, modified by the requirements of the environment. It is here where the inheritance of acquired characteristics, the idea for which he is best known, played a role in his theory, for this is the mechanism he thought caused specific modifications. But ironically "use inheritance" was not *his* idea or discovery. Lamarck took the inheritance of acquired characteristics for granted, as it was widely accepted at that time.[26] Lamarck applied his theory to human origins. For him, both physical and mental characteristics had an organic basis. Thus, mental or behavioral evolution was possible and subject to the same vital forces driving increasing complexity. Suppose, he speculated, there was an arboreal, quadrupedal prehuman race. If it left its arboreal habitat and began to use only its feet for locomotion, Lamarck argued it would eventually learn to stand and walk upright. Mental changes would follow, he suggested, for instance, as the need to communicate resulted in speech. This was, perhaps, his greatest break with his contemporaries, as Lamarck was relegating to natural process the difference between humans and the rest of the world that Descartes had identified as most fundamental. There could be no inferior human races in Lamarck's view:

> Hindering the great multiplication of races closely related to it, and keeping them relegated to woods or other uninhabited places, this race will have stopped the progress of the perfecting of the faculties of the other races, while itself, master to expand itself everywhere, to multiply itself unhindered by others, and to live there in numerous tribes, will have successfully created new needs that will have excited its industry and gradually perfected its means and faculties . . . finally, this preëminent race having acquired an absolute supremacy over all the others, it will succeed in putting between it and the most perfect animals a distance, and, in a way, a considerable distance.[27]

But ultimately Lamarck was not an important influence. Ironically, what took hold and lasted, to finally be associated with and attributed to Lamarck until the present, was the inheritance of acquired characteristics, the idea of use inheritance that he shared with many of his contemporaries. The implications of his theorizing for human origins and the relations of the races were lost. In part this was due to the end of the Enlightenment that framed his thinking. In part it was due to a younger contemporary, and rival, Georges Cuvier (1769–1832).

Cuvier, at the National Museum of Natural History in Paris, was one of the foremost anatomists of all time. Not one for schemes of universal explanation, he was a careful and meticulous observer who believed that species were fixed in form for all time. He realized that the past differed from the present; the fossil record being discovered below the streets of Paris showed that the differences were quite remarkable. But *why* did the past differ? Cuvier found his answer in the concept of extinction. Although the existence of extinct organisms had been accepted by several naturalists (for instance, Blumenbach), Cuvier's work on the extinct fauna in the Tertiary of the Paris Basin made the idea of extinction incontrovertible. For the first time, abundant remains of megafauna (mammoths and mastodons, for example) were described that clearly had no living survivors. In a sense, Cuvier supplied more evidence for evolution than any other person thus far, because he could explain how change took place. Accepting the notion that all forms of life were present at the act of creation, he had evidence that the cause of change was extinction. In each era different species became extinct, and new species migrated in from other regions to replace them, changing the composition of faunas. Cuvier posited a series of past catastrophes to account for the extinctions, the last of which was the great flood of biblical times. What might appear as a series of transformations was actually a series of replacements. Cuvier did not live long enough to see that while winning the battles with transformationists such as Lamarck, he had lost the war with evolution.[28]

Before Cuvier, most natural historians were primarily physicians, whose interest in other animals was motivated by their interest in human physiology. Their work was necessarily quite anthropocentric. Cuvier's primary interest differed because it was in morphological structures, which transcended species boundaries. He was one of the first prominent anatomists to focus on the function of physical structures; he believed their forms could only be understood through a study of their functions. For him comparative anatomy was everything, not just a way of gathering knowledge, but a way of organizing it. Function, rather than relationship, was the basis of his organization.

Cuvier's opposition to any form of evolution stemmed from his commitment to essentialism and the concept of unity of type. He recognized two kinds of variation. The first kind separated types. It was immutable and represented major, structural differences between animals. This kind of variation distinguished different products of creation and could not be ranked in linear order. Cuvier also recognized a more superficial, form of variation that occurred within species and was very plastic, superficial and changeable. The question is, what kind of variation do the human races fall into? He believed humans were all members of the same species. He was primarily interested in human anatomy, believing intellectual differences were reflected in racial anatomy, and whites were biologically superior to other races. His views on these issues changed, however, even as his society and the world views it reflected were changing. In 1790 he expressed views that the "intellectual" differences between groups were environmental, but by 1817, he had come to believe differences between human groups were racial in origin.[29] This change can be related to his whole view of environmental influence on variation and the kinds of variation he considered racial. His major disagreement with transformationists such as Lamarck was that evolutionary change could not transcend the unity of type. There might be a lot of superficial variation among organisms, but major, structural differences were more conservative and not subject to alteration.

Actually, Cuvier largely avoided work on humans and is in many ways self-contradictory on the issue of race. While he believed in the intellectual and physical superiority of Europeans and came to consider cultural differences to be a product of race, he was not a polygenist. He did believe that there were immutable cranial differences between races that corresponded to differences in intellect and behavior.[30] Therefore, even if racial differences were typologically superficial because they existed within a species, they were also large and intrinsic. Cuvier ultimately resolved this conflict by postulating a long separation of the races. While he believed that all humans descended from the same creation, he thought they had been separated for almost the entire time since then. So if not actually a polygenist himself, polygenism as it was understood in the 19th century could never have flourished without the natural history perspective he contributed.

Cuvier, it may be supposed, never bothered rebutting Lamarck's theory because he didn't consider it worthy of his time or effort; Lamarck's views were "a source of entertainment"[31] for him.

FROM PHILOSOPHY TO SCIENCE

European expansion was critical to the rise of polygenism because it made Europeans aware of human diversity in the most intimate ways and presented them with the necessity of explaining "the other" philosophically. But it did more, exposing Europeans to biological diversity in all species, and so forever changed the nature of biological science by creating the conditions in which an understanding of evolution could emerge.

However, long before the advent of Darwinian theory, the new practitioners of biology were attempting to understand variation of life on earth and the place of civilized and savage humans within it. This quest was necessarily essentialist in nature, as was all pre-Darwinian biology, which resulted in a major emphasis on classification. Thus, the purpose of the scientific study of the newly discovered life-forms in the 17th and 18th centuries, including humans, was to determine and name the types they represented and determine their usefulness. The natural philosophers (later the scientists, as distinct disciplines emerged) who systematically collected this information may have been satiating their curiosity, but others were interested in what merchants could buy and sell at profit and what farmers could grow.[32] Europeans' systematic surveys were not just abstract quests; they sought useful plants and animals, and their efforts had unexpected scientific consequences for biology and the emerging new discipline of anthropology. Collecting this information necessarily brought focus on differences rather than similarities. When native populations were classified in the pre-Darwinian world, it was a small step to consider different types as products of separate divine creations. This platonic cast to the study of human biological diversity was linked with attempts to understand the status of "the other" in the new discipline of anthropology.

Polygenism, to some extent, was made acceptable as a natural consequence of applying the typological principles of the fledgling field of biology to human beings. It was developing scientific legitimacy as it became the initial major theory of the new anthropological discipline, a theory immediately embroiled in controversy.[33] At the same time, the discovery of such undreamed of, seemingly infinite, diversity fostered increased interest in natural history, which resulted in dramatic changes in European biological perspectives. The culmination of this was Darwinism in the second half of the 19th century and the true emergence of biology as a science, separate from the physical sciences, but equally valid, with necessarily different methods and paradigms.[34] The merging of anthropology and evolution, though, was not at all immediate, and marred by many false starts.

Evolution never came easy to anthropology. Even before Darwin it was not Lamarckian evolution but the essentialist natural history tradition that became a major part of early 19th-century anthropology, supplanting more humanistic approaches of the Enlightenment to understanding cultural differences. Although it was heretical, polygenism found a home in the new discipline, and there were major supporters in America and all countries of Europe. This was also a time when intellectual traditions were disengaging, separating further from natural philosophy, with positivist-inductivist traditions in the spirit of Francis Bacon emerging in the sciences. Polygenism had no real power as a theory until it moved from the realm of philosophy to anthropology. As a social theory it was largely disliked, and until it was adopted by the natural historians of this discipline, polygenism remained a social theory. Philosophers like Lord Kames could ruminate on the differences between races, anatomists who had never seen a nonwhite cadaver could speculate as they pleased, but major racial differences had never been "scientifically shown." In fact, most anatomists had come down on the side of monogenism. But in the hands of the new anthropologists, polygenism became something science could "prove." Certainly not all anthropologists steeped in the natural history tradition became vocal promoters of polygenism; many did not. But the polygenists included some of the most preeminent scientists in France and America. While polygenism was adopted by naturalists from other countries (England and Germany in particular), it did not achieve the prominence it enjoyed in 19th-century French and American anthropological communities.

4

SLAVERY AND ITS REVERBERATIONS

MULTIPLE ORIGINS THEORIES about human races are disturbing for a number of reasons, perhaps the most pressing of which for Americans is the association of polygenism with the justification of slavery. This might seem like an issue of ancient history, but the consequences of the past century echo today and remain powerful influences on the shape of social and political life in our country. Public science is not immune from these influences, and indeed cannot afford to ignore them.

In the 19th century, before Darwin and before the acceptance of biological evolution, polygenism was particularly popular in America because of its implications for the status of African slaves. The theory "proved" whites were qualitatively different from the so-called colored races and therefore was seen as potential scientific justification for slavery. As we saw in the last chapter, racial thinking, the attribution of human differences to something deep-seated and unchanging, was part of the link between the birth of anthropology, the rise of polygenism, and the change in European attitudes toward "the other." This change in attitudes was related to the slave trade, where interactions by their very nature were dehumanizing and polygenism represented the ultimate dehumanizing scientific theory, since it in effect denied humanity to nonwhites.[1] This chapter explores the relationship between polygenism and slavery: while the acceptance and uses of polygenism were clearly political, the relationship between polygenists and the politics of slavery is less straightforward than it first appears, and bears closer examination because of its reverberations today.

Although several loud voices supported the idea of multiple origins earlier in the century,[2] polygenism did not enter the American scientific mainstream until 1839, when the Philadelphia physician and natural scientist Samuel George Morton (1799–1851) published *Crania Americana*. Although he initially did not state his support for polygenism in so many words, the results of his research backed no other position.[3] Morton was one of the new breed of positivist scientists who believed that objectively derived data and analysis would lead the intelligent scientist to the truth. Morton's new, scientific anthropology focused on craniometry, the measuring of heads, to understand the differences between races. Although it was never demonstrated, and is now widely believed incorrect, it was assumed in those days that larger heads always meant larger brains, and larger brains meant greater intelligence. This link was then tied to civilization: anthropologists contended that their research showed nonwhite races had smaller brains, were less intelligent, and therefore incapable of developing advanced civilization.

These beliefs fit nicely with the 19th century's version of a longer, historic tendency to see "savage" cultures as stages in the evolution of civilization that naturally culminated with the Europeans. This emerged from traditional, well-accepted views of life on earth that were dominated by the concept of the "Great Chain of Being." The "Great Chain of Being" was a natural (meaning at that time a Newtonian) law, holding that all life on earth was ranked on a scale of superiority from the lowest to the highest, the highest being most godly. Some early challenges to the fixity of species were based on the great chain concept, accepting the ranking but going beyond the great chain with the presumption that the rungs on the ladder to perfection could be ascended. Such a position was developed by Robert Chambers (1802–1871) as part of his idea of a progressive unfolding of a divine plan he put forth in his immensely popular and influential book, *Vestiges of the Natural History of Creation*.

But in the context of the times Chambers' suggestions that species could be changed were quite controversial. Lamarck's earlier ideas about the transformation of species had been dismissed, along with other explanatory creations of the Enlightenment. The idea that species could change by ascending a kind of ladder of life from simple to more complex forms was unlike Lamarck's phylogenetic tree in any event. A more direct influence on the times was Cuvier's dismissal of the great chain, let alone the ability to ascend it, which directly contradicted biblical doctrine and essentialist ideas about the fixity of species. Therefore, because his ideas about the mutability of species were extremely controversial, Chambers had to publish anonymously in 1844.

The ranking of species from lowest to highest forms was intrinsic to the European world and moreover was regularly being applied *within* the human species. A small group of other pre-Darwinian progressionists, following Chambers, went even further. In accord with a more "evolutionary" view, race differences were sometimes explained by a series of directed progressive developments in which the most primitive race gave rise to a more advanced one, and that to an even more advanced form, and so on. The different races, which some considered to be different species, represented stages of the white man's progress.

Most ethnologists of the latter part of the 19th century bought into this form of evolutionism. Progressionists, who believed that modern primitive cultures represented stages of European culture's evolutionary past, reasoned there was some sort of biological reason for the "retardation" of "primitives." In spite of this, some prominent ethnologists increasingly recognized the complexity of society and its elements in the natives they were studying. The American ethnologist Lewis Henry Morgan (1818–1881), for instance, argued that civilizing influences were a direct cause of biological change and that intelligence *and* brains were plastic enough to respond to these influences.[4] He thought the cultural environment could affect biological change, and therefore it could be inferred that savages could improve their biological capacities through contact with whites. Most anthropologists, however, believed the brain and the intelligence that resided in it to be immutable and a cause rather than an effect of the difference between savage and civilized races.

Whether one saw races as evolutionary steps on the ladder toward perfection or as unrelated products of creation, the focus was increasingly on the brain (or more directly on the head) as a reflection of the intrinsic racial differences 19th-century science believed existed. Scientists questioned how different behavioral and mental abilities could be measured and studied in a systematic way. One answer was provided by anthropologists, who measured the volume of skulls and other features of cranial anatomy, from which they inferred a variety of behavioral capacities, from intelligence to integrity. These were the days, after all, when it was widely believed that strength of character could be read from the prominence of the chin. Cuvier had claimed earlier that Europeans had larger heads than other races, in accord with their greater intelligence.[5] But his sample sizes were very small. Samuel

Morton wanted his discipline to be more objectively scientific. He strove to use large samples and prove beyond question that racial differences existed in skull size. Craniometry was scientific; it made no aesthetic judgments, and differences in osteological features were considered intrinsic—not the result of climate. Differences in the bones were much more than skin deep. These differences could separate species. Morton had friends around the world who collected heads for him. Many of them were physicians in colonial positions in a variety of countries who had access to bodies. Skulls were sent to Philadelphia, often with known information about the individual: sex, age, occupation, cause of death, and that all-important attribute, race. Some were the skulls of 19th-century people, victims of executions, poverty, or natural causes; others were from ancient burials. A huge collection resulted, numbering well over a thousand crania at the Philadelphia Academy of Natural Sciences, now housed at the University of Pennsylvania. Morton was only interested in the relative sizes of these crania; they constituted an entire collection whose sole purpose was dedicated to proving inferiority.

PHILADELPHIA CONNECTIONS

As an undergraduate student at the University of Pennsylvania in the late 1970s, I knew nothing about Morton and even less about the sordid history of physical anthropology his collection represented. I was attracted to physical anthropology before I really knew what the field encompassed or how to ask meaningful questions within it. When I first met the Morton collection, I had not yet learned to be a "populational thinker" and I lacked an intuitive understanding of the importance of variation. I knew variation was important and evolution could not proceed without it, but I had not yet attained an appreciation of how variation was manifested in bones. I did not have an intrinsic "feeling" for what it means skeletally to be human. I began to learn all this from the Morton collection, and it may have been the most significant element of my education, because different experiences with, and views of, variation are the most important influences on how one treats the evidence of human evolution.

I was able to learn and practice by cleaning, reconstructing, and cataloging skeletons from archaeological excavations run by the University Museum, but this was not a scientific analysis. I wanted a project and my adviser, Alan Mann, a well-known paleoanthropologist, suggested doing something with "nonmetrics" on the Morton collection. It was an excellent idea. The focus on nonmetrics is the perfect way to "see" skulls. Nonmetric data are just that; they are systematic observations you don't measure, but describe in some sort of discrete way by putting them into categories. Sometimes osteological nonmetrics are really discrete—they include traits like the presence or absence of different holes for vessels or nerves or the presence of extra bones. But often nonmetric studies involve continuous features like brow ridge shapes or degrees of expression of a feature that need to be somehow fit into a set of discrete categories like "small" or "large." You are placed in the position of trying to pigeonhole variation of a continuous feature into set divisions, and this can be fraught with difficulty. What do you do with the examples that aren't really large or small? Make an intermediate category? How many intermediate categories do you make? It makes you understand the problems of treating variation typologically. In studying cranial nonmetrics you look at variation all over the skull, and from attempting to describe it gain an appreciation of the pattern of morphology and its diversity.

Alan Mann, now our good friend, is an excellent teacher, and he used the Morton collection in the only way it makes sense, as a vehicle for understanding variation. It is not a population, there are no related individuals, it is not a community that shared social practices, it is not a collection that makes biological sense—but what it can do is provide insight into the vast range of variation of the human species. He knew that after studying that collection in a focused way it would be impossible for me to view humans typologically again; after systematically looking at the skeletal remains of close to one thousand people, it is impossible not to gain an appreciation for human unity and diversity. My approach to human evolution, which involves placing a temporal dimension on that enormous variation, owes everything to this experience. How ironic that while Morton used this collection to try to construct typological categories of human variation, in keeping with the idea of multiple creations, the value finally attained by the collection lies in its ability to demonstrate how *continuous* and broad human variation can be.

But Morton was up to more than typology. It was not enough for him to demonstrate differences between races: he wanted to rank them. This was the door that Blumenbach opened; whether the races could be arranged from an original form to a degraded one, or from the ungodly to the godly along a Great Chain of Being, they could be arranged in *an order*. Morton used his collection to support the prevalent idea of a linear scale of increasing racial superiority: Africans were lowest, Aboriginal Americans and Asians were an intermediate level, and Europeans were of course on the top. Morton's cranial data reflected this by showing that nonwhite races had smaller heads than whites, and of these, Africans had smaller heads than Indigenous Aboriginal Americans.

His premises and observations were badly flawed.[6] Far from the objective science he hoped to bring to craniology, it is clear that Morton's preconceived ideas biased his data collection and analysis. One of his offenses was using different materials to determine the volume of the inside of skulls, a measurement that approximates the size of the brain. Cranial capacity can be determined by filling a skull with some kind of pourable nonliquid material and then measuring the contents in a volumetric flask. Common materials for filling the skull included seed and lead shot. Samuel Morton used both of these, although he knew that shot is preferable since there is less room for observer error. Seed is compressible; you can press a lot of it in, or you can pack it quite loosely. The difference in measured capacity between a tightly packed skull and a loosely packed one of the same size can be significant, and the potential is there for a researcher's bias to take over.

This represents a major problem in inductive research programs, one that is not just limited to Samuel Morton's work or to antiquated science. In fact, with the modern data explosion and packaged methods of analysis afforded by our rapidly growing technologies, we are seeing the emergence of a whole new realm of inductive science. Morton's case is only one example of why it is important to recognize that science is not objective, and scientists are not without their own prejudices and preconceptions; they are quite human. Scientists accepting the Popperian protocol of refutation take advantage of this by packaging their prejudices and predilections in the hypotheses they try to disprove. But inductivists can fall victim to it no matter how hard they try to attain objectivity. It is often very true that we don't see things as they are, but as *we* are. In Morton's case we will never know exactly what happened, but somehow his conclusions came to fit his expectations better than the data allowed and it seems possible that his use of the compressible seeds played a role in this.[7]

But there was more to this than the use of seed for measuring capacities, and it seems certain that in some cases Morton fudged the conclusions he drew from his measurements. Because he published his raw data, it is possible to show that Morton's summary tables do not accurately summarize the data he collected. Gould noted:

> During the summer of 1977, I spent several weeks reanalyzing Morton's data. In short, and to put it bluntly, Morton's summaries are a patchwork of fudging and finagling in the clear interest of controlling *a priori* convictions. Yet—and this is the most intriguing part of the case—I find no evidence of conscious fraud; indeed, had Morton been a conscious fudger, he would not have published his data so openly.[8]

Morton's finagling provides an example of how easy it is for scientists to fool themselves.[9] Human races, however significant or insignificant the *average* differences between them may be, are astonishingly variable. For most features the variation within a race is far greater than the differences between them. Morton's data actually indicated this quite accurately. For instance, within his Caucasoid racial group the *average* cranial capacities of different populations ranged from 75 cubic inches for Hindus to 96 cubic inches for English.[10] These differences between populations in a single race (Caucasoid, in this case) are much larger than the differences between the different races (for instance, between Caucasoids and Negroids). But this would be difficult to see in the data he summarized, where the differences he reported between races are *much* bigger than the *actual* racial averages one could calculate from his data. Why the discrepancy? The answer is that he picked and chose as he constructed the samples to summarize the study, and he ignored or even took advantage of the different sample sizes. For instance, he eliminated all but 3 Hindus from his Caucasoid sample (he had 17) before he calculated the Caucasoid average, thereby artificially elevating its magnitude because the crania he omitted had smaller capacities. At the same time he retained in his sample all of the Aboriginal Indigenous Americans in his sample who had the smallest cranial capacity (Peruvian Incas). Because their sample size was much larger than most of the other native American groups with larger capacities in the collection, this artificially depressed the Aboriginal Indigenous American average. The result was that Aboriginal Americans (manipulated average too low) and Caucasoids (manipulated average too high) appeared much more diverse in Morton's summaries than they actually are.

Ironically, the strength of Morton's claims was in their perceived objectivity. His conclusions were considered objective and scientific because they were based on actual measurements and the statistics calculated from them. If Morton were alive today, no doubt he would be doing multivariate analyses of the crania on a high-speed computer, and his results would probably be positively received in the same way and for the same reasons: the allure of numbers and the widespread perception that objectivity can be found in statistics.

The major support Morton brought to polygenism was not simply his "demonstration" of inequality, which actually could be accommodated under many origin theories of the time. He made another point. His work on ancient Egyptian skulls was important in the context of the broad acceptance of Archbishop Usher's date for creation: 4004 B.C. This put the ancient Egyptians close to the time of creation. In studying them, Morton established an antiquity for modern racial variation that showed there was not enough time for the races to have diverged from the common ancestor that monogenism requires, or for them to have progressed toward perfection in any significant way.[11] In his 1844 publication, *Crania Egyptica*, he provided "evidence" against both monogenism and progressionism, the idea that history revealed a march toward progress. In this important publication he identified Negroes and Caucasians of "modern type" both living in ancient Egypt. This was especially significant because Morton's colleague and fellow polygenist, G.R. Gliddon (1809–1857), once the United States vice consul in Cairo, provided him with evidence that the Egyptian remains were much older than three thousand years. Morton's analysis indicated racial divergence already existed in modern form only a short time after creation. Therefore, to polygenists of the day Morton's cranial analysis showed that races did not progress but were static and unchanging for most of their 6,000-year history.

The polygenism/monogenism debates, while international, became centered in America, and their basis was established long before Morton reached his explicit conclusions.[12] Although embraced by students of human diversity from all over the world, polygenism lay at the foundation of the "American school" of anthropology. It found acceptance in America by addressing a major moral and political dilemma—the reconciliation of the treatment of people of African descent, both within the institution of slavery and without, with the tenets of the Declaration of Independence. Polygenism could justify both the unequal treatment of blacks and segregation. Politicians who preyed on fears of miscegenation often used polygenist arguments to justify the separation of the human races.

It is no accident, then, that polygenism took off in pre-Civil War America. America was the first modern country that brought together members of different races in sufficient numbers to affect its demography. Vast numbers of Africans were brought in as slaves, and the indigenous Native Americans were still represented in significant number. Unlike most Europeans, white Americans were *in contact* with members of different racial groups, and miscegenation was a real issue in the minds of many. Moreover, polygenism provided a tool for solving the "Jeffersonian dilemma." America was unique in having a Declaration of Independence that proclaimed that all men were created equal. This was more than a legal or social equality.[13] By tying equality to the creation, Thomas Jefferson had invoked a biologically endowed, God-given equality. Clearly, slavery was in direct opposition to such an equality doctrine *if* Africans were products of the same creation—*if* they were, in fact, "men" in the meaning of the Declaration. Jefferson himself was unsure of the equality of races, although he tended to favor it. In the end, he abdicated responsibility for this judgment to science, assuming with the optimism of the times that the issue would be settled in that arena. Jefferson had turned to a discipline that, in the second half of the 18th century, was strongly influenced, as was he, by the humanitarian and egalitarian principles of the Enlightenment. Although very few anatomists studied race—there were few cadavers and skeletal materials for comparison—most who did emphasized the unity of humankind. It is unlikely that Jefferson would have considered nineteenth-century polygenism as the scientific resolution to the equality problem.

Thus the concept of the humanness of different races became more than a moral or philosophical debate in America; it became a matter of science, and the face of science was changing from that which Jefferson knew. If science showed human unity, the more popular moral position, then the groundwork would be set for abolition. But, the proposition of separate creations raised the possibility of science providing a different solution to the Jeffersonian moral dilemma—one that could uphold slavery as an institution. And while most anatomists of the late 18th century were strong unity advocates, by the mid-19th century many prominent American anatomists and natural scientists had come to endorse polygeny.

This brings us back to Morton. Whatever one thinks of Morton and his motives, he played an important role in the development of polygenism. While there is a modern tendency to portray the American school and polygenism as pseudoscientific, prior to 1859 and the publication of Darwin's *Origin of Species*, the theory was well accepted and in many, perhaps most, scientific circles became *the* prevalent theory of human origins. In America, the most influential natural scientists were polygenists. The most prominent of these was Louis Agassiz (1807–1873), the eminent Swiss-born Harvard zoologist.

Before coming to America in 1846 to study fossil fishes, Agassiz had espoused monogenism, but weakly, relying solely on religious grounds to support it. A "spiritual disciple"[14] of the monogenist Cuvier, he envisaged the history of the world in terms of a succession of catastrophes. In fact, monogenism was antithetical to his ideas on the diversity of the nonhuman part of the natural world. Originally when Agassiz discussed the place of humans in nature, he claimed that they were a departure from the pattern of natural diversity. The human exception, he felt, provided testimony of the hand of God.

In the last set of lectures he gave in Switzerland, Agassiz spoke on his ideas of multiple creation for the natural world but not for humans, whom he then regarded as a single species deriving from a single common origin. His studies of natural diversity indicated that the flora and fauna of different regions were created in the environments where they occur today, environments where they were meant to flourish, and that species were limited to these geographic ranges. He also thought the geographic limits of particular groups of animal species corresponded to the geographic ranges of human races, but noted that because similar animals in different environments were generally different species whereas humans were not, the lack of human speciation could be taken as evidence of their superiority:

> And here is revealed anew the superiority of the human genre and its greater independence in nature. Whereas the animals are distinct species in the different zoological provinces to which they appertain, man, despite the diversity of his races, constitutes one and the same species all over the globe.[15]

Agassiz did not directly address human diversity, and while he asserted a single origin for humans, he was quite aware of, and in fact provided detailed scientific accounts of, the flora and fauna of different regions that supported his idea of multiple origins. With humans the exception to the rule, Agassiz's belief in human unity was not stalwart.

Arriving in America at a time when many things in his life were changing, Agassiz changed his mind about human races and very quickly adopted polygenist views. On an intellectual level this is not surprising, since this brought his view of humanity more closely in accord with his views on the rest of nature, which were very "antimigrationist." Additionally, since he discounted literal interpretations of the biblical story of creation as applied to other animals, he laid the moral foundation for his own acceptance of what was to many a heretical doctrine. This was an important requisite in the context of his own personal beliefs, as Agassiz was a religious man, the son of a Neuchatel pastor.

Agassiz's convictions contradicted more than his church. Later in his life he adamantly opposed Darwinism, and remained in opposition until his death, characterizing it as "a scientific mistake, untrue in its facts, unscientific in its method, and mischievous in its tendencies."[16] One of the reasons he opposed Darwinism so strongly, we believe, was Darwin's emphasis on the biogeographic evidence for common descent. Agassiz himself was convinced biogeographic diversity showed quite the opposite, the *lack* of connections between different organisms—it was the foundation for the multiple creations he perceived. His earlier monogenist views on the origin of human races, held long before the publication of the *Origin of Species*, were compatible with Darwin's notion of common descent, but he considered this an exception to the rest of nature and in any event moved away from this stand even as Darwin was finishing his book. Because Agassiz considered the biogeographic variation of the nonhuman world discrete and the consequence of different creations, he was convinced historical processes and migration played a minimal role in variation and used the patterns of biogeographic diversity to support this. Of course, he also opposed Darwinism because it challenged his entire creationist, essentialist world view—but in a sense it added insult to injury since the strongest evidence for common descent was the same variation he interpreted entirely differently. As an egotistical man described as "something of a showman,"[17] he may have had as much difficulty admitting his interpretations were wrong as modifying his world view to incorporate common descent.

For precisely this reason it was easy for Agassiz to mix polygenism with his religious views: as a scientist faced with evidence against the literal interpretation of Genesis, he already had to reconcile his religious and scientific views. Agassiz was an expert in fossil fish and faced incontrovertible evidence of extinctions and the emergence of new species. He knew the world was not static and unchanging, yet he believed in the fixity of species and that change was a direct result of the hand of God. To explain this, he had embraced a catastrophist approach, believing in multiple creations[18] and extinctions. In fact, Agassiz was just about the last catastrophist holdout in the United States. At the same time he was a progressionist because he felt there was progress through time as the catastrophes that wiped out entire faunas allowed replacement by more advanced organisms. This sort of serial creationism over time was not at all incompatible with the idea of multiple creations separated by geography instead of time, and Agassiz promoted both.

Agassiz's conversion to polygenism has been compellingly portrayed as racism overcoming religious views. He described his repugnance at initial encounters with the black waiting staff in his Philadelphia hotel at length in correspondence to his mother, in which he claims that, while human, they are of a different species from the white man. All things considered, his vehemence is surprising.

> It was in Philadelphia that I first found myself in prolonged contact with Negroes; all the domestics in my hotel were men of color. I can scarcely express to you the painful impression that I received, especially since the feeling that they inspired in me is contrary to all our ideas about the confraternity of the human type (genre) and the unique origin of our species. But truth before all. Nevertheless, I experienced pity at the sight of this degraded and degenerate race, and their lot inspired compassion in me in thinking that they are really men. Nonetheless, it is impossible for me to repress the feeling that they are not of the same blood as us. In seeing their black faces with their thick lips and grimacing teeth, the wool on their head, their bent knees, their elongated hands, I could not take my eyes off their face in order to tell them to stay far away. And when they advanced that hideous hand towards my plate in order to serve me, I wished I were able to depart in order to eat a piece of bread elsewhere, rather than dine with such service. What unhappiness for the white race—to have tied their existence so closely with that of Negroes in certain countries! God preserve us from such a contact![19]

It was also during this Philadelphia trip that he first met Samuel Morton, with whom he quickly became fast friends, maintaining a close relationship until Morton's death. No doubt Morton's support also influenced his conversion to polygenism, but a strong case can be made that the racism and abhorrence of black people apparent in his correspondence were major factors for Agassiz's adoption of the doctrine. Perhaps he needed some kind of emotional "evidence" that different races did represent species in order to overcome equally emotional religious objections to polygenism. But it bears pointing out that he didn't have very much to overcome. He no longer had any commitment to a single source of human creation since multiple creations were critical for his explanations of biogeographic animal diversity. Agassiz, a splitter, treated small amounts of difference as evidence for species distinctions and defined racial variation out of existence by changing it into species-level variation. Monogenist views of humans were, if anything, a thorn in the side of these scientific beliefs. Agassiz must have been very happy to shed his ideas of human monogenism for those reasons alone, and while there is absolutely no doubt he held very racist views, it is probably equally true that any excuse would serve. Yet, Agassiz opposed slavery.

POLYGENISM, SOCIETY, AND POLITICS

Agassiz was one of the American polygenists whose views were infused with a sociopolitical agenda, significant in the ongoing controversies about slavery and the future of interracial relationships.[20] There is little doubt that internal American conflicts over slavery spawned the warm reception polygenism found there. Indeed, no other country except France, and there for reasons stemming, not from slavery, but from the social and intellectual changes associated with the French Revolution, accepted polygenism so fervently. However, *justification* for slavery is only one of many reasons polygenism took hold in America; most of the others are related to slavery but far less directly, and the relationship between slavery and polygenism is not as simple and straightforward as it may first appear. Polygenism obviously served to dehumanize nonwhite races and therefore justify their exploitation, but this does not solely account for polygenism's popularity. Because of its heterodoxy, polygenism never found favor in the deeply religious slaveholding South, and conversely, many of its strongest supporters were not supporters of slavery. In fact, many vocal supporters of polygenism disliked slavery because they were afraid of miscegenation, something they viewed as an inevitable consequence of slavery, and used polygenism as scientific support for their opposition to racial interbreeding. Additionally, antireligious/proscientific movements adopted polygenism for reasons that had little to do with the status of the races in America, but instead because it challenged the doctrines of Christian orthodoxy.[21]

Although he was one of the most influential figures in the American school, it is not clear that Samuel Morton himself had an active political agenda. His Quaker upbringing was in strong opposition to slavery, but on purely scientific grounds, characterized by positivism and a splitting mentality, he thought racial differences were primordial. If these were uninfluenced by "climate," he considered them unchanging and therefore indicative of species.[22] However, many others, from Charles Cauldwell, a pompous and outspokenly racist phrenologist, to Josiah Nott (1804–1873), a Mobile physician and scientist who was Morton's close colleague, used Morton's work in overtly political ways.

Most American polygenists in the scientific community claimed to have no political agenda, often accusing the monogenists of being blinded by political or ethical concerns instead of being guided by scientific principles. Using the age-old argument that politics or social values should not thwart science, Agassiz wrote:

> We disclaim, however, all connection with any question involving political matters. It is simply with reference to the possibility of appreciating the differences existing between different men, and of eventually determining whether they have originated all over the world, and under what circumstances, that we have here tried to trace some facts respecting the human races.[23]

This assertion of objectivity was a smokescreen for the clear social agenda Agassiz held. In his 1850 *Treatise on Race*, the same document in which he disavows political interest, Agassiz specifically emphasizes the relationship between the scientific knowledge he is imparting and judicious social policy:

> We entertain not the slightest doubt that human affairs with reference to the colored races would be far more judiciously conducted if, in our intercourse with them, we were guided by a full consciousness of the real difference existing between us and them, and a desire to foster those dispositions that are eminently marked in them, rather than by treating them on terms of equality.[24]

Agassiz explicitly stated that while blacks should have "legal equality," they should be denied social equality. Clearly his definition of legal equality was pretty narrow, apparently applying only to eliminating slavery. It seems never to have occurred to him that blacks, whom he considered feebleminded and childlike, should be accorded such legal rights as the vote on a par with whites. He was such a segregationist that he probably could imagine these rights being manifested only within the black community or within black states. Indeed, he states, "Let us beware of granting too much to the Negro race in the beginning, lest it become necessary to recall violently some of the privileges which they may use to our detriment and their own injury."[25]

Agassiz opposed slavery for a number of reasons. First, he was a religious man who may well have considered it immoral, in spite of embracing the idea of separate creations. More important, he was a separatist who was actually opposed to slavery for racist reasons. He was repulsed by the idea of miscegenation, and considered it a precursor for the moral and social decline of the United States. He thought that the disgust he felt in contact with blacks was commonly held, on both sides of the racial barrier. Accepting the reproductive isolation definition of species, he assumed that mutual revulsion at social/sexual contact was what kept species separate in the natural world. Thus, he believed this natural distaste would keep the races separate, were it not for slavery.[26] He contended that intermixture under the conditions of slavery was prevalent because there was a loss of the natural repugnance for blacks held by young men of the South due to initial sexual contact with mulatto house servants. With convoluted reasoning he explained these contacts as the "white half" of mulattos attracting these young men, as did the women's easy availability. However, once consummated, the relationship with mulattos caused the southern men to become habituated to black contact and led the southern men to "gradually seek more spicy partners."[27] Sexist as well as racist, Agassiz considered the women to be willing partners, with the mulatto women being more receptive than white women since, as Gould describes Agassiz's view, their "black heritage loosens the natural inhibitions of a higher race."[28] Agassiz felt that an end to slavery would put an end to this unnatural interbreeding, which he considered the most horrible thing about interracial contact. Left to their own devices, after the unnatural institution of slavery was abolished, races would reestablish a natural geographic segregation, he believed, and blacks would stay in or migrate to the southern lowlands, as they were created for hot humid conditions, leaving the "bracing North" and the coastal regions for the white species. He expected that "the New South will contain some Negro states. We should bow before this necessity and admit them into the Union; we have, after all, already recognized both Haity and Liberia."[29] This natural segregation should be rigidly enforced, in his view, although given the different geographic inclinations of the races, by and large nature would be "the accomplice of moral virtue."[30]

Even if many of the polygenist scientists themselves didn't advocate slavery, polygenism was used by others to do so. In some instances, the advocates were close associates of the researchers. The best-known case is the alliance between Morton and several who popularized his views as part of the new anthropology, such as G. Gliddon and J. Nott (trained, in part, in Morton's home). While Morton did not directly address the sociopolitical ramifications of polygenism, his cronies certainly did. In the 1857 publication *Indigenous Races of the Earth*, Gliddon and Nott promoted a polygenic theory in which blacks are inferior to whites. Even before the publication of *Crania Egyptica*, Gliddon promoted polygenism in the South from the lecture circuit, publishing pamphlets about slavery in ancient Egypt and claiming Morton's research proved "the Negro races had ever been *servants and slaves*."[31] Nott was one of the founders of the American school of anthropology, and in 1843 was among the first American scientists to publicly espouse the position of separate creations for the human races. His opinion of the Negro race, a distinct and separate species in his scientific framework, was very low, and he declared slavery to be the only reasonable alternative for their existence. Abolition could not possibly succeed, as freed Negroes could not be expected to compete successfully with whites and couldn't survive in Africa. Race mixture was an "insulting and revolting" alternative that would lead to the white race being "*dragged down* by adulteration and their civilization destroyed."[32]

The sociopolitical aspects of the monogenism/polygenism debates involved many issues, but predominant among them were the conflicts between slavery and abolition and between science and religion. Monogenists, often the liberal abolitionists, were tarred as unscientific—affected by blind adherence to dogma—so polygenism picked up some supporters purely on the grounds that it was antireligious. Morton's publicizer Gliddon, for example, was most interested in polygenism because it flew in the face of the dogmatic religion. Most American monogenists who argued against polygenism came from religious, rather than academic, circles—for them the battle was clearly one of religion *versus* science. Since it was touted as "objective science," polygenism was challenged by those who felt that materialism in general was the downfall of society. Consequently, monogenist views were also held by those whose main interest was antimaterialism, who gave little thought to the issue of racial origins. The American polygenists used this as well, appealing to the materialism of many liberal thinkers to override their social concerns. The social and religious issues were inexorably intertwined, and it is not just politics that makes strange bedfellows.

POLYGENISM IN EUROPE

The polygenists in pre-Darwinian Victorian England, even more so than in America, were directly involved in the proslavery side of the abolition controversy and were most vocal in the 1860s. Post-Darwinian polygenism arrived back in England and wound up influencing major students of human evolution well into this century. But although polygenism later became a major part of international anthropology, as a pre-Darwinian "separate-creationist" school, it was never embraced in England or in Germany as it was in America and in France.

In Britain, as in Germany, the "science of man" was dominated by monogenists, with polygenists, though vocal, not strongly represented in the scientific establishment. The strongest polygenist voice was that of the Edinburgh anatomist Robert Knox (1791–1862), who was indeed a leading figure internationally in the polygenist "movement," in many ways anticipating Gobineau[33] in his assertions of the basic inequality of races. Knox's was a classic polygenist argument, emphasizing the innate differences between races and minimizing the effects of environment. But polygenism was not well accepted by most of the scientific community, and the polygenists actually became stronger in the post-Darwinian years (this is discussed in more detail in the next chapter). They formed the Anthropological Society of London, countering the more progressive Ethnological Society that included the Darwinians, among others, in its membership. The Anthropological Society was responsible for the publication of polygenist articles and papers and the English translations of several powerful polygenists from Germany and France. But the major figures in British natural history and anthropology were monogenists; scientists were too religious, too Lamarckian, or too committed to the unity of type to recognize species-level differences between groups of humans.

Richard Owen (1771–1858), the dominant Victorian figure in natural history and anatomy, was an essentialist much like Cuvier. Yet while recognizing large differences between races, he did not see them as conforming to different archetypes. In fact, as was argued in a recent analysis of the Huxley-Owen debate,[34] Owen could even be considered antiracist because, like Camper, he saw a clear division between all humans and apes. He did not view different races as stages in between Europeans and apes as Huxley did. But the situation is more complicated. Egalitarianism probably had nothing to do with Owen's scientific stance; the sharp divisions between apes and humans he envisaged more likely stem from his devotion to essentialism and the importance of archetypes. In fact, Owen recognized large differences between races, and his nontransformationist views, if politically motivated at all, probably resulted from reluctance to change the *status quo*,[35] rather than from any egalitarian feelings he had toward different races or social classes.

The British Lamarckians, whom Owen vehemently opposed, were the evolutionists he fought long before he fought the Darwinians. The British Lamarckians have been described as political radicals whose evolutionary philosophy went hand in hand with their social philosophy.[36] This is ironic because while Lamarck himself was quite political, he did not link his views of life to any philosophy of social reform.[37] The British Lamarckians opposed the rigidity of the traditional English class system that closed access to professions to members of nonelite social strata, and they believed that chances to succeed in society and entrance into professions should be based on merit, not class. Lamarck's idea of direct environmental causation of individual change appealed to the idea of individual betterment within society and the notion that in an equal environment all people could potentially succeed. The Lamarckians were much more progressive than later Darwinian evolutionists—the idea of individual advancement responding to the demands of the environment is much more compelling to progressive social causes than harsh models of competition and relative fitness—and it may be that Lamarkism was retained for so long (until the modern evolutionary synthesis of genetics and Darwinism in the 1940s) because of its progressive social implications. On the other hand, it is also true that the political implications of a malleable, modifiable person were important as Soviet communism developed (malleability was required if there was to be a "New Soviet Man"), and the power and influence of that infamous Russian peasant D.T. Lysenko spread. Lysenko's brand of Lamarckism and his hostility to scientists were to sweep away a generation of biology in Russia and the evolutionary synthesis it had achieved.[38]

In contrast with Britain, but as in the United States, in France very influential figures in the scientific community adopted polygenism. Even after Darwin's principle of common descent appeared to peal a death knell for polygenism, Paul Broca (1824–1880), a physician, anatomist, and craniometrist, was an ardent polygenist. Broca is probably most famous for his work on the brain—even the freshest student in the biological sciences has heard of Broca's area, one of the language centers of the cortex. Through his work with aphasics Broca is responsible for the discovery that functions were localized in specific areas of the brain. Broca, an excellent anatomist, influenced anthropology, biology, and medicine in countless ways. He is considered the founder of French anthropology,[39] starting the *Société d'Anthropologie de Paris* in 1859. But his influence extended well beyond France.[40] He had a major impact on all of physical anthropology and was responsible more than any other individual for establishing it as a science. He developed an anthropology laboratory, school, and society in France and influenced similar developments in other countries as well. Broca originated many of the techniques and instruments that are still in use today for taking measurements on human crania and other bones.

Broca was also a polygenist and a racist who was strongly influenced by Samuel Morton—his "hero and model" according to Gould[41]—and although he outclassed Morton as a scholar and an intellect, he had strong ties to the Philadelphia physician, whom he admired and with whom he shared data. Much of his craniometry work was dedicated to the establishment of the inferiority of nonwhite races through "objective" inductive methods that included showing that whites had larger and better organized brains (i.e., larger frontal lobes) than nonwhites. His most important contributions to polygenism, however, were his attempts to show that races did not successfully interbreed and were therefore different species. He researched hybridity in general and finally concluded the human races were not truly interfertile, but crosses of different races had varying success in producing fertile offspring, depending on their degree of relationship. More closely related races were interfertile; more distantly related ones were not. He contended the hybrids of the different races were inferior to pure races, and even that some crosses, those between very "disparate" races, were virtually inviable. Yet in spite of his focus on interfertility as a defining characteristic of species, Broca never adopted Darwinism. Like many others in the 19th and early 20th centuries he could not understand how selection could go far enough to produce species.

While other anthropologists in France, in particular Broca's monogenist contemporary Jean Louis Armand de Quatrefages (1810–1892) and his own protégé, Paul Topinard (1830–1911), disagreed with his polygenism, Broca's commitment to polygenism made it an influential part of the French anthropological tradition, and in this the French and American anthropological traditions resembled each other. Also like the American scientific community, the dichotomy between the "racist" polygenists and the "egalitarian" monogenists was far from universal, and often mischaracterized the Europeans just as it did the Americans. For example, Joseph Arthur Count de Gobineau (1816–1882), one of the world's most notorious racists, was a confirmed monogenist. He was a diplomat by profession who published books on politics, art, and history, wrote an epic poem but also four volumes on the inequality of the human races (1853–1855) that were taken very seriously by Nazi and other racist "scholars" earlier in this century—hence, the basis of Gobineau's infamy. Gobineau, hardly a liberal or an abolitionist, believed the races were quite different biologically in spite of their common origin, and that most of them were incapable of civilization. He was interested in explaining the downfall or decay of great historic civilizations, attributing this to changing racial constitutions resulting when "pure" races mixed or were engulfed by inferior peoples. The application to 20th-century politics, from "Aryan superiority" to "ethnic cleansing," couldn't be clearer.

A PARADOX

Much of polygenism's support came from its being considered a science, as opposed to the religious dogma associated with monogenism. Many scientists who supported monogenism were attacked for being unscientific, while polygenism was touted as science free from emotional social and religious constraints. If the results of scientific study affected the sensibilities of the "bleeding hearts," should those results be repressed?

Long after the extinction of slavery, these arguments reverberate through the 20th century. Most of the polygenists were racist, certainly in today's terms, but the interrelationships were complex. Many monogenists were also racist and may well have supported polygenism if they could have—there is little doubt a large number of monogenists were such solely on religious grounds. So if the polygenist/monogenist debates were a 19th-century pre-Darwinian version of the age-old science *versus* religion controversies that continue to the present day, the reflection of those debates in the issues of today is muddied by the intervening century, and by Darwin's profound influence on all aspects of biology.

It is ironic that in defending Genesis from the arguments of the polygenists, monogenists were forced to take positions that foreshadowed the tenets of Darwinism, the theory that, later, really shook up the biblical foundations of origins. Monogenists postulated human descent from a single source, which is fundamental to Darwinism. Furthermore, in order to explain how human variation emanated from that single source, many adaptive principles, precursors of natural selection, were invoked. This placed monogenists in the paradoxical position of having a theory more supportive of Darwinism, while having religious philosophies that rigorously opposed it. Darwinism, of course, provided firm scientific grounds for monogenism, but these were not always appreciated by the monogenists, as a 180-degree shift in dogma was required to accept them.

As the controversy burned, and in America was to boil over into the Civil War, Darwin's *Origin of Species* appeared. Although some parts of Darwin's theory were slow to gain acceptance, common descent was rapidly acknowledged by the scientific community. With the acceptance of Darwinism, polygenism lost some of its support simply because evolution was recognized rather than creationism. But while it may appear that the recognition of common descent for all species was incompatible with polygenous ideology, polygenism did not die. It was reworked and has appeared and reappeared as an influential theory in different guises since the mid-19th century. Instead of depending on independent creations, scientific separation of the races came to involve long-standing segregation by isolation and in its most recent manifestation in the 1960s the temporally staggered achievement of *Homo sapiens* status. This kind of polygenism has directly affected the development and acceptance of Multiregional evolution.

5

POLYGENISM AFTER DARWIN

OUR VISIT TO ZAGREB during the spring and summer of 1988 was not our first. It was my second and most important since I was doing my dissertation research that year; Milford had been coming regularly since 1976. Given the history of the curation of the Krapina remains, it now seems like a miracle that anybody was doing paleoanthropological research there at all. This is because when Gorjanović closed the books on Krapina with his retirement in 1923, his office and the Neandertal collections he worked on were virtually frozen in time. Both his office and the exhibit of Krapina fossil remains stored near it on the second floor of the museum, were kept just as he left them, actually not very easy to see.[1] Only a few scholars came to study the Krapina Neandertals over the years, in part due to the fragmentary nature of the specimens. All that remained of some 70 individuals were 279 teeth and over 400 bone fragments. The value of this collection lay in its disclosure of the *variation* of many skeletal elements literally from head to foot. However, one could *not* reconstruct a complete Neandertal skull or skeleton, and thus the collection was not amenable to the types of analysis generally preferred by paleoanthropologists. The same question resounds that characterized Milford's old disagreement with John Robinson: which is the better source of information, a large sample of a few elements, or a single but mostly complete specimen? Of course,

Gorjanović-Kramberger at about the time the Krapina Neandertals were published.[2] *His study of the fragmentary Krapina Neandertals led to a 1906 monograph, but was soon eclipsed by the fewer but more spectacularly complete Neandertal burials found in France.*

both sets of data are valuable for different types of problems, but Milford's interest in the patterns and causes of variation made the Krapina collection particularly attractive to him.

However, many paleoanthropologists have taken Robinson's view that completeness is the most relevant criterion for a successful study, and the importance of the Krapina collection was eclipsed by discoveries of complete Neandertal burials in France at Le Moustier and La Chapelle in 1908 and the La Ferrassie adults in 1909 and 1910. The most intact Krapina pieces, like the C skull and J mandible, were widely publicized when they first were found, but these paled before the completeness of the French specimens, and the faraway Croatian site played only a minor role in western European and even in central

European paleoanthropological thinking (Croatia, of course, was part of the Austro-Hungarian Empire). Inadvertently, the influence of Paul Broca contributed to the problem. Broca had defined how certain measurements should be taken on human bones. His goal in standardizing these was to allow observations from different scientists to be validly compared. If two researchers measure the length of a cranium, but define it differently, their results are incomparable and useless as far as the comparative method is concerned. The search for the standard set of measurements Broca demanded continued through the earlier part of this century. Different schools of measurement definitions and techniques arose, often along national lines; and while none became universal, they all shared the assumption that measurements were to be taken on complete specimens. It is easier to duplicate measurements on complete specimens, and much is to be gained by examining them. The measurements were developed mostly for studies of modern humans (i.e., races) not fossils. Therefore, virtually no standardized measurements existed that could be applied to the large sample of fragmentary Krapina specimens, and this made it difficult to compare the Krapina Neandertals with published material.

There was another reason for the historical eclipse of Krapina. Gorjanović was highly revered, almost deified, by his successors at the museum. While this was entirely justified—history has provided more reasons for deep respect for Gorjanović than even his successors knew—this reverence had a negative consequence. At the museum, the view came to prevail that Gorjanović had done everything possible with the collection.[3] If nothing could be gained by further study, the preservation of the Neandertals was the museum's predominant (if not sole) concern, and while scholars were never kept away from the collection, they were also not encouraged to come.

But by the late 1960s, paleoanthropology was changing in a direction that emphasized questions about variation, and a major rethinking of the place of Neandertals in human evolution was well under way. Even as the importance of the Zagreb collection grew, it became more inaccessible than it had ever been, especially for Americans, because of an unfortunate incident involving an American the decade before. Milford, an almost-Ph.D. in his first job at the just-merged Case Western Reserve University in Cleveland (Case Institute of Technology + Western Reserve University), was completely unaware of this.

Milford was hired in the fall of 1968, just a few months before finishing his doctorate degree. The next year a graduate student from the University of Tennessee came to work with him. Fred Smith, now one of the leading Neandertal experts, was a young man from Lenoir City, just outside of Knoxville, with what Milford thought of as a thick, charming Tennessee accent and an enthusiastic desire to study the Krapina Neandertals. Milford was familiar with this site—while working on his dissertation on dental evolution, he had read virtually everything Gorjanović had written on the Krapina teeth—but Fred had a much clearer vision of the promise the human remains from Krapina still held. Fred had been there on an undergraduate visit to Yugoslavia and saw the Krapina Neandertals. He was overwhelmed with what was there, and learned German to be able to read Gorjanović's publications in the German language. A scholar in the Austro-Hungarian empire, Gorjanović wrote his major publications in German, using the less accessible Croatian language to present his more speculative (and often more interesting) interpretations and explanations.

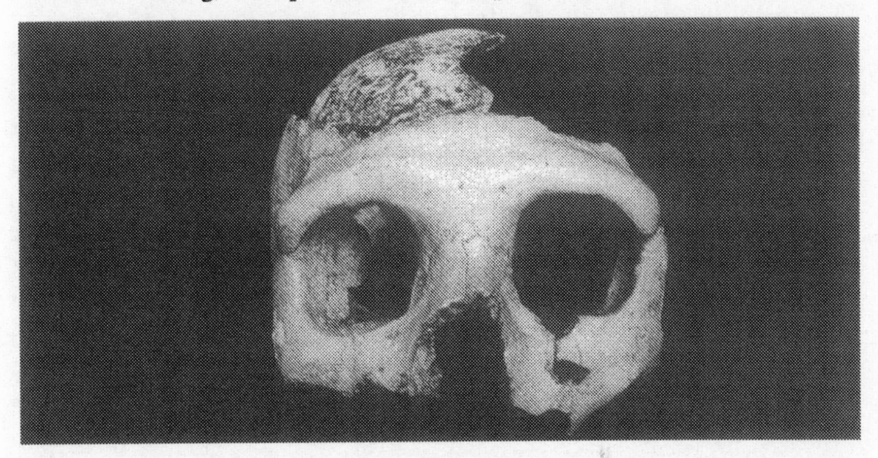

The Krapina C woman's face[4] is the most complete of those in the Krapina collection. Its "completeness" is the result of many attempts at reconstruction over the years—when the fragments were first discovered and described there was less to the cranium, and even today more of its pieces may still be found. Many researchers visiting Zagreb hope a fresh look will result in gluing a new piece of bone on this or other specimens. The C skull was given its letter designation in order of discovery, but identifications became confusing. For instance, the C skull is the 3rd to be found (following A and B), and the 3rd mandible to be discovered is the C mandible. But the similarity in letters is completely accidental and does not mean that the C mandible is the same individual as the C skull; in fact in this case it couldn't be, as the C mandible belonged to an individual not yet even a teenager, while the skull is that of an adult (her cranial bones had knit together internally, a sure sign that she was 25 years or older at death). There is also a C maxilla, and this may *be the same person as the C mandible. Confusing enough? It was to everyone who had worked there, and Milford encouraged the museum to renumber all of the specimens with a single set of consecutive numbers so that each would have its own unique ID. The C skull, thus, became Krapina 3. But renumbering in 1988 does not change papers and books written before.*

Although just beginning his graduate training, Fred was anxious to establish himself and pursue his interests, and was to leave for Zagreb the following spring when classes in Cleveland were over. However, only his native charm prevented him from becoming an unknowing victim of the Natural History Museum's reluctance to let foreigners, especially Americans, study the Krapina remains. After Fred left, Milford received a letter from the director, Ivan Crnolatac, asking Fred *not* to come. Milford decided not to contact Fred, who was still at home in Lenoir City but on the verge of departure for Zagreb. He thought Fred's earnest exuberance and southern politeness would afford him access to the collection. And indeed they did; after the unsuspecting young man arrived in Zagreb, he quickly made friends of the museum staff and never learned until returning home that his request to see the specimens had been initially denied.

Starting work, he soon found the collection was difficult to examine for the very advantage it had over material from other Neandertal sites—its variation. Specimens had been given museum numbers, but only relative to the box they were stored in, so that several different bones or teeth kept in different boxes might have the same number. The numbers of some had worn off over the years, and these would be unassociated if they were separated from the box they were kept in. Fred was restricted from studying the bones in more than one box at a time, and as a young student, there was never a possibility for him to attempt reconstruction of the fragmentary remains. Yet there was much he could do with this extraordinary collection that had remained virtually untouched for more than a half century. He accomplished a mission impossible by just being there. He opened the collection for future studies by convincing director Crnolatac (soon known by his nickname Crni) that there was much more to learn about the Neandertals he curated. Fred's dissertation was ultimately based on this and subsequent studies of the Krapina Neandertals.

When Milford first studied the collection in 1976, Crni was prepared to see more done with it. He was deeply impressed with the results of Fred's studies, his dissertation, and now realized much more could be accomplished. One part of Milford's work focused on an attempt to determine the age-at-death for as many of the Krapina individuals as possible. The teeth are the most useful single element for doing this. He applied a technique learned from his friend Alan Mann, who developed a means of estimating the age-at-death for skeletal remains that include juvenile and adult jaws with at least some teeth.[5] Mann's technique lets one estimate age from the amount of wear on teeth. Eruption times for the deciduous and permanent teeth can be estimated for these Neandertals. With the exception of the last molar (wisdom tooth) they are within the range of modern human populational averages. Age-at-death estimates for an individual who died at the time a tooth was erupting are pretty accurate. Someone who died when the permanent upper incisors just erupted was dead at the age of 8 years or so. But what about people who died between tooth eruptions, or after all the teeth had erupted? This is where Mann's technique is needed. The idea is that, for instance, if the first permanent molar tooth erupts at 6 years of age and the second molar erupts at 12, the amount of wear on a first molar *when the second molar is just erupting* represents 6 years of use. A second molar with that amount of wear, 6 years, represents an individual who died at 18 years of age (eruption at 12, plus 6 years wear). These are approximations and not real birthdays, but Mann showed they could be reasonably accurate (give or take a year) for large samples.

To estimate ages this way and determine the age distribution it was necessary to get all of the teeth on the table at once, and this required

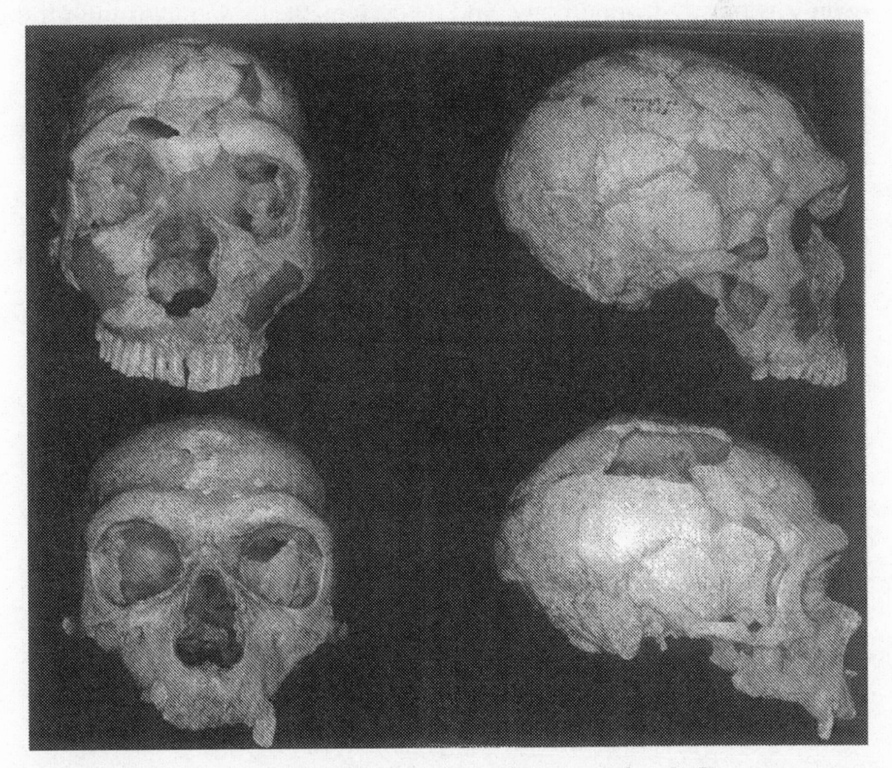

Complete crania of two French Neandertal men found just after the Krapina remains were published, La Ferrassie (above) *and La Chapelle. These specimens are burials, with complete associated postcranial skeletons.*

mixing up teeth from different boxes. They had to be renumbered, and this was a difficult request because it meant changing something Gorjanović had left and, more important, admitting he hadn't done everything that could be done. After Fred, Crni was *almost*, but not quite, ready for this.

It took a fortuitous accident to convince Crni to promote new reconstructions. Milford noticed many of the skull fragments appeared to be freshly broken. Their edges were sharp and clean without signs of erosion. Breaks like this are common in bones from European caves because of trampling by cave bears and the motion of the cave floor as parts of it freeze and thaw during a year's weather cycle. Milford reasoned that if Gorjanović had carefully collected all the human bones, some pieces might fit together. Working evenings, Milford sat trying systematically to match pieces while Crni, who did not want to go home either, looked on. One night Milford was working with the C skull, the most complete and famous of the Krapina partial crania, looking for fragments to fill in the top of the vault. The missing portion was mostly frontal bone, and he systematically went through the

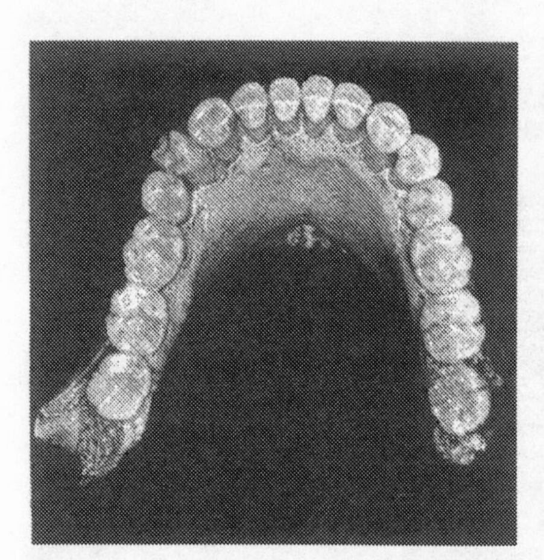

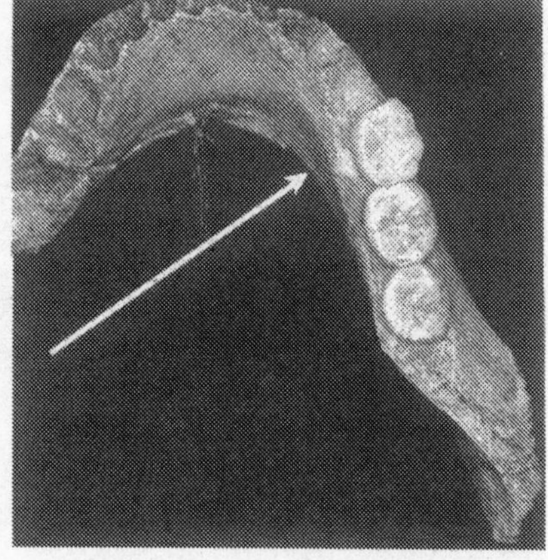

Krapina mandibles G (right) *and H. The wear on the last molar of the H mandible is about the same as on the first molar of the G mandible. The enamel is flattened from being worn away, and some dark spots can be seen, which is where the darkened, underlying dentin of the tooth is exposed. This should mean that at death the H mandible was some 9 or 10 years older (the last molar of H erupted at 15 or 16, less the time of first molar eruption, 6 years, giving the difference). Now, looking at the younger specimen, G,* its *last molar just erupted. If this was also at 15 or 16, this would be the age of the specimen at death. We could estimate that G died at about 15, H at about 25, give or take 2 or 3 years in each direction.*

frontal bone boxes trying to find a fit. Failing in this, he then went to the other cranial pieces. The chances of a fit were much poorer, but it seemed too early to go back to his hotel room, and Milford worked on. The last piece in the parietal box was a funny shape and the wrong color—it was speckled while the C skull was a cream color. He almost didn't try it for a fit, but then remembered lecturing his own osteology students to ignore color when reconstructing specimens. So he picked up the piece. It was not a parietal but much of a frontal bone, the flat part, missing the orbits and the browridges above them, which is why it had been misidentified as a parietal bone. The piece fit on the C skull perfectly, joining at two different contacts. The shared excitement of this particular reconstruction solidified the relationship between Crni and Milford and represented a turning point in Crni's attitude toward the collection. Because the reconstruction added something significant to the best-known cranium from Krapina (the C skull is an icon of the collection, for many), the experience convinced Crni to let the collection be renumbered and allow its systematic comparison and analysis. First the teeth, and later the other bones could be laid out and compared, in some cases for the first time. Throughout the rest of his life,

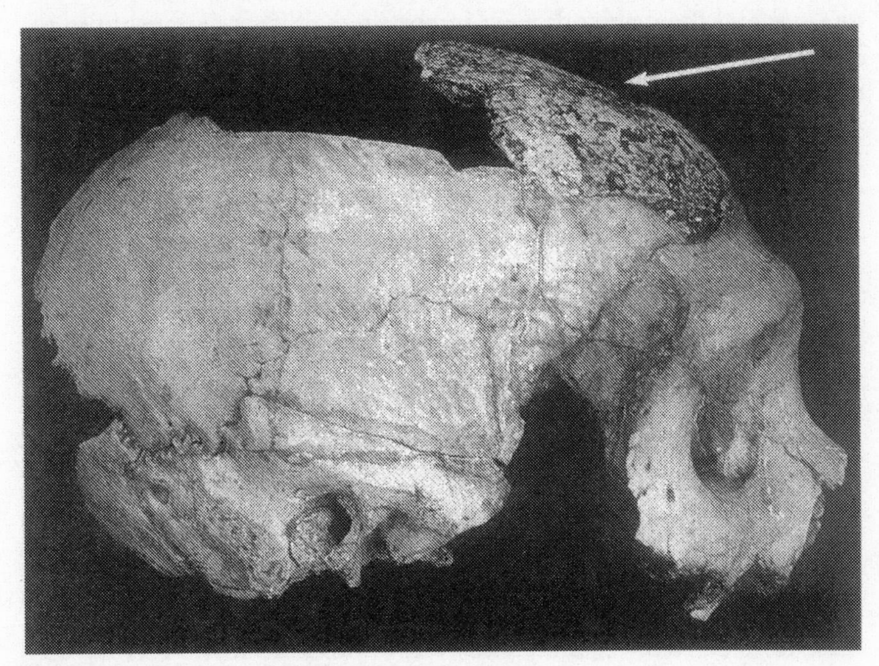

Krapina C, with the colored frontal piece attached.

Crni was the most enthusiastic of converts and the most helpful of curators.

During this first visit a tall, very skinny, young man walked into Gorjanović's office and sat, watching Milford work. This is how he came to meet Jakov Radovčić, soon a close friend.

Twelve years later, in the summer of 1988, Jakov was finishing work on a popular biography of Gorjanović and the development of paleoanthropology.[6] He wrote in the Croatian language and a bilingual friend (who knew nothing of paleontology or anthropology) translated it into English. The published book included both languages, every page with a left column in Croatian and a right column in English. Jakov wanted help smoothing out the English translation; he wanted to make sure that it was accurate, especially in its use of English jargon. We stayed with the Radovčić family for several weeks while we were looking for an apartment, and during that time every evening was devoted to reading, translating, and discussing Gorjanović. A tremendous amount of information about his life and times and writings, previously only found in Croatian, was appearing in English for the first time, and we were privileged to be the book's first English-speaking readers. (Milford ultimately wrote the foreword for the book.) There was a lot of new material here, much to learn about the history of the site and the man and, most interestingly, their relationship to the whole of central European science. The position of central Europe in the history of paleoanthropology is usually eclipsed by a strong western European bias that marked our own education and is present in most treatments of the subject. The writings of Gorjanović and Jakov emphasized the centrality and prominence of central Europe in the active theorizing about human origins taking place around the turn of the century and beyond.

POLYGENISM AND THE FOSSILS:

VIEWS OF RACE AND THE NEANDERTALS

In 1891, just before the Krapina Neandertals were found, there was an explosion in the knowledge of human evolution, with Eugène Dubois's discovery[7] of the much earlier *Pithecanthropus* skullcap and femur from the Trinil site on the faraway island of Java. This added a much more archaic specimen to the exclusively European fossil record then known. By the turn of the century several Neandertal fossils had been discovered:

- The Engis Cave (Belgium) child's cranium found in 1829
- The Forbes Quarry (Gibraltar) cranium discovered in 1848
- The 1856 Neander Valley discovery, giving the Neandertals their name, of a skullcap and skeleton found in the Feldhofer Cave (Germany)
- The two Spy (Belgium) burials, including skulls and skeletons, unearthed in 1886

Most of them were recognized as Neandertals, although Engis remained unidentified until much later, and Forbes Quarry seems simply to have been forgotten[8] for a half-century. But did these Neandertals demonstrate human evolution? Evolution was rejected by many, some who saw the Neandertals as pathological and others who saw them as no different from other human races.[9] Some holdouts held considerable influence, such as Rudolf Virchow (1821–1902) in Germany, who thought the Feldhofer Cave Neandertal skeleton's differences could be explained by pathology, and quite a few French anthropologists. These men were academically powerful; for instance, at the 1892 Anthropological Congress in Ulm many of the German experts denied there were human fossils,[10] especially European ones.

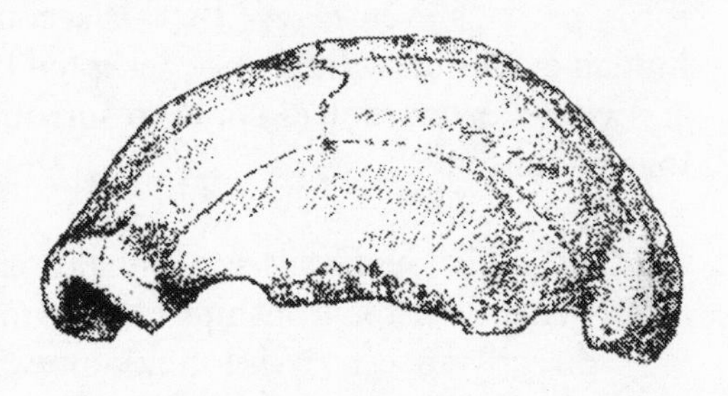

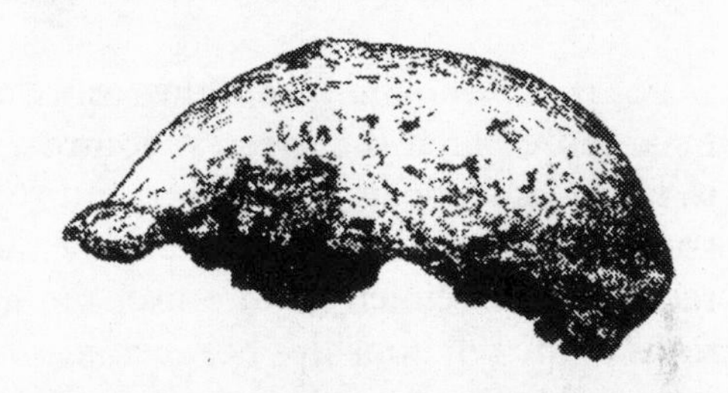

These two faceless crania[13] are from Trinil on the island of Java (the first "Pithecanthropus" skull, below*) and the Feldhofer Cave Neandertal from Germany* (above). *With the 1891 Trinil discovery, their comparison epitomized the problems of understanding the course of human evolution, even as they provided firm evidence of it.*

Trinil provided time depth to the fossil record. The top of the Trinil skull was decisively more primitive than the European remains,[11] and theories specific to human evolution could be developed. They soon were, and Krapina quickly came to play a role in them. Yet, ironically, this was also the time when Darwinian evolution was being questioned.[12] Partly this was because of the growing number of non-Darwinian evolutionary theories, partly as a reaction against Darwinism, but as we discuss in Chapter 8, partly because it was increasingly realized that the theory as expounded by Darwin didn't explain everything.

When Dubois returned to the Netherlands from Java, just having published his first detailed description of Trinil,[14] he began a campaign for acceptance of his contention that *Pithecanthropus* was a link between humans and their anthropoid ancestors. He attended several international congresses in Europe and by the end of the century had published extensively. Other scientists responded with almost 80 publications.[15] They were far from unanimous, though. Of these authors,[16] only 14 accepted Dubois's interpretation. More rejected it, 9 outright and 16 because they considered *Pithecanthropus* a human form, perhaps even *Homo.* By this account, at the turn of the century human evolution was still not accepted by many, perhaps most.

Two major issues had come to surround the fossil evidence for human evolution:

1. Whether the fossils actually did represent prehistoric humans
2. What their relationships were to modern humans (i.e., were they directly ancestral, did they represent extinct side-branches, how were they related to different races, etc.?)

For those who did accept the reality of human evolution, the second issue represented the greatest controversy of the time, and continues to be an issue today. As physical anthropology added time depth to its study of human variation, effectively, the frameworks used to interpret races were extended to interpret the human fossil record, and questions of the relationships between fossil and modern humans were tied to ideas about race. The views various workers held on race inevitably influenced their views of human fossil remains, and in turn, the fossil record became a source for establishing racial histories. We believe this might have been on Gorjanović's mind when he decided where and how to first announce the Krapina discoveries.

On October 18, 1899, eleven days after he finished the first excavations at Krapina, Gorjanović wrote to Johannes Ranke (1836–1916), a Munich anthropologist and author of papers on races and fossil humans, about his discovery of the Neandertal fossils from Croatia. He may have known Ranke from his student years at the University of Munich when he studied under the paleontologist Karl Zittel, but he wrote to Ranke in his capacity as editor of the preliminary reports section of the *Archiv für Anthropologie*, journal of the German Society for Anthropology, Ethnology, and Prehistory (ironically the Society was founded in Berlin by Rudolf Virchow, a polygenist who had no place for Darwinism and was perhaps the leading *anti*-Darwinian of the time). The Society published the letter in the next year's journal and it was one of the first announcements of the Krapina Neandertals to the wider European scientific community. The Krapina Neandertals represented the largest single Neandertal collection then (and now) known, and they became new ammunition in many of the developing controversies concerning the fossil record.

Ranke, the first person outside Croatia contacted about the Krapina finds, was a leading authority on racial variation and had just published, in 1896, on the relation of human fossils and the living races.[17] Like Gorjanović, he was a confirmed monogenist, inheriting the old Blumenbach anthropological position. Ranke embraced the ideas of Adolf Bastian (1826–1905), a German monogenist who rejected Darwinism and focused on human unity rather than diversity, especially the "psychic unity" of all humankind. Bastian ignored physical differences and considered cultural diversity as environmentally induced local variations of universal mental processes shared across humanity. Ranke was more anatomically oriented, contributing to the development of some of the instruments that are used to measure crania, and he classified races on the basis of physical characteristics such as head shape, skin color, and hair form. Influenced by Virchow, he believed the human races were unchanged through prehistory, certainly since glacial times, and like Bastian thought that from the first there were representatives of modern humanity already differentiated into races but that these differences were quite minor. Ranke, in fact, contended there were no true fossil races taxonomically different from living ones, and Neandertals were true human beings, members of the species *Homo sapiens* because "its characteristic peculiarities do not remove this race from the series of forms already known to us from modern races."[18] Ranke's views of race and the unity of man were very close to those of Gorjanović, who felt sure Ranke would change his mind and come to accept Neandertals as more different from the living than he had realized, as true human fossils, after he saw Gorjanović's initial interpretation of the fragmentary Krapina remains.

But other anthropologists who would soon come through Zagreb had very different ideas about race; many belonged to a new breed of polygenist, differing from Virchow by incorporating evolution into their ideas of racial origin and interrelationships. In 1859 the major controversy in what now would be called anthropology was the monogenism/polygenism issue. The Darwinian revolution, far from resolving or overshadowing this controversy, was absorbed into the preexisting debate. But being a Darwinist means you believe in common descent. How could one postulate different origins if we all stem from a single common ancestor? Darwinism should have spelled the death of polygenism, but surprisingly it not only failed to herald its demise, but many new Darwinian forms of the doctrine were soon to emerge and the new polygenism was more popular than ever. This was possible because anthropologists accepted common descent, one aspect of Darwinian theory, but rejected natural selection as the only mechanism of change. This allowed polygenists to invoke non-Darwinian evolutionary mechanisms such as directed evolution (orthogenesis)—evolution controlled by internal regulation that unfolded through the progression of life[19]—to account for parallel changes in separated races. If the races shared the same inner mechanisms, their evolution could be expected to proceed the same way. This way the new polygenists explained how the races could have evolved separately, in parallel, to each independently become human (at least to varying degrees) after they diverged from a very ancient, often prehuman common ancestor. In pre-Darwinian polygenism, when races were thought to have separate origins, there was an easier explanation: it was assumed they were simply created in their present form.

Racial politics played an important role in how evolution was accepted and interpreted, but evolution played an equally important role in how races were conceived. Two-dimensional schemes of racial relationships came to involve a third, temporal dimension, and various fossils ultimately were placed on racial lineages. But the situation was not simple. The frameworks of monogenism and polygenism were much more complex after Darwin and can be viewed in relation to several different factors and issues, including:

- How variation is partitioned in terms of lumping or splitting
- The importance of ranking races and the need to provide an explanation for it
- Whether humans are considered unique or an integral part of the animal kingdom
- The acceptance or rejection of Darwinism as an overarching explanation for variation

Perhaps the most important of these was the way variation was treated—the issue of "lumping" or "splitting" that threads its way through the history of biology. Anthropologists who were splitters saw human variation as discontinuous and acknowledged large differences between modern human groups. They also recognized major gaps between themselves and fossil humans. All polygenists were splitters.[20] However, Darwinism provided a mechanism whereby a monogenist view could also encompass significant racial variation—for instance, from "primitive" to "advanced." The races could be derived from a single source but reflect different evolutionary directions or, more important, different amounts of evolutionary change, and post-Darwinian monogenists could be splitters as well. It is also true that some monogenists, like Gorjanović, tended to regard human variation as continuous, recognized the unity of the human species, and extended this notion into the past. But others who recognized the unity of modern humans did see a major typological distinction between fossil and modern humans.

GORJANOVIĆ'S HISTORY AND INTELLECTUAL POSITION ON NEANDERTALS

Gorjanović-Kramberger[21] was born in 1856, the year of the Neander Valley discovery, into the Austro-Hungarian empire, of which the Croatian capital of Zagreb, his city of birth, was still considered a rather peripheral part. This was more than political marginalization, as it translated into the standing of non-German scientists in the wider European intellectual community. Nevertheless, cultural and scientific institutions were formed during a Croatian national revival period in the mid-19th century that continued to grow. Scientists who were educated in other parts of Europe returned to Croatia; the Yugoslav Academy of Arts and Sciences was formed (in Croatia), and scientific institutions including the Croatian National Museum were founded. Zagreb University, founded much earlier in 1669 and once well established, was renovated by the latter part of the 19th century, but it took some time for Zagreb and other southern Slavic capitals to be recognized as important centers of learning. Although Croatian scholars spoke and read German, published in German language journals, and had close ties with Vienna, there was a sort of linguistically based isolation of the Slavic parts of the empire. The "mainstream" intellectual community did not read Slavic languages, and many important Slavic works were not published in German. Croatian intellectuals were often ignored by the German-speaking central European community. It is to Gorjanović's credit that he was not ignored. He entered into and held his ground in debates with central European authorities in his various specialties of paleontology, and he was well known in the central European academic world. With the Krapina finds, Zagreb could not be ignored, although it is safe to say Krapina would have received much more attention over the years had it been a western European site.

Gorjanović's father, Matija Kramberger, was from a German family that settled in Slovenia in 1648 after the Thirty Years War, and his mother, Terezija Dusek (née Vrbanovic), was Croatian. Gorjanović was christened Karl Kramberger. Years later in 1882, as a Croatian patriot, Gorjanović adopted the Croatian versions of his German names. While he dropped Karl completely in favor of Dragutin, he retained Kramberger in hyphenated form. Gorjanović studied in Tubingen, under Karl Zittel, a renowned vertebrate paleontologist, and finished his degree in 1879 with a thesis on fossil fish of the Carpathians. He returned to Zagreb and, after a brief period that included further study in Vienna, became curator of mineralogy and geology at the National Museum. In 1884 he took a position at Zagreb University as well, as assistant professor of paleontology. Although he published on geology, some of Gorjanović's best-known work and early publications were on fossil fish and reptiles. These studies illustrate his early commitment to Darwinian thinking, emphasizing adaptation and continuous variation of fossil groups over time and space. He used stratigraphy to reconstruct paleoecological conditions that could account for evolutionary changes in certain groups of reptiles as he traced phyletic lineages and noted the differences and similarities between fossil and more recent forms.

When Gorjanović began work on the Krapina remains his focus shifted to humans. Gorjanović arrived in the town of Krapina on August 23, 1899, after receiving extinct rhino bones from the local schoolteacher. These had been sent four years earlier, but Gorjanović was so preoccupied with his other duties that he had no earlier chance to visit the site. He was taken to the place where the bones were found, a rockshelter in a 9-meter sandstone cliff that had been quarried for sand. The stratigraphy of the deposits in the shelter was apparent, with horizontal bands containing ash, charred sand and charcoal—evidence of hearths—along with the remains of extinct animals. He looked closer and saw there were stone tools in the sediments as well, and then he found a single human molar. He realized at that moment that he was on the threshold of a most important discovery—the skeletal and cultural remains of prehistoric humans. On September 2 he began excavating. Although hurried, the excavations were very methodical and careful, especially considering the standards of the day. The Spy Neandertals, discovered in 1886, had been so quickly excavated and received so little field documentation that when the bones of the two bodies were mixed up with each other in moving them to the laboratory there was no field diagram or cataloging (for instance, putting numbers on the bones as Gorjanović did) to straighten out which bones went together. At Krapina the stratigraphic layers were labeled and drawn in Gorjanović's notebooks, he kept meticulous records, and all human and most animal remains were labeled with their stratigraphic provenience (which level they came from).

The site ultimately yielded the remains of over 70 individuals, most of whom were children. It was the largest Neandertal collection from a single site in the world, and one with virtually limitless information about the life, times and death of the Croatian Neandertals. Most of the remains were from a short time span[22] and many of the individuals may have been related. Numerous bones were covered by artificial cut marks that were taken by some researchers to be evidence of cannibalism but by others as evidence of bundling in preparation for secondary burial.[24] Some of the animal bones in the cave also had cut marks, usually in different places than the human cut marks,[25] and seem to have been Neandertal prey.[26] Although he did not then know it, the Krapina Neandertals were early, over 100,000 years old. Their bones were taken back to the National Museum where Gorjanović studied them. By the end of the year and the turn of the century, he had announced the find, given brief descriptions of the morphology of the Neandertals in lectures in Zagreb and Vienna and in his letter to Ranke, and taken a clear position on their place in human evolution. He considered them directly ancestral to modern humans.

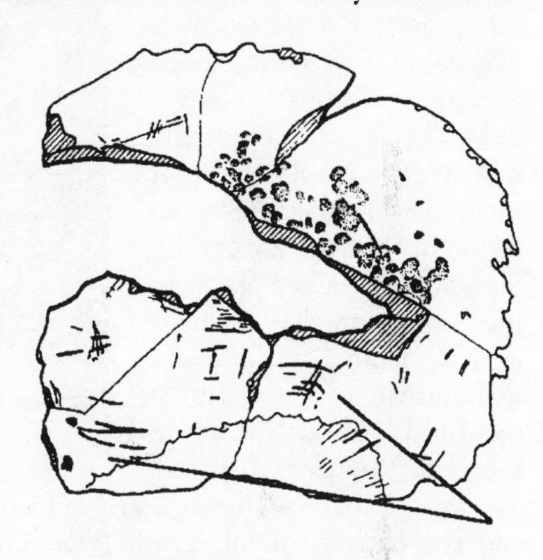

Parietal 16 from Krapina,[23] with the position of some of the cut marks indicated. These long, parallel cuts that reached the bone through the flesh might be the consequence of removing the mandible from its articulation with the skull, but more probably result from scalping.

In the field of human evolution Gorjanović most admired Gustav Schwalbe (1844–1917), a German anatomist and human paleontologist at Strasbourg. He was the primary authority in German, indeed central European paleoanthropology, the author of definitive descriptions of several of the most important fossil human remains known then. Gorjanović met Schwalbe in 1903 at the Congress of the German Anthropological Society in Kassel. He disagreed with Schwalbe's treatment of Neandertals in his introductory lecture but developed an appreciation for his ideas so deep that he dedicated his 1906 monograph on Krapina to him.

Schwalbe, in 1914.[27]

Following Ernst Haeckel, whose importance in the development of polygenism and other matters could never be overstated (he is discussed at some length below), Schwalbe set up a hierarchy of living and fossil forms that more or less represented steps in human evolution. Schwalbe was a key player in the argument that humans evolved from the apes and that human fossils were direct links between them. He founded the *Zeitschrift für Morphologie und Anthropologie* and in its first issue published an analysis and comparison of the Trinil *Pithecanthropus* skullcap[28] that put to shame the description by its discoverer, Eugène Dubois.[29] Schwalbe's publication, in that fateful year of 1899, firmly established the intermediate position for the skull, just as Haeckel had envisioned when he coined the name that was ultimately applied to it, *Pithecanthropus* = ape man. Perhaps it was this date, as much as his progressionist idea that fossils could represent sequential stages leading to humans, that linked Schwalbe so closely to Gorjanović.

Schwalbe followed this paper with two more substantive works comparing the Neandertals from Spy to the Feldhofer Cave specimen and effectively laid the groundwork for naming them *Homo primigenius*, an appellation for the extinct prehuman group first coined by Haeckel in 1868. This was not Haeckel's name for Neandertals, however, but for the prehuman, speechless, common ancestor of all human races that he hypothesized. It was Ludwig Wilser (1850–1923) who in 1898 first used it to refer to Neandertals.

Neandertals, of course, are no longer among us, at least as a distinct race. They are gone—extinct—and this poised a problem for Schwalbe: how could Neandertal extinction be reconciled with their status as the transitional species he envisioned? Actually there are two ways. The first is "extinction" through evolutionary change. Here, the Neandertals evolved into modern humans and are extinct because of this transformation. The second is extinction by replacement, wherein the Neandertals were replaced by modern humans. How could they still be ancestral then? This would be possible if the known European Neandertals, the ones who were replaced, were the basically unmodified but later surviving offshoots of the Neandertals who were the human ancestors.

Schwalbe recognized both of these possibilities but was effectively unable to make up his mind about which was the correct interpretation. He vacillated throughout his career about the position of Neandertals in human phylogeny. In his most detailed statement,[30] Schwalbe put forth two potential human phylogenies: in one, he arranged the known human fossil groups in a linear progression leading to modern humans. The Javan form *Pithecanthropus* gave rise to *Homo primigenius*, which in turn gave rise to *Homo sapiens*. The other scheme, which was more popular at the time, depicted *Pithecanthropus* and *Homo primigenius* as collaterals to the modern human line, each representing different species whose terminal members included the known fossils.

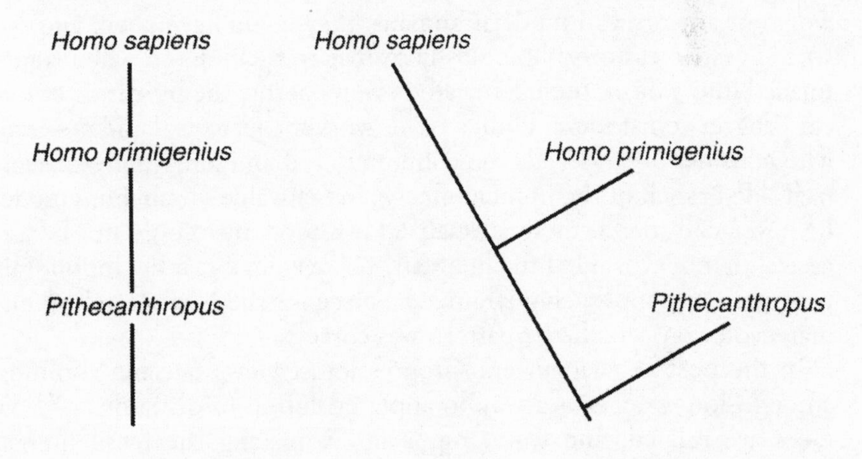

Schwalbe's Alternative Phylogenies.

Schwalbe emphasized this scheme in his 1903 Kassel lecture, which Gorjanović understood (or perhaps misunderstood) to mean that the Neandertals were being excluded from human ancestry. Gorjanović's objection was that

> we do not know of a single non-adaptive organ that would force the developmental course of that species to die out. Instead, all the most observable atavisms [features inherited from remote rather than immediate ancestors] on later fossils and recent humans instruct us as to a developmental continuity between the old diluvial and modern man.[31]

Schwalbe, however, was making a distinction without a difference as far as the ancestry issue was concerned. What Schwalbe's schemes actually contrasted, apart from the relatively minor issue of whether real ancestral fossils or only their close cousins had been found, was the issue of the *pattern* of evolution. He was grappling with what still is a controversial and unsettled question—whether species appeared gradually with one changing into another (anagenesis) or whether new species were always the result of an ancestral species splitting into two descendants (cladogenesis). It is an early version of today's "Ladder *versus* Bush" debate.[32] Which of his two schemes was correct was of little importance to Schwalbe, since in either case there were transitions, filling the gap between apes and humans. Even if the second, offshoot pattern is the better description of prehistory, as he believed toward the end of his career, he did not consider the Neandertal and *Pithecanthropus* limbs specialized side branches. He thought they were unchanged retentions of the ancestral forms. If the known Neandertals were not ancestral to modern humans, they could have been, and the true ancestor was morphologically similar to them. Even when transformed into a bush, then, Schwalbe's views betray the influence of linear progressionist ideas. Unlike other workers, historical and modern, who consider Neandertals the culmination of an increasingly specialized side branch of the human lineage, in Schwalbe's branching model he never considered them specialized offshoots that could not be ancestral to moderns. He fundamentally believed in linear evolution. His progressionist approach guaranteed a place for the Neandertals in human evolution, whichever pattern was correct.

In the post-Darwinian era, progressionist views became common among monogenists as a way to apply evolution to questions of how races are related, and were important in placing the fossil human record in that racial framework. The ideas of either separate or progressive, serial, creations were now dismissed. Instead, in many post-

Darwinian linear progressionist schemes, the different statuses of the "lower" races could be explained by freezing each at a different place in their evolution along the way to the "highest" race, Europeans. In this context Darwin viewed races as the links between humans and their anthropoid ancestors. This form of evolutionary monogenism depicted different races as different stages on an evolutionary ladder, each evolving from a "more primitive" ancestor. Thus some of the Darwinian monogenists ascribed an apelike status to certain races, even as they affirmed human unity. Other monogenists, though progressionists, were decidedly more liberal.

Schwalbe, like many other paleoanthropologists of the time, treated races and fossil humans together in his evolutionary schemes. While Schwalbe definitely considered Neandertals a different species based on anatomical criteria, they differed mainly because they were more primitive than any of the living races. Because he accepted the basic precepts of linear progressionism and Schwalbe believed neither Haeckel's polygenism nor his precept "ontogeny recapitulates phylogeny" (the "biogenetic law"). He also argued against Julien Kollmann (1834–1918) and others who believed that neoteny (the retention of the juvenile characteristics of ancestors in the adults of their descendants) explained evolution, and that therefore the most primitive human form was the most neotenic, African Pygmies. Schwalbe rejected this "Out-of-Africa" theory and asserted that Neandertals were much more primitive than Pygmies or any other race. In his view of human evolution the races diverged from a common ancestor that descended from Neandertals, a classic application of evolution to the monogenist position.

But *why* the change occurred between the Neandertal and modern populations is a question Schwalbe never asked, and this more than anything else distinguished him from Gorjanović, who was a classic Darwinist. Gorjanović expected natural selection to account for evolutionary changes and tried to understand the details of why and how it happened. And Gorjanović found more reasons than this to be at odds with the Strasbourg anatomist. Disagreement stemmed from Schwalbe's vacillations about the role of Neandertals in human evolution. Without the advantages of hindsight in viewing Schwalbe's entire career, Gorjanović focused on Schwalbe's emphasis on the nonancestral status of Neandertals, and not on his overall linear theoretical position. At professional meetings and in correspondence with Gorjanović in 1903 and 1904, Schwalbe espoused his nonancestral interpretation, considering Neandertals a side branch in the evolution of modern humans, which sparked rebuttal by Gorjanović. At later times they appeared to be in agreement; and for good reason. Increasingly, Schwalbe stood out in contrast with Marcellin Boule (1861–1942), the other great paleoanthropologist of the day, who firmly believed that the Neandertals were contemporary with humans and they, and anything much like them, were too specialized to represent human ancestors.

Gorjanović's views of variation were unusual, although perhaps not surprising given the nature of the sample he worked with. He considered an understanding of the full range of modern human variation critical to an accurate interpretation of the fossils, and one of the first things he did in his analysis was to focus on the huge collections of human skeletons in Vienna. He was interested in the range of variation in features that seemed unusual in Neandertals. For example, one of his first foci was the mastoid process of the temporal bone, an area that is different in Neandertals than it is in modern humans. He sought to understand how the mastoid varied in modern humans as a whole and then how the fossils fit into this variation. He was not interested in seeing how Neandertals compared with "primitive" races, often a goal of progressionists who reasoned that such races are more closely related to them, but he rather wanted to have a grasp on the variation of humans as a whole to see whether and how his Neandertals fit in. In fact, despite the large numbers of non-Europeans he studied, the individuals that he found shared many Neandertal features were Europeans.

Gorjanović saw the Krapina fossils as primitive, but not qualitatively different from modern humans. The modern skeletons that retained "Neandertal features" were evidence of continuity for him, and he considered the shifting frequency of variations an indication that moderns and Neandertals were all part of the same variation. He emphasized the importance of adaptation and the significance of variation within the Neandertals and modern humans. Neandertal "primitive" features such as large browridges and faces were, in Gorjanović's view, the adaptive consequences of mechanical forces minimized later due to technology. Therefore he interpreted Neandertal morphology as variations that could be explained adaptively, a theoretical view very different from Schwalbe's who regarded them as reflecting a "primitive phase" to human evolution. Gorjanović reasoned that the moderns evolved from Neandertals as their adaptations changed, a classic selection argument. As early as 1899, he wrote in the Croatian language:[33]

> All the osteological features in the jawbones of diluvial man are apparently remnants of the original adaptation to the natural circumstances of life. With a powerfully muscled jawbone he was doing all those tasks that

> living man today does by cooking, cutting, mowing, either using a knife or other aids. The disappearance of these robust muscle attachments is clearly related to the considerable reduction of the action of said muscles, due to the use of artificial aids.[34]

Coming as he did from a geological background rather than an anthropological or medical one, Gorjanović never wrote specifically about his ideas of race, but his views about human variation are clear from the way he interpreted human evolution. His view of "Culture" as a major factor in the natural selection on humans fits very well with the historical geographic school of racial variation promoted by Bastian and recognized by Ranke, which envisaged human differences as a response to different ecological conditions and their associated cultural practices. Gorjanović considered the Neandertals members of the human race, even though he later called them *Homo primigenius*, after Schwalbe's use of the term, to reflect the magnitude of their differences from the other races. We believe this doesn't reflect a change in his thinking about their relationship to moderns but indicates his recognition, following more discoveries from the cave (most of the specimens were not found until 1905), that Neandertals vary more from modern races than the modern races differ from each other. Therein we can see Gorjanović's broad view of human variation; he did not think that the differences between even the most extreme of modern races were sufficient to consider any of these humans different species.

DIFFERENT VIEWS OF EVOLUTIONARY MECHANISMS

After the announcement of the Krapina fossils, many European paleoanthropologists communicated with Gorjanović, and one of the first to come to Zagreb was Hermann Klaatsch (1863–1916), a comparative anatomist and professor from the University of Heidelberg.

Klaatsch was a student of the preëminent anatomist Carl Gegenbauer, a close friend of Haeckel's. In 1899, the year of Gorjanović's first visit to Krapina, Klaatsch gave a talk before the German Anthropological Congress, meeting in Lindau. Although a Darwinian, he opposed the views of Darwin as well as those of Haeckel (and others) by denying an apelike ancestry for humans. He propounded a theory in which not only were the human races quite distinct from each other, but all of humanity was distinct from the other primates.

In his speech[35] Klaatsch argued that humans form a separate branch of the primates, using comparative anatomy to support his view that humans and the other higher primates (apes and monkeys) were different evolutionary lines, independently derived from prosimians. This phylogeny set him apart from other Darwinians, including Schwalbe, who was perhaps the most prominent supporter of an ape ancestry for humans, not only writing on this point but managing to discredit most other polygenist interpretations, like Klaatsch's, in the process of detailing the specifics of human ancestry.

Hermann Klaatsch.[36]

During the next decade Klaatsch's views changed dramatically, as he developed a scheme of evolutionary polygenism that incorporated different ape species *within* human racial lineages. Yet he never saw apes as *ancestral* to humans. In spite of this, Klaatsch was strongly influenced by Schwalbe's detailed study of Neandertal anatomy, and came to appreciate the significance that fossils could have in supporting his views. Klaatsch regarded the Neandertals as a distinct species of the genus *Homo* and had tried to contribute to their understanding by contending that the shapes of the Feldhofer Cave limb bones were not the result of rickets, as the antievolutionist Virchow had proclaimed. Equally erroneously, though, he thought they were a reflection of arboreal adaptation, a primitive feature linking them to other primates. With his focus on the importance of fossils to support evolutionary scenarios, he immediately recognized the significance of the Krapina discoveries.

Klaatsch spent a lot of time in Zagreb; he contributed to some of Gorjanović's Neandertal reconstructions although Gorjanović never gave him permission to either reconstruct or describe the Krapina fossils.[37] However, Klaatsch nevertheless published descriptions of several, particularly occipital and temporal, which he compared with other Neandertal bones from Spy, and this, no doubt, contributed to some of the irritation that developed between the two men. Gorjanović and Klaatsch had very different views on the Neandertals' place in human evolution; Gorjanović saw them as ancestral to modern Europeans and Klaatsch most certainly did not. These differences and Klaatsch's disappointment at not being permitted to work more broadly on Krapina gave rise to rather heated exchanges in print later on and prevented any professional collaboration between the two men.

Klaatsch's views on Neandertals were closely tied to his views on race. As he became involved with Neandertal fossil remains, and studied race more fully with a three-year visit to Australia (1904–1907), he developed his polygenic model of human evolution. Klaatsch was therefore a Darwinist and a polygenist. How could he be both? He could, because for him like so many others, Darwinism only entailed the tenet of common descent.

The full impact of Darwinism on anthropology was not widely realized at that time. Most anthropologists were quick to accept the principle of common descent and its implications. They recognized that Darwin's works meant the constituents of species and other taxonomic groups were bound together through a shared common ancestor. Innumerable theories about the identity and nature of man's last common ancestor with other anthropoids and/or other human groups were created. But the second part of Darwin's theory is the mechanism of natural selection. The full title of *The Origin of Species* is, after all, *On the Origin of Species by Means of Natural Selection: Or the Preservation of Favoured Races in the Struggle for Life.*

The natural selection mechanism was not well received, and as presented it was difficult to understand how it worked. In fact, only a few dealing with paleontological problems adopted natural selection as the mechanism that caused most evolutionary change until the modern evolutionary synthesis in the 1940s unified Darwinian selection and genetic theory. Ernst Mayr, perhaps the most famous evolutionary biologist of the latter part of this century, points out that all fields of biology were slow to accept natural selection as an evolutionary mechanism,[38] but students of human evolution, perhaps for obvious reasons, were particularly loath to abandon notions of progress or direction in evolution which are incompatible with selection. They still considered themselves Darwinists because they accepted common descent with change as the basis for diversity, but the cause of that change remained a problem and is still one of the most hotly debated issues in evolutionary biology.

Darwinian selection was most often understood in its role as pruner: the "survival of the fittest" is also the demise of the unfit. It was difficult to understand how this pruning process could create the kinds of major changes seen throughout the evolution of life. Pruning can get rid of inadequate or harmful variations, but how can it create new ones? While many of the new breed of evolutionary biologists understood that selection could account for small, inherited changes or variations within populations, they did not believe that this process could explain the kinds of differences noted between species and even higher taxa, both living and extinct. In fact the German word then widely used to describe these changes due to natural selection was *kinker*, or trifles.

How selection could lead to innovations and major adaptive change was unknown, but most, including Darwin, realized there was a serious problem. Without an accurate genetic model it was not possible to understand how natural selection, and all that it implies, could act as a major agent of evolutionary change. Moreover polygenists in particular had a second problem—they *could not be* selectionists. It was actually easy to adjust polygenism to ideas of common descent instead of separate creations; races were then recognized to have had a common ancestor but so long ago that their humanity was acquired separately, and for some so long ago that their ancestries were found in different primate species. Those who found the origins of human races in different primate species generally thought the human races themselves were different species. But the Darwinian mechanism, natural selection, could not plausibly account for the kind of parallelism these polygenist theories required when explaining the independent acquisition of human traits—even of humanity. Different mechanisms were sought to explain this evolutionary pattern.

All polygenist approaches after Darwin required major parallelism or convergence in the evolutionary process, and this trapped them as they had to rely on some kind of non-Darwinian force to explain it. External causes, in this case meaning Darwinian selection, could not account for the independent evolution of the same things, since "parallel" and "convergent" have very particular meanings in the context of evolution. Features (like feature **A** below) may be similar or the same because of descent (the ancestor had the same feature) or independent acquisition (separate parallel development from a different feature **B** in the ancestor). Convergence is quite different and an even more unlikely idea. In convergence, similar or identical features evolve independently from different sources, as features **C** and **D** in the common ancestor shown below. While these processes occur due to natural selection, the odds that they would occur for all human features are astronomical.

The assertion that the evolution of humanness in races was parallel or convergent means that the same changes happened in different races independently of each other, or perhaps (in the case of conver-

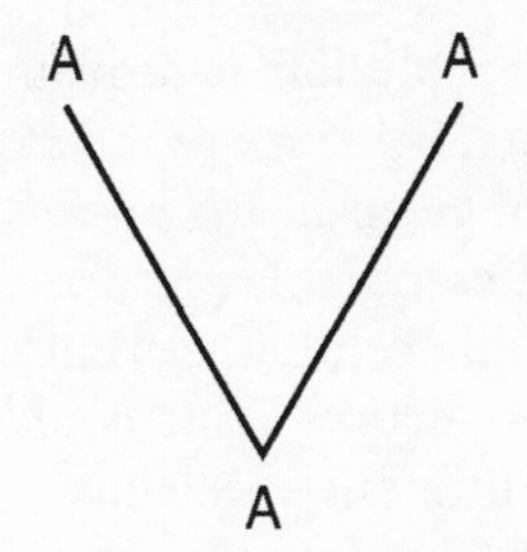

Similarity by descent (homology).

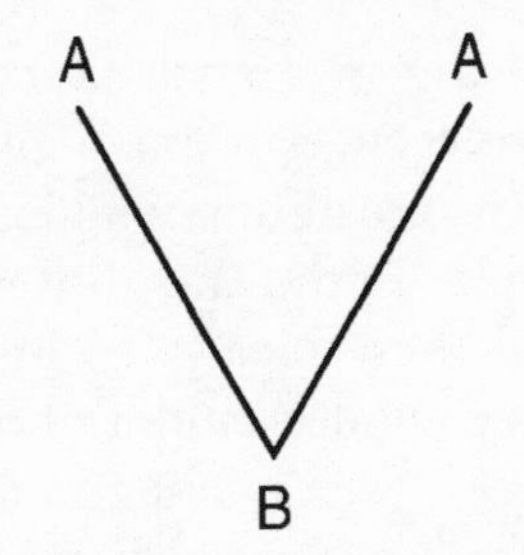

Similarity by independent development from a single source (parallelism).

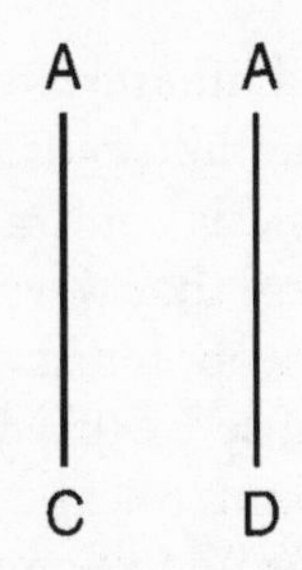

Similarity by independent development from different sources (convergence).

gence) evolved from different sources. Yet, such assertions did little to undermine polygenists' credibility since most anthropologists, as well as many others in the natural sciences, invoked several different forms of non-Darwinian mechanisms to explain human evolution. Virtually all were Lamarckians, orthogeneticists, or both, relying on some kind of externally or internally directed evolutionary process.

Lamarckism, as it developed into neo-Lamarckism in the 19th century, came to focus on an assumption and minor issue in Lamarck's writings,[39] the notion of *externally* directed evolution, with direct environmental influence causing change through "use inheritance" or the inheritance of acquired characteristics. Many scientists of the time believed this form of external direction took place in the evolutionary process and accepted the notion that developmental changes could be inherited, especially if they persisted for several generations. We discuss this further in Chapter 8.

In contrast, orthogeneticists invoked *inner direction* as an explanation. The cause of change was often considered a form of heterochrony (changes in developmental timing, such as neoteny, caused by a slowdown in rate of growth, or recapitulation, an extension of the growth and development period). Orthogenic change was sometimes seen as a form of preformation, a gradual unfolding of a plan from some preprogrammed "seed."[40] Its direction was established by vital forces or other causes of internal predisposition. Although orthogenesis has been described as a pessimistic reversal of the old progressionism, with many orthogenic trends seen as ultimately leading to the degeneration and/or extinction of a lineage,[41] anthropologists applying orthogenesis often sounded very progressionist indeed (as did the neo-Lamarckians), believing in a goal-directed evolution to explain the parallelisms and convergences so many of their schemes required. This was epitomized by the recapitulation ideas of Haeckel and his followers that were so popular in the late 19th and early 20th centuries. Anthropologists retained these ideas well into this century, partially because of the difficulty in understanding how selection alone could account for the origin of species. Numerous different mechanisms continue to be invoked for "species level" evolution.

The point is that it was, and among the lay public often still is, difficult to conceive of human evolution as *only* a consequence of natural selection. The process, based on varying genetic contributions to the next generation caused by differences in reproduction and survivorship, is opportunistic. It depends only on differences in ability of varying anatomy and behavior to promote survival and successful reproduction in specific environments. Selection is, in effect, a reactive process rather than a directed one, and this is difficult for people to associate with human evolution. Progress is a part of our culture; today, as in the past, the Western world thinks in terms of progress: in society, civilization, technology, and biology. It is more appealing to think in positive terms of ancestors striving forward, conquering hardships, and moving onward toward humanity and increasing perfection. Today's understanding of selection dashes these romantic images and, at the same time, makes parallelisms of the magnitude called for by polygenism difficult to account for by removing directive forces from the picture. But this understanding was more than a half-century away when Klaatsch first visited Zagreb.

FOSSILS IN THE POLYGENIST SCHEMES

Avid Darwinians such as Klaatsch embraced the "new polygenism" which had flowered into many more guises than its nonevolutionary counterpart prior to Darwin. In fact, well into this century most anthropologists embraced some sort of independent evolution of the different races, and polygenism was strongly supported in Germany, France, England, and America. Most forms of "evolutionary polygenism" recognize monophyly, a common ancestry for the human races extending far back into antiquity. These polygenist theories differ from monogenism that just postulates a distant common ancestor *in that they suggest that the races were separated for so long they acquired their humanity independently*, either through parallelism or some kind of adaptive convergence. While races were no longer considered products of separate creations, according to these polygenist theories the races are products of independent evolutionary lineages, albeit lineages that shared a common ancestor. Moreover, a powerful minority of the polygenists saw races as different species,[42] and some even postulated different anthropoid ancestors for them, pushing the common ancestor back further still. A few like Klaatsch came to consider different human races ancestral to different anthropoid species! They accepted a polyphyletic origin for humans, an origin from different ancestors in which humans were not all the descendants of a single ancestor.

Klaatsch's polygenism became more extreme over time, especially after his 1904–1907 visit to Indonesia and Australia where he spent much time with the natives. He appears to have liked and even respected them, a factor that probably influenced the elaborate evolutionary scheme he developed. Klaatsch contended that the human basal stem was to be found in Australia, a stem also giving rise to the other anthropoids. The great apes, he believed, derived from different human lineages in various areas. From the basal Australian stem branched the two main human lineages, an Asian one and a European one. The Asian branch gave rise to the orangutans and also the humans of Asia and later of Europe. The Aurignacians who replaced Neandertals were members of this lineage. Aurignacian was the name he used generally for people of the Upper Paleolithic, a period after Neandertal times when European cave dwellers had a modern-type anatomy, painted the walls of their caves, and made sophisticated tools and

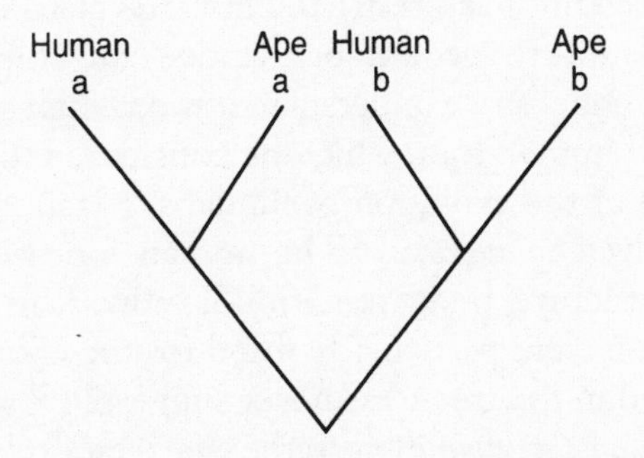

A polyphyletic interpretation of polygenism. It is polyphyletic because the two humans (a and b) do not have a unique common ancestor.

weapons. The earlier European lineage gave rise to the Neandertals and the African races, whom Klaatsch argued, share a number of degenerate characteristics. It also gave rise to the African apes. Thus, modern Europeans are actually more closely related to orangutans than they are to African people, according to Klaatsch. Australians are the little-modified descendants of the basal stem and are not associated with the degenerate lineage and thus were seen favorably, as ancestral to Europeans. Here he combines his polygenist views with linear progressionism; within this lineage, Australians represented the primitive, ancestral condition. They were the European past, frozen in time.

Klaatsch believed adaptive convergence caused the common human attributes. Evolution produced the same results in similar environments, especially in similar species which, meeting the same challenges, would be expected to evolve the same adaptations. Neandertals and other degenerate races independently developed tool use and large brains. This outlook linked him to Gorjanović in an important way. They both focused on adaptation as the consequence of evolutionary changes. Both approached variation with questions about how particular anatomical features functioned and what advantages they brought. But what was the cause of change? Was adaptation not just the consequence but also the goal? On this issue he was never clear. In different writings Klaatsch invoked virtually every mechanism to account for this part of his model, clearly its most controversial aspect.[43] The mechanisms were united only by the predominant effect of environment in his thinking. Although he did not invoke a purely orthogenic explanation, some kind of vitalism in the guise of inherent trends played a role in his explanation of some evolutionary directions; for instance, Klaatsch believed there was an intrinsic tendency for all human and humanlike primates to undergo brain size expansion—ape brains would have reached human size had they not been restricted by large facial muscles.

Yet other attempts to account for adaptive convergence seemed to fit the Darwinian model as we understand it today. In his discussion of the evolution of the human foot, which he considered to be the most important change in the evolution of humans, Klaatsch contends: "Primates always lived in trees. . . . The human foot first appeared as a 'spontaneous structure' [read 'mutation']. Individuals who had this accidental variation were particularly fitted for the erect posture."[44] But this foot evolved in the trees. He never suggested it appeared in order for it to be *used on the ground* but rather gave an explanation we today call "exaptation"—an adaptation that became prevalent because of the advantages it provided for one activity or behavior, but is now used for another purpose. Klaatsch proposed:

> The upright posture of man is one of those features without which we can scarcely imagine him. . . . As long as his ancestors were supposed to be "quadrupeds," it was a hopeless puzzle how an upright animal could be evolved from an animal running on all fours. . . . The starting point was a climbing posture. . . . —the capacity for climbing thick trees [too thick for a grasping foot to be useful] . . . the foot is applied to the trunk as a whole and the body is pushed forward.[45]

Klaatsch's ideas were poorly accepted for several reasons. It is easy to focus on the disbelief that the theory of a human ancestry for ape species was met with,[46] but, looking deeper, he was burdened by the same the problems that most polygenists contended with. None of the mechanisms he postulated, even jointly, were enough to account for the amount of adaptive convergence his thinking required. For instance, in the evolution of the foot, whether the basal ancestor had a humanlike foot or an apelike grasping foot, spontaneous structures (mutations) would have to occur independently in different lineages for this explanation to work, since each lineage gave rise to both humans and apes. Thus if the basal Australian foot were assumed to be human, grasping feet needed to independently arise in both lineages (ape a and ape b in the figure above). Moreover, ironically given his background in comparative anatomy, Klaatsch's theories also were met with disbelief because colleagues disagreed with their anatomical basis.

Klaatsch was one of the first polygenists to include the fossil record in a model of racial origins. He used the fossil record to support his polygenist view, and one way he could do so was to show that the Neandertals and the Aurignacians (races which he believed had separate origins) coexisted. Then he could argue that Neandertals were not ancestral to modern Europeans (descendants of the Aurignacians), and this would support his view they were different species. Klaatsch was therefore anxious to find Aurignacian fossils that were temporally close to Neandertals. He collaborated with and acted as scientific adviser to an unscrupulous fossil collector named Otto Hauser.[47] Hauser was extremely mercenary, primarily interested in selling fossil remains to various museums for high prices. For the fossils to be valuable, they had to have provenience—the museums needed to be sure that they came from ancient sediments. So Hauser, who was instrumental in discovering two sites in France, Le Moustier (a Neandertal site) and Combe Capelle (an Aurignacian one), reburied and "rediscovered" fossils several different times in the presence of the potential buyers, until his specimens were sold. Linked only by this unsavory background, it is ironic that the fossils from the two sites shared the same unfortunate fate. They were among the very few from western Europe to be partially or fully destroyed during the Second World War.[48]

Klaatsch was involved in the discoveries and analysis of both of these fossils. Although the actual ages of these sites, particularly Combe Capelle, were suspect, he considered the Combe Capelle fossil, supposedly from the earliest Upper Paleolithic strata, particularly important since it was fully modern and yet lived, he believed, right after the Neandertals. Therefore, he felt it was an important demonstration that the two were different species with different origins.

In spite of their similar predilections to explain evolutionary changes as functional responses to the environment, Klaatsch and Gorjanović disagreed about too much. They were clearly at odds over the place of the Krapina Neandertals in human evolution. Klaatsch regarded them as degenerate while Gorjanović included them in humanity and considered their notable differences to reflect different adaptations. Klaatsch and Hauser tried to involve Gorjanović in the French "discoveries" described above, but Gorjanović refused to deal with them. To Gorjanović's irritation, Klaatsch also tried to use the Krapina remains to show the coexistence of the two human species his French discoveries were supposed to represent, *Homo primigenius* and *Homo aurignacensis*. In the large amount of variation at Krapina, Klaatsch recognized the two species who he believed fought each other for the possession of the Krapina rock shelter. Gorjanović quickly refuted Klaatsch's claims, writing that although there was great variation at Krapina, it was not unusual but normal (this is one of the things he learned in Vienna), and only Neandertals were present.[49] This has been born out by all subsequent studies. One upshot of Milford's age-structure analysis was the recognition that many of the Krapina specimens are youths. Differences in the growth and development attained before death is an important element in the Krapina variation that Klaatsch interpreted so wrongly.[50]

EVOLUTIONARY POLYGENISM

While Klaatsch's view that apes were degenerate progeny of different human groups may seem bizarre, it was part of a popular scientific movement. The idea of an "evolutionary polygenism" had already been accepted and promoted by many Darwinians, although Darwin himself was a monogenist who scorned polygenism as a justification for slavery. As early as 1864 Alfred Russell Wallace (1823–1913), codiscoverer of Darwinian evolution, attempted to resolve the monogenist/polygenist debate as it then existed, in the context of Darwinism. In a paper[51] presented at the Anthropological Society of London, a conservative proslavery group,[52] he suggested a compromise. He argued the races shared a common anthropoid ancestor, but so long ago that it was before the final evolution of the human brain. This implied that *humans* descended from a single ancestor and were phylogenetically united, but that the races were formed before *humanity* was attained—the different races became human separately. This interpretation became the keystone for uniting polygenism and Darwinism. The criterion was changing from one of separate origins to a separate attainment of humanity.

Wallace was trying to please both the Anthropological Society and members of the Ethnological Society, a rival group consisting of opponents to slavery including many of the famous Darwinians such as Darwin himself, Thomas Henry Huxley (1825–1895) and others, who were partly responsible for the demise of "old-style" polygenism in England. Wallace proposed that before intelligence was important in human evolution, geographic groups evolved racial features in response to climatic factors that remained unchanged even after selection for intelligence. He thought that after intelligence became important, selection only acted on the brain, and it did so in all human groups. This common selection was responsible for the similarity of human groups, and later in his life he came to believe that supernatural forces were responsible for this mutual evolutionary direction. But it was not really a compromise at all. In a sense Wallace's arguments presented the thesis that all "evolutionary polygenism" came to embrace; since it postulated the separation of races prior to their humanity, it is essentially a theory of the independent acquisition of human status, a polygenist theory because it postulated separate human origins.

The year 1864 was a busy one for the proslavery Anthropological Society, very likely because of the political situation in America. In addition to publishing Wallace's talk that year, they published two translations of major polygenist works from the French and the German. These were Georges Pouchet's (1833–1894) *Plurality of the Human Race* and Karl Vogt's (1817–1895) *Lectures on Man*. By the end of the year all of the variations of evolutionary polygenism were present, if not actually presented before the Anthropological Society of London. However, Pouchet's polygenism, though published after Darwin, was not evolutionary. Like most of the French polygenists of the last century, Pouchet was not an evolutionist. He equated evolution with linear progressionism, which he thought to be unrealistic and associated with the monogenist position. In spite of Louis Pasteur's experiments, Pouchet still relied on a nebulous spontaneous generation theory to explain the origins of the separate races.

Vogt, in contrast, was a German evolutionist who tried to link the pre-Darwinian notions of polygenism and Darwin's theory. He did not recognize a single common anthropoid ancestor for humanity, but instead traced the origins of the races to different primitive anthropoid progenitors, ape lineages in different geographic regions of the world that later converged and were bound together by shared humanness. He rejected the idea of common descent, instead arguing that there were many original types, each with its own potential for further development, that unfolded during the evolutionary process. Vogt broke with much of the earlier German egalitarian tradition; for him, the races were quite unequal. He applied recapitulation theory (the theory that there was an internal drive to evolution that resulted in the repetition of the adult form of ancestors in the juvenile or embryonic growth stages of the descendants) to this issue and reasoned that as a "primitive" race, Africans are like the children of the "advanced" form. What better for justification for everything from slavery to colonialism could there be? What better scientific explanation for the "white man's burden" could one ask for?

Vogt was a scientific radical, a materialist who linked the mind to the brain and denied the existence of the soul. Trying to promote the cause of descent from apes,[53] he became known as the "monkey keeper" ("keeper" being the German meaning of "vogt,"[54]) and was ultimately thrown out of Germany for his belligerent materialism and vocal positions taken against religion. But he was promoting more than human descent from an ape. The precept of multiple descents was an important element in his thinking, and the common trends in the different human lines, created by recapitulation, resulted in a form of progressionism as all races, to a greater or lesser extent, approached the advanced form. More than most, Vogt felt there was progress in evolution and the separate racial lines were essentially preprogrammed to evolve toward humanness. This form of orthogenesis was too extreme for most anthropologists, who wanted to explain polygenism with more explicit material mechanisms. For many, these mechanisms were developmental and linked to some form of "heterochrony" (as mentioned above, like recapitulation), a term coined by Haeckel that has come to refer to changes in the onset, or timing, of development. Heterochrony links ontogeny and phylogeny (two more of Haeckel's terms). The time of appearance of a feature or its rate of development in a descendant may be retarded or accelerated relative to the appearance of a feature or its rate of development in its ancestor. This could lead to the adults of the descendant resembling the children of the ancestor, or the children of the descendant resembling the adults of the ancestor. If there was a heterochronic cause to evolution, changes would only proceed in a certain direction, and the heterochrony mechanism in two closely related species would cause the same, parallel, changes.

In his famous 1828 work on animal embryology, Karl Ernst von Baer (1792–1876) contended that individual development proceeds from the general to the specific. His view was that the earliest embryonic stages of all organisms are the same, later stages are only similar for related organisms, and the latest stages are unique. This was a highly influential interpretation and quite different from the *scala naturae* of the essentialists which saw all life arranged on a single ladder extending toward perfection, with the development of an individual animal mirroring the development of an animal series. The *scala* does not imply evolution, and one of its strong supporters was the anti-Darwinian Agassiz, who wrote that "the phases of development of all living animals correspond to the order of succession of their extinct representatives in past geological times . . . the oldest representatives of each class may then be considered as embryonic types of their respective orders or families among the living."[55] Ironically, Darwin favored a variation of Agassiz's interpretation of ontogeny to that of von Baer. But in a second irony this variation was soon transformed by Haeckel into an evolutionary mechanism that, for many, replaced Darwin's natural selection as the cause of evolutionary change.

HAECKEL:
POLYGENISM ENTERS THE 20TH CENTURY

Ernst Haeckel (1834–1913), Darwin's German apostle, was arguably the most influential of the post-Darwinian polygenists,[56] having a profound international effect on later physical anthropology and evolutionary thinking in general. His most popular work, the 1899 *Welträtsel* (*The Riddle of the Universe*), was one of the most widely read books of its time.[57] Over 100,000 copies of the first edition were sold in one year, and there were 9 additional editions, translated into 23 languages. His works can be directly linked to 20th-century British polygenism, especially as expressed by Sir Arthur Keith and R. Ruggles Gates, and the formation of the polygenism that became a major part of 20th century American anthropology. While the French also had a strong polygenism school—the principal doctrine of Broca's *Société d'Anthropologie de Paris* was polygenism—the French polygenists never accepted Darwinism, and this variant of the doctrine did not have lasting influence. Only very late in Broca's career did he accept the likelihood of transformation, but while his views on the fixity of species were changing, he never accepted the notion of natural selection. Paul Topinard (1830–1911), who succeeded Broca and wrote influential textbooks in anthropology,[58] agreed that human races were ancient, so ancient, he thought, nothing could be said about their origin. He was not a polygenist, and the strong influence of the polygenism movement in France died with Broca. Haeckel, in contrast, was a *Darwinian* polygenist, who like Huxley in England was responsible for Darwinism's popularization and its adoption into German public and scientific thought.

Haeckel was an extraordinary scientist whose biological insights gave us the concept of phylogeny from Darwin's common descent, the identification of a science of "ecology" (the science of the household of nature), the recognition that the nucleus of the cell "has to take care of the inheritance of heritable characters,"[59] and the reduction of a century of unclear and contradictory embryology to what became for a short time a tenet of Darwinism, the biogenetic law. He saw Darwinism as fitting in a larger scheme: monism, a universal theory of development that encompassed "the whole domain of human knowledge."[61] For him matter and spirit were the same, in a larger view drawing on the works of Goethe and the romantic tradition (he dedicated the second volume of *General Morphology*, his first major work, to Darwin, Lamarck, and Goethe). "While championing a cosmic mechanism, Haeckel talked in ecstatic tones of nature being 'alive,' and was soon to dispense to each atom an individual soul; the holy of holies within the cosmic tabernacle which was to form the basis of his religion of nature-worship."[62] Thus, his treatment of evolution was in some ways a form

Ernst Haeckel.[60]

of vitalism, but was also mechanistic, Huxley at one time showing trepidation at Haeckel's "mechanistic excesses."[63]

Haeckel took Agassiz's interpretation of ontogeny, the one Darwin favored, and from it developed the biogenetic law. If Agassiz and others were right, the history of evolution was written into the development of individuals. All of it should appear in the ontogeny of the highest form of life, humanity. *Ontogeny*, in other words, *recapitulates phylogeny*. This was an important element in his evolutionary reconstructions, for he realized there were three sources of information about the path that evolution took: comparative anatomy for determining relationships, fossils for the details of history, and embryology that could be used to fill in missing data because of biogenetic law.

This law revealed Haeckel as a progressionist who envisioned life unfolding along a sequence leading to man—apes were failed attempts to attain humanity. He viewed lower forms of life as only slightly modified relics of earlier evolutionary stages and is no doubt the source of the woes suffered by today's budding evolutionary biologists taking a comparative anatomy class—the reconciliation of evolution with what is presented as an evolutionary progression in the dissections from frog to cat to pig on the pathway to human. But the human races were different species too, according to Haeckel, and they also could be arranged in an evolutionary progression. In his scheme, which combined the new polygenism and progressionism, the races evolved independently, and some races evolved further and lost more apelike characters than others (guess which!).

Through the biogenetic law Haeckel brought an important realization to evolutionary biology—its distinction, as a historical science, from the other derivatives of natural philosophy: physics and chemistry, the purely mechanistic, mathematics-based sciences.[64] Phylogenetics is the study of how genealogical relationships can be determined. It is arguably the most important of his conceptions. He recognized that phylogeny must represent history, by showing descent (he was an artist and the trees of relationship he drew were picturesque). He came to believe genealogical relationships between organisms could be ascertained by amount of shared ontogeny.

But his biogenetic law was more than simply a tool for reconstructing past evolutionary relationships. Haeckel was perhaps the strongest advocate of natural selection as a cause of evolution apart from Darwin himself, and this he used to explain differences between the human races. But natural selection couldn't act alone. Haeckel faced the problem of all polygenists in having to explain the parallel development of the human races, perhaps more than many because of his insistence that they are distinct species. He concluded this because he believed racial differences among humans were equivalent to species differences in other animals. Haeckel turned recapitulation into an active evolutionary force, which through "natural laws" pushed adult ancestral features into the earlier ontogeny of their descendants. Therefore, modern juvenile traits were adult traits of the past. This, combined with the progressionist bent of his thinking, convinced him that the various human species (read "races") represented different evolutionary stages. For some, the evolutionary process did not extend the species far beyond the adult form of the ancestor. These species, the inferior ones, are childlike because they represent the juvenile stage of the ontogeny of advanced species, according to his logic. Others evolved further. Thus the biogenetic law explained parallelisms between the races, while natural selection joined it to account for the differences.

Natural selection was responsible for the different evolutionary progress achieved by the different races. It was a powerful force to be reckoned with in all aspects of evolution and not just the biological ones. In Darwinian evolution, selection acts on the *individual*. Haeckel saw *groups* as another target of selection. He applied to species Darwin's understanding of the pruning action of selection acting on individuals, and he argued human evolution was directed by constant competition between the different human species. Evolutionary models based on species competition are not unknown today,[65] but Haeckel viewed races as species and combined this competitive mechanism with his polygenic-progressionist precept that some human species were decidedly inferior to others: "That immense superiority which the white race has won over other races in the struggle for existence is due to Natural Selection . . . That superiority will, without doubt, become more and more marked in the future, so that still fewer races of many will be able, as time advances, to contend with the white in the struggle for existence."[66]

Haeckel was a powerful supporter of common descent from a single ape ancestor for humans and did not derive races from different ape species as Vogt did (let alone apes from different human species like Klaatsch). His polygenism was an extreme version of Wallace's. The human species each evolved from a different species of ape-man (*Pithecanthropi*[67]) living in different regions. Each attained human status separately, as their human attributes were independently acquired through competition between them. The ability of these species to interbreed did not serve to shake his faith in polygenism. Like many other evolutionary polygenists, he claimed cross-species fertility was common in other animal species. He was more disturbed by the absence of discrete racial boundaries, a bigger problem to polygenists than interfertility.[68]

Haeckel, in the later decades of his life, had a nationalist social agenda. His monism provided an interpretation of Darwinism within the Romantic framework of Naturphilosophie, a popular pantheist German school of thought, romanticizing links between the human spirit, the land, and nature. Links between elements in the natural world, similarities between organisms, and progress were explained developmentally; ontogeny was the unfolding of a plan that explained progress. Haeckel interpreted human actions as a part of nature, justified by the laws of evolution. He saw Darwinism as a social theory: society and nature were, in effect, one to him.[69] Strongly influenced by Goethe, Haeckel believed in a transcendental racial unity of the German people (Deutsche Volk) that held a common spirit that bound them to the fatherland, and through Darwinism he found the mechanism explaining their natural racial superiority. Contrasting with the way that social Darwinism was used in Britain to justify laissez-faire capitalism by showing that individual competition was the natural way, Haeckel applied Darwinism through his theory of the competition between racial groups (species, for him) to explain why the extermination and exploitation of other racial groups were the inevitable and desirable consequences of natural selection. Haeckel believed through natural selection European superiority would result in the extermination of other races and therefore progress for both the species and industrialized nations.

Haeckel's interpretation of Darwinism, combined with the already prevalent idea of Naturphilosophie and Völkische, was adopted by the public and the scientific community. He was an apostate of evolution, and throughout his career he wrote popular books to make evolutionary science available to the masses. His romantic, idealized view of the German Volk and the superiority of the Aryan race was adopted by much of the intellectual community, and especially after the unexpected defeat of the First World War, with the great rips it left in the social fabric, it became well established in German political culture. In 1919, the year of his death, the German Workers' Party took special note of his racial theories and eugenic views with their promise of racial improvement. The next year they changed their name to the National Socialist German Workers' Party.

There is a direct link between Haeckel's interpretation of Darwinism and his version of polygenism, and the biopolicy of the Nazi regime.[70] The biopolicy of the National Socialists arguably formed the central theorem of Nazism, underlying the most infamous political ideology and social policies of 20th century Europe. After all, "National Socialism is nothing but applied biology."[71] It was "the only element of the Nazi era which was neither modified nor manipulated in response to strategic or tactical requirements."[72] There was virtually nothing in the Nazi doctrines that was not put forth by Haeckel and well known and accepted by educated Germans when Hitler was still a housepainter.[73]

Nazi Germany had the first political organization based on an explicit biopolicy.[74] The National Socialists considered this policy to be fully congruent with Darwinism as they understood it from the writings of Haeckel.[75] In his Darwinism, combined with his notion of polygenism, Haeckel provided the scientific basis to support the contentions that

- the human races were different species
- some were far more advanced than others
- competition between them was the main mechanism of their evolution

This was no mere ivory-tower academic principle without significance for everyday life in the real world. Natural selection was not just a matter of upholding the Spartan ideal of smothering weak infants.[76] Haeckel asserted, "The lower races—such as the Veddahs or Australian Negroes—are physiologically nearer to the mammals, apes and dogs, than to the civilized European. We must, therefore, assign totally different value to their life."[77] Not one to shrink from the implications of his scientific views (in this case the view that progress comes from competition), he presaged by decades Hitler's conclusion (in *Mein Kampf*) that war was the only workable mechanism for race competition today, arguing "melancholy as is the battle of the different races of man, . . . might rides at all points over right . . . and in the end . . . the nobler man triumphs."[78] Others were listening carefully.

6

THE LAST STAND

THE ANTHROPOLOGY DEPARTMENT at the University of Pennsylvania, one of the oldest anthropology departments in the country, is housed in the University Museum. The department and the museum library now have their own wing, built in 1971, but until then, anthropology at Penn was in the old building, a Greco-Roman structure designed by Wilson Eyre, a prominent Philadelphia architect, at the end of the last century. It is a beautiful building that conjures up the very essence of Victorian anthropology—romantic explorers' images of exotic peoples and places. The new wing is an imposing edifice of poured concrete and glass incongruously attached to the older structure, a marriage that, like the merging of old and new anthropology, somehow seems to work in spite of its opposing elements.

When I was a student I spent many hours in the anthropology lounge on the third floor of the new wing. It is a large and open room at the junction of a hall of faculty offices and labs and the hallway to the elevators and library. The long hallway to the elevators is narrow and its one concrete wall is bare, but adornment is unnecessary since the other wall is glass from floor to ceiling. Through this one can see the old museum building a few feet away, separated from the new wing by a small courtyard of grass and statues, intersected by the concourse connecting the two structures. The old facade, the best example of Greco-Roman architecture in the city, is also lovely. Its dark bricks of various shades of brown and red and its Italian terra-cotta tiled roof provide sharp contrast for the colorful mosaic tiles that line its windows and architectural features. From the lounge one has an easy view through double glass doors, across hallway and courtyard to the old building where so much of the history of anthropology was enacted.

We both know this lounge well; even twenty years ago when I was studying the Morton collection bones, Milford often visited Alan Mann, his colleague and friend. The lounge's Danish modern chairs that we sat on then[1] are still there, their orange leather now reupholstered in a more fashionable brown, some facing a poured concrete wall, apparently designated the department's memorial gallery. On it hang the photographs of dead or otherwise retired anthropologists once affiliated with the department, forced to look, mostly unsmiling, out at a bank of metal lockers in the concrete room. I remember many days sitting on those orange chairs looking into the faces of old men whose names I had heard of but who meant little to me at the time—knowing that what attracted me to anthropology was different from what attracted them. Now, years later when we go back to visit Penn, we look at that wall with different eyes: it is now invested with more meaning, and there is among the portraits a face that compels more of our attention than the others. It belongs to Carleton Coon (1904–1981), whose very name has come to symbolize one of the major controversies in anthropology, a controversy that we find ourselves reluctantly immersed in even now, years after his last publications and death.

Coon's career spanned a time of many-faceted change in physical anthropology as it grew from a discipline primarily concerned with race and racial typologies to an evolution-based modern biological science. The installation of Darwinian principles as the cornerstone of biological thought was arguably the most important change to affect the biological sciences,[2] altering them from largely descriptive fields to ones with well-grounded theories and methods of addressing the how's and why's of variation in the natural world. Yet for many reasons it took physical anthropology a long time to incorporate the evolutionary synthesis into its paradigms; even now, half a century after the "modern synthesis," the merging of Darwinian selection, genetics, paleontology, and field biology, was first proposed,[3] many anthropologists continue to think typologically. Certainly in the early 1960s when Coon's most influential book[4] was published, few physical anthropologists were yet thinking along populational lines, although many were changing. Notably, Sherwood (Sherry) Washburn was vocally bucking the typological race studies of his Harvard mentor, Earnest Hooton, and first at Columbia, later at Chicago and then at the University of California, Berkeley, he was leading the way to a new expanded, evolutionary, physical anthropology, producing primatologists, behaviorists, and prehistorians.[5] On the other hand, many other workers like Coon remained wedded to the race concept, and Coon, for one, tried to incorporate evolutionary thinking into anthropology in his own way, by forcing a reluctant union of two opposing concepts—evolution and race, as he understood them. But what he came up with was another polygenic model of human origins, probably the last expression of polygenism in 20th-century mainstream anthropology.

Coon's polygenism sometimes appears ambiguous, precisely because it is incompatible with the Darwinism he tried to embrace. In spite of his Darwinian premises, Coon viewed races as lineages with virtually separate evolutionary histories. Coon embodied the end of an era; his was the last generation of an older, descriptive anthropology based on race. Coon's ideas about the origin of races, in part inherited from Harvard and Hooton, his mentor there, can only be understood within this broader context. But the transition between the old and new was not easy and gradual; the new arose and coexisted in tension with the old for decades, reflecting rifts within American physical anthropology that exist to this day.

Many major changes in American anthropology took place in the period following the Holocaust and reflected changes in both evolutionary biology and public attitudes toward race. Two trends were of particular importance: first, the entire concept of race was coming under fire, largely from anthropologists at Columbia University, and second, the modern synthesis promoted a new understanding of Darwinism and natural selection that undermined the validity of the mechanisms of the old polygenism. With the modern synthesis, the progressive evolutionary forces that were necessary to explain parallelisms were increasingly recognized as invalid. Consequently, the modern synthesis should have marked the end of polygenism in scientific anthropological thought, particularly when coupled with the growing disenchantment with the race concept. But once again, Darwinism failed to end polygenism.

HARVARD

That American polygenism did not end was, in large part, because it was so well ensconced at Harvard, seat of the major tradition of physical anthropology in the United States since the field's inception. Earnest A. Hooton (1887–1954), foremost among the Harvard physical anthropologists for the first half of this century, was not an evolutionary theorist; evolutionary mechanisms were not very important in his thinking, and when the new synthesis was formulated, it was not incorporated into his view of science. He also never embraced the view that races did not exist, promoted after the war by Ashley Montagu at Columbia. Hooton's was the intellectual antithesis to the Columbia school, focusing on the definition and study of race instead of its demise. Coon represented the newer generation of Harvard anthropology, incorporating elements of evolutionary theory, particularly adaptation, but maintaining the typological and polygenic views of his mentor.

The movement to professionalize American physical anthropology was a slow one and, in fact, a single individual was largely responsible for the development of the entire field. Had things been different, we might have had two founding fathers. But although Aleš Hrdlička, a Bohemian-born, French-trained anthropologist with the Smithsonian,[6] tried hard to create a Physical Anthropology Institute like Paul Broca's in Paris, he was never successful. When he originated the American Association of Physical Anthropologists in 1928, he envisioned an associated institute that would include museum work and research, but also would be a center for advanced training.[7] However, its funding never materialized, and the training of physical anthropologists remained in the hands of the established universities. Instead it was Hooton, Coon's mentor, who proved to be the wellspring for American physical anthropology.

Earnest A. Hooton.[8]

Hooton, professor at Harvard from 1913 until his death, was responsible for training virtually an entire academic field.[9] Either directly or indirectly he contributed to the education of almost every American-trained physical anthropologist. He spawned several generations of students at a time when few other universities offered physical anthropology as part of their curricula, and while most of these workers are now retired, their students and their students' students constitute most of the profession. Before 1925, only six American Ph.D.s in physical anthropology had ever been awarded, and five of these came from Harvard. They were granted to candidates trained by specialists in other disciplines. In 1913, Hooton was hired by Harvard to replace William C. Farabee, a Harvard-trained physical anthropologist who specialized in genetics and was research-oriented, producing no students. Hooton's Ph.D. was in classics; he had little prior training in anthropology and it took some time to get his training program off the ground. But starting in 1926, a flood, virtually one a year, of his Ph.D.s in physical anthropology emerged from Harvard, including the most influential figures in the discipline. Within a very few years, physical anthropology became a major component of American anthropology.

Hooton formed his scientific opinions about race, as well as his emphasis on race as a fundamental question of physical anthropology, after he arrived at Harvard. There, polygenism was incorporated into his thinking and he disseminated these ideas throughout the burgeoning field. Race studies developed into one of the foci of Hooton's career. As a typologist who came to apply Karl Pearson's biometric techniques to the study of race,[10] he was interested in the definition of races as typological categories. He believed that races could be defined and delineated and that they could be distinguished statistically: it was an anthropologist's job to do so.

This kind of work had a typological premise and was the antithesis to the emerging school that races had no validity as biological categories (socially, of course, races are real entities and one's race, as socially defined, has had and continues to have considerable consequence). Hooton's thinking on race was adopted by some of his students and rejected by others, but in either case, whether accepted or rejected, it strongly influenced subsequent generations of scholars. Polygenism thereby remained fundamental in the thinking of the field whether or not it was agreed on, and this legacy shaped subsequent considerations on the antiquity of human regional variation. The Harvard framework from this time still exists. Within this framework, regional antiquity is automatically understood to mean racial antiquity, and racial antiquity, since old type concepts of race have yet to be truly abandoned, is simply a new manifestation of the old polygenism. Hooton's Harvard students were not incapable of changing, but typological thinking was a consequence of what they were studying. One of the drawbacks of skeletal biology is its emphasis on description, identification, and classification. In other words, it is typological. While description will always be important in skeletal biology, its preponderance is a part of the discipline's historical, pre-Darwinian legacy. In effect, it has been harder for skeletal biologists than it has been for most others to become populational thinkers.[11] It was for the students of Hooton's students to accomplish this last important change.

Hooton was much admired by his students and by now has attained almost mythical proportions. Stories about him abound. In them he emerges as intelligent, humorous and witty, and although his jokes often bordered on the ribald and racist, they were appreciated by his mostly young white male audiences. He drew cartoons and wrote funny poems, some of which he published to illustrate his many books, that kept his friends, students, and colleagues in stitches. He was also very benevolent, interested in the welfare of his students. He nurtured them personally and professionally, helped them find funding for projects, and his friendships with them continued through their professional lives. Indeed, he maintained such a following that it is difficult to discuss his now discredited polygenic and eugenic ideas, although they are common knowledge. His treatment of race was ambivalent. While he spoke out against discrimination, on the one hand, what we now consider racism is apparent in his publications, and even more so in his semiprivate writings and drawings still discussed in the hallways and common rooms of anthropology departments around the country.

Hooton's views on race were clearly influenced by another Harvard anthropologist, Roland Dixon (1875–1934).[12] Hooton's 1931 classification divided races into primary races and secondary races, composite ones that were created by the crossbreeding of the primary races, just as Dixon wrote in 1923. Along with this thinking about pure "types" came a shared commitment to polygenism. This was explicitly put forth by Dixon. According to him, the primary races have great antiquity, dating back to the divergence of humans and apes. Dixon saw the primary human races as equivalent to other anthropoid species, although less variable as they could interbreed.

> The acceptance [of the hypothesis that] the existing varieties of man are to be explained . . . as developed by amalgamation of the descendants of several quite discrete types, places us squarely in the ranks of the long discredited polygenists. But, quite apart from the results of the present inquiry, the whole trend of recent anthropological investigation, together with the archaeological discoveries of the last decade, can have . . . no other outcome than the abandonment of the monogenist position and the frank acceptance of polygenism.[13]

The discredited polygenism Dixon refers to is the creationist polygenism of the "American school" that was supported by Morton and Agassiz in the early and mid-19th century.[14] The "new" evolutionary polygenism he bought into and promoted was even more widely accepted and publicized than its earlier counterpart, largely through the work of Haeckel early on, and it became a major part of American anthropology through the dissemination of these ideas at Harvard. It is perhaps here more than any other place that we find the clearest links between Haeckel and 20th-century physical anthropology.

Hooton accepted Dixon's ideas of polygenism, but more so than his colleague he elaborated on specific polygenic schemes. Like the evolutionary polygenist Vogt of the previous century, Hooton derived human races from different fossil anthropoid species (Hooton called them different genera). By Hooton's time these fossil anthropoids had been discovered, and were, as a group, known as dryopithecines.

> During the Miocene period a family of giant generalized anthropoid apes, the *Dryopithecus* family, which was spread over a wide zone in the Old World, evolved into the ancestors of the existing and extinct forms of anthropoid apes and those of several varieties of man. . . . Certain of the progressive *Dryopithecus* genera took to the ground in several parts of the anthropoid zone. Some of these became the ancestors of extinct human precursors, while others were the progenitors of the lines leading to present day races of man.[15]

In accord with Dixon's views, Hooton believed the races were different enough to be separate species, writing "the differences between the several races are quite as marked as usually serve to distinguish species in animals."[16] Furthermore, he considered the racial lineages as different experiments toward becoming human,[17] but unlike Haeckel, he did not contend that competition between the different lines was the cause of their evolution. Instead, Hooton regarded the course of human evolution leading to the present races to be marked by parallelism, and he played with a Lamarckian explanation for adaptive changes, contending the inheritance of acquired characteristics must be an integral part of any evolutionary explanation.[18] Many of the parallel changes, however, were in nonadaptive features that, to a large extent, were caused by orthogenesis (the inner-directed evolution we discuss in the last chapter). For instance, in his book *Apes, Men, and Morons*, he wrote that the reduction of the jaws in different human lines was due to the sharing of "certain progressive and non-adaptive hereditary forces making for jaw reduction."[19]

HOOTON AND KEITH

Hooton's ideas on race were thus virtually identical with Dixon's and not surprisingly so. They were formed after Hooton arrived at Harvard and faced the problems of teaching. Dixon, a senior member of the Harvard faculty, mentored Hooton.

Hooton earned a Ph.D. in classics at the University of Wisconsin, granted in 1911. He went on to Oxford in 1910 to continue his studies as a Rhodes scholar, but instead of continuing in classics he earned the newly created Diploma in Anthropology. At Oxford his mentor was Robert Marett (1866–1943), a noted classical scholar who had just begun archaeological excavations at the Neandertal site of La Cotte de St. Brelade, Jersey, Channel Islands. Originally a student of philosophy and ethics, Marett also read anthropology. He wrote and presented several ethnological papers on the development of religion[20] and helped found the degree-granting school of anthropology at Oxford. Marett introduced him to anthropology, but the strongest anthropological influence on Hooton came from Sir Arthur Keith (1866–1955). Keith, who played so prominent a role in the Piltdown affair,[21] was a major influence on Hooton's career[22] and they shared mutual admiration for each other. Undoubtedly Keith affected Hooton's thoughts about human fossils and anatomy—they were quite close to Keith's. However, it is unclear whether Keith's very strong, very racist views influenced Hooton at all. On the contrary, Keith's late adoption of polygenism may have been due to Hooton's influence. Keith did not become a polygenist until 1936, well after Hooton was publishing his polygenic model. But Hooton never shared the fervent racism that played an important role in Keith's polygenism when it finally emerged.

A lifelong friend, Hooton was loyal to Keith, sometimes supporting his positions even beyond Keith's own patience with them.

> For many years I have adhered to Sir Arthur Keith's view that *Homo sapiens* is geologically very ancient. . . . Sir Arthur has been wavering from that opinion for the past few years, and now he seems to be in full retreat from this paleontological Verdun. It is not for me to raise again the standard and say, "They shall not pass"; but I think that Keith's former position is still tenable, and in a humble, unheroic way, I am sitting on it.[23]

T. D. Stewart, who replaced Hrdlička at the United States National Museum, observes: "As a teacher and as the author of 'Up from the

Ape,' Hooton has been largely responsible for the tolerant reception" of Keith's views.[24]

That Keith was a *late* convert to polygenism is in itself quite surprising. He should have been a polygenist all his life because of his strong racial views and his notion that racism was a valuable human attribute. He had very romantic views about race and the superiority of the northern European "types" and contended that racial aversions were actually beneficial and responsible for human evolution. He saw races as evolutionary units and held many views similar to those of Haeckel, whom Keith much admired. Haeckel had interpreted Darwinism to apply to an interracial competition for survival and dominance that formed much of the basis of the German biopolicy of the 1930s and 1940s. Such notions of competition between races thrived in a polygenist paradigm, and the idea of racial competition as a motivating factor in human evolution probably influenced Keith's ultimate adoption of polygenism, but, curiously, not until later in his career.

Earlier, Keith had argued vigorously with Hermann Klaatsch about the improbability of as much convergence between races as Klaatsch's theory of independent race evolution required. There was an exchange in the British journal *Nature* in which Keith concluded that the adaptive convergence Klaatsch advocated "passes somewhat beyond the limits of rational speculation."[25] In 1910 Keith felt that a multiple origin of a single species was highly unlikely and all humans constituted one species. Although polygenism was perfectly compatible with his racial views, he could not accept the mechanism Klaatsch proposed. But the disagreement, in retrospect, was perhaps more over degree than kind. Klaatsch, after all, associated the human races with different primate species, and Hooton derived them from different species of dryopithecines, fossil apes. Both of these schemes required parallelisms, or convergences, and ultimately Keith's did as well.

Keith, writing later, also came to believe the races were diverged from different ape species, but more recently. For him the ancestors of the human races were species of anthropoids on the human line who existed *after* the dryopithecine, the orangutan, and later the chimpanzee and gorilla lines branched off. The ancestral human ape evolved into different branches which independently became australopithecines ("Dartians," as Keith called them, after the insightful scientist who first recognized them as human ancestors, Raymond Dart). The different Dartians independently evolved "in parallel" into the main races. Keith reported that in East Asia, the intermediate form of "Peking Man" from the Chinese site of Zhoukoudian was the remote ancestor of today's Chinese—a relation discovered by the German paleoanthropologist Franz Weidenreich,[26] who was then working in China. In Australasia, "Java man" was ancestral to Australian aborigines, and according to Eugène Dubois,[27] discoverer of the first specimen, the links could be seen in fossils such as Wadjak from Java and Talgai from Australia. The Caucasians of Europe, he argued, evolved independently from a yet-undiscovered population from Western Asia.

Keith realized adaptive convergence didn't work as a reasonable explanation of how different human branches could independently evolve into interfertile races, and this was the basis of his disagreement with Klaatsch, but he became so wedded to the polygenic model that he had to rely on orthogenesis as an explanation. He posited a nebulous inbuilt genetic mechanism that created the parallel direction to the evolution of the races. In a 1936 paper, publishing an address he presented to the British Speleological Association, Keith noted: "As in the past, the future of each race lies latent in its genetic constitution. Throughout the Pleistocene period the separated branches of the human family appear to have been unfolding a programme of latent qualities inherited from a common ancestor of an earlier period."[28]

But there was more to the evolution of races than the forces of orthogenesis. The genetic program provided direction, but the engine that drove evolution, he thought, was competition. He posited that the races evolved as they competed with each other, another idea he borrowed from Haeckel. Competition between races was good, in an evolutionary sense, and anything promoting this competition was desirable. For Keith, racism was the main impetus for the kind of racial competition that creates evolution, and it is not at all surprising that he described Hitler as "a naked nationalist, racialist, and evolutionist."[29] Keith also had his romantically derived dislike for other races and romantically elevated opinions of his own.

Racism, in fact, was no stranger to Keith's writing, although he wrote explicitly against it when focused on individual people. But its tie with the evolutionary process provided a veneer of acceptability, if not inevitability, for his attitudes about races as groups. The interwoven elements of racism and evolution can be seen in his discussion of the Jews and anti-Semitism. Keith believed the Jews were a distinct biological race. At the same time, it was his theory that

> human evolution is carried out by group contending with group . . . groups are kept apart and isolated by their mutual antagonisms or aversions. Isolation is a condition that must be preserved if a group is to evolve. It is to the dislike or animosity which separates evolving groups

that I attribute the evil feelings which are so apt to arise in Gentile nations towards their guest communities of Jews.[30]

Anti-Semitism is the fault of the Jews, he reasoned. It does not come from their numbers but because of their separatism (albeit a necessary consequence of the fact they are an evolving race). The proof is that anti-Semitism does not "break out where Jews are most densely planted."[31] Keith argued that while there was anti-Semitism in Poland, where 3.3 million Jews composed 10 percent of the population (this was written after the Holocaust), "in the city of New York Jews now form nearly 20 percent of the population, and yet the city is free from organized outbreaks of anti-Semitism."[32] Keith did not mention that the Jewish community in Germany was the least separate, most assimilated one in Europe.

The Holocaust had no effect on Keith's views of racial competition. He became a polygenist during the Nazi rise to power, and his views did not change even after the atrocities of the Holocaust became widely known. He adopted and promoted Haeckel's polygenism, not Hooton's, and repeatedly published works attempting to justify racial hatreds by incorporating them in an evolutionary context. Like Haeckel, Keith had a romantic view of the relationship between humans and nature. As a romantic in line with the German monists, he believed that the laws of nature applied to humans and that they should be embraced. Therefore, racial prejudice was not something to eradicate—it was something to celebrate. He was "troubled" by the Nuremberg war trials, asserting "hitherto mankind has drawn a sharp distinction between crime and war. The men whom the Allies hanged as criminals will live in the German national memory as patriots and as heroes."[33]

As for the Jews, there was after all one way that they could release themselves from the burden of anti-Semitism—they could assimilate. This was an idea Keith shared with his friend, Hooton.

JEWS

Earnest Hooton was not by any stretch of the imagination a Nazi sympathizer. Still, he didn't necessarily believe in racial equality[34] and was a vocal eugenicist. But his approach to eugenics was quite different from many others, as he focused on the individual and not the race. Hooton believed it was possible, and desirable, to improve each race through negative eugenics,[35] preventing genetically transmitted handicaps from being passed on. "Every tree that bears bad fruit should be cut down and cast into the fire."[36]

His complex approach to race and racism is embodied in his attitudes about Jews; they might border on anti-Semitism by today's standards, but they were really not so much racist as condescending. In *Twilight of Man*, Hooton's essay on the Jews was titled "Noses, knowledge, and nostalgia—the marks of a chosen people." In it, he mixes culture and biology in his argument that "Jewish intellectual preeminence, the hothouse flower of Jewish genius, the virtues of Jewish religion, the peculiar merits of essentially Jewish culture can be preserved only through a maintenance of the traditional policy of inbreeding and social exclusiveness."[37]

But if these merits are maintained by inbreeding, so are the negative traits. Other Jewish stereotypes, which Hooton also considered partially biological in their nature, helped make Jews responsible for their own problems "The Jew possesses as part of his heritage, perhaps reinforced by the traditions of his people, a certain emotional intensity which expresses itself in modes of behavior alien to certain northwest European stocks—especially to Anglo-Saxons."[38]

The answer to Jewish problems was biological assimilation. It would do the species good—the Gentile gene pool would become smarter, and Jews would become nicer, even as they disappeared from the face of the earth.

While critical of Jews—as Hooton was indeed critical of all groups he could put a label on, French, British, and especially Germans—these views were not racist in Hooton's mind because they did not consign all *individuals* to the curse of their blighted ancestry. Combined with his eugenic outlook, this created a mix of ideas not unlike that of *The Bell Curve*,[39] where individuals would be affected by the biological attributes of their social group as a whole. For instance, he once wrote, "We have no proof that racial differences in psychology and in behavior actually exist,"[40] but if they do "they should be studied and explored so that each racial element could be taught to realize its fullest possibilities of success capitalizing its strength, and to avoid certain endeavors which through its peculiar weaknesses lead to failure."[41] To make his position clear, he added, "No race has a monopoly of virtues or vices. . . . there are no hierarchies of mankind by total racial ability, ranging from something a little lower than the angels to something a little higher than the apes."[42]

Hooton was the single American physical anthropologist to rise to the occasion by conducting a campaign against racism. It came about when Franz Boas (1858–1942), the German-born Jewish founder of the American Anthropological Association,[43] became concerned about the fate of Jews in the Nazi state soon after Hitler was appointed chancellor.[44] He organized the Lessing League to combat "the anti-Semitic agitation which is being carried on in this country,"[45] but realized that to be effective a scientific denouncement should not be made by a Jewish scientist.

The scientific community was not at all united on this issue. In the 1934 London meeting of the International Congress of Anthropological and Ethnological Sciences, the great British evolutionary biologist J.B.S. Haldane spoke out against racism, warning his audience against the abuse of science in support of race theories. He was not alone; an antiracist tract was published a year later by Julian Huxley and Alfred Haddon.[46] (It was the harbinger of the later American movement and questioned whether races actually existed, suggesting that "race" be replaced by "ethnic group.") Nevertheless, Boas was unable to convince the Congress to pass a resolution on the issue. Returning empty-handed to the United States, he asked two prominent figures to take public antiracist positions, Livingston Farrand, president of Cornell University, and Raymond Pearl, editor of *The Quarterly Review of Biology* and *Human Biology*. Both refused. He then turned to Hooton.

Hooton held a prominent position and might be listened to as a Harvard professor. But he was a third choice, and in describing his relationship with Boas, E. Barkan terms it "the frustrated antiracist campaign of an odd anthropological couple." The oddness reflected an old tension manifested along the Columbia-Harvard intellectual axis. Hooton represented biological determinism, and his commitment to eugenics reflected how little he thought the environment shaped an individual. Boas was responsible for the unlinking of biological and cultural change,[47] a link that was assumed in the theories of older American anthropologists such as L.H. Morgan.[48] The accelerating changes of the 20th century showed Boas that biological evolution could never be the cause of culture change—it could not possibly even keep up. Boas argued against the stability of many biological features. He examined head form,[49] expressed by a simple index (calculated by dividing the breadth of the head by its length) that was commonly used to distinguish races. He showed that it changed significantly in the first American-born generation of immigrants who came to America. Relatively long-headed people became broader, and relatively broad-headed people became longer. Because the change was rapid and converged on the same head form from different starting points, Boas argued it was environmentally induced—an important trait to distinguish races could easily change from one generation to the next. Hooton had minimized the significance of these results, arguing that the populations were small and might not be stable, but two of his own students, Harry Shapiro (Hooton's first Ph.D. who was, himself, Jewish) and Fred Hulse, soon showed other immigrant populations (the Japanese in Hawaii) underwent similar dramatic first-generation changes.[50]

Hooton often seemed ambivalent on race issues, but he was the only game in town for Boas, and when faced with the German program of race-hygiene (this was in 1935, just at the time when the Nuremberg Laws were passed) he tried to address the problem, proposing that American scientists make a "dispassionate and impartial statement." Hooton wrote the draft of such a statement and sent it to seven leading American physical anthropologists. Only the Bohemian-born Aleš Hrdlička of the United States National Museum[51] would sign it, and a second draft was unsuccessful. Hooton finally published his own "Plain Statement About Race" in the widely read journal *Science*[52] and continued to speak and write on the issue, often combining it with his eugenic ideas. This encapsulated his ideas about human variation; the human species could be bettered by improving each race, not by eliminating one of them.[53]

These and other public positions had little apparent significance, even in the scientific community. In the 1939 meeting of the American Association of Physical Anthropologists, a resolution was proposed that disassociated human racial variation from differences in psychology or culture, discredited "Aryan" and "Semitic" racial categories, and denounced racism. It was not passed, but referred to the executive committee of the association, which took no action. Barkan reports that a year later Hooton's student, William Howells, asked his mentor what could be done. Hooton replied: "Not only has the horse been stolen, but the barn has also been burnt."[54]

HOOTON'S POLYGENISM

Hooton's polygenism mellowed only somewhat over the years. He retained from his earlier ideas a focus on the distinction and degree of difference between the races, which he believed were originally "pure" and unmixed. And he could not give up the precept of separate origins and subsequent convergence, writing in the 1946 revision of *Up From the Ape:*

> The probable cradle of humanity, the site of our prehuman ancestors' first ventures on the ground, is not some single hallowed spot or limited Garden of Eden, but the whole broad area through which the progressive great apes ranged. A terrestrial habitat and an upright gait were not God-given attributes of a single Adam and Eve among the great anthropoids. It is difficult to avoid the conclusion that, of the diverse families and genera of giant anthropoids developed in the Miocene period, several may have taken to the ground in different areas and at various times. Some of these attained a semi-human status, and some achieved complete humanity; some have survived to the present day. . . . If the finds of fossil man hitherto brought to light mean anything at all, they mean that nature has conducted many and varied experiments upon the higher primates, resulting in several lines of human descent.[55]

It is unclear whether his perception of the races and their evolution really changed that much between the editions of his famous book that spanned the implementation and fall of Germany's biopolicy. He still theorized that the white races evolved from the soon-to-be-discredited Piltdown chimera,[56] and that the Australians evolved independently from a separate *Pithecanthropus* stock, but not quite as far—evolution stopped for them.

As a polygenist, Hooton was a moderate. Yet, in the late 1940s, he supported the most ardent and unabashed 20th-century polygenist, R. Ruggles Gates,[57] for instance, writing the introduction to his 1948 book *Human Ancestry from a Genetical Point of View*,[58] although disassociating himself from some of its contents. Gates wrote this after his retirement, while a research fellow at the Biological Laboratories at Harvard (across the street from Anthropology). Gates was a Canadian plant geneticist who for most of his career worked at King's College, London, where he developed the idea of parallel evolution in plant groups. Increasingly, toward the end of his career, he applied these ideas to human ancestry. In doing so, Gates, like Keith, accepted wholeheartedly the fossil evidence of regional continuity, and agreed it meant there was a separate evolution of the races. He in fact argued vehemently that they were different species. Gates accounted for their parallel evolution using the concept of directed, parallel mutations. He elaborated on Klaatsch's scheme, resurrecting an orangoid European ancestor, linked to Europeans through Piltdown, and a gorilloid line giving rise to Neandertals (now extinct) and the African races. Like the French polygenists of the last century, he considered the sterility criterion for species to be invalid and believed, as did Hooton, that the racial differences were of the magnitude of species differences in other mammals (Hooton, however, had not concluded the races were different species). Gates was an extremist and although not alone (Keith and the well-known primate anatomist and evolutionist Osman Hill[59] had written much the same thing), he was not very influential. He had no students (at least in anthropology), and it was hard to take him very seriously, at least in part because the mechanisms he proposed to explain parallel evolution were so weak. Nevertheless, some scientists did believe his work should be addressed. Franz Weidenreich at the American Museum of Natural History argued vociferously against his position in several papers.[60] It probably wasn't worth the effort. Gates was at the periphery of evolutionary science as it was being applied to race and human evolution. At a time when many biologists were embracing the union of genetics and evolutionary thought, Gates was suggesting genetic mechanisms that would even strain most orthogenetic models. There was no place for selection in his scheme, and he certainly never adopted the modern evolutionary synthesis, although he wrote after it.

Unlike Gates and most other polygenists, Hooton is considered an unabashed racist only with difficulty. He recognized the likelihood that there were racial differences in character or ability, but would not rank them. Hooton was unwilling to attribute racial differences to different evolutionary stages. He regarded morphological traits as being more primitive (meaning similar to) or advanced (meaning different from), relative to the anthropoid condition. For instance, he considered the lips and noses of Negroes advanced, while their prognathism was primitive. He did not impart much value to this, however, and made it clear he thought the balance of primitive and advanced features did not make one race "more evolved" than another. Although he was a eugenicist, even after the Holocaust, he never held the ideas of racial hygiene and racial purity that earlier polygenists and the defenders of German biopolicy promoted, because he applied eugenic principles *within* races, not *to* them. In a sense this reflected what was becoming of polygenism. Races were still treated as discrete entities (because of the essentialist world view) and were believed to have by-and-large separate evolutionary trajectories, but it was recognized that they were no longer "pure" and the differences between them reflected the consequences of their intermixture—they might differ in degree but not in kind, as there was ample opportunity for all skills and abilities to spread widely. But Hooton's position went a step further, and unlike most other polygenists, he was able to separate polygenism from blatant racism. It was not a position that prevailed.

Hooton's career spanned the creation of the modern synthesis, but it is unclear whether it had any major effect on him. At the famous 1950

Cold Spring Harbor Symposium on Quantitative Biology, held on the topic "The Origin and Evolution of Man," Hooton's role was restricted to discussing the papers on constitutional types. Chairing one session in which his student Sherry Washburn used the word "population" in every second or third sentence, Hooton complained, "I hate the word *population*."[61] Hooton never theorized about cause or process in evolution. The modern synthesis may have led to the construction of more adaptationist models and evolutionary scenarios, but the treatment of fossil specimens and views on race didn't change that much[62]—they were both still treated typologically by Hooton and many of his contemporaries and students. The essentialist legacy compatible with the non-Darwinian evolutionary models was difficult to shake. As a consequence, while pre- and post-synthesis views of human evolution and the relationship between the races are quite different, the change did not happen overnight, and many writers, while overtly selectionist, do not seem to have fully incorporated populational thinking into their models. And once again, as with the original Darwinian revolution, polygenism did not really die. The face of polygenism changed once more with the modern synthesis, but for most its inherent racism remained unchanged.

Hooton's students incorporated his view of human evolution and race into their thinking, whether or not they agreed with it, as exposure to a world view always affects the invention, perception, and interpretation of ideas. Some, such as Sherry Washburn, openly disagreed with Hooton's views;[63] others such as Stanley Garn and William Howells disagreed more carefully. And still others modified and built upon Hooton's views, incorporating the fossil record into evolutionary schemes, in a way he never had. Although polygenists such as Keith and Gates used observations made on fossil human remains of distinct regional evolutionary patterns to support their polygenic models, Hooton was much more cautious: "Bridging the gap in racial history between dry bones and living persons is a very precarious business. . . . My long and extensive experience in the fields both of skeletal raciology and of the racial classifications of living peoples has made me very critical of my own efforts and those of other anthropologists."[64]

Hooton, as described above, was an extremely influential figure in American physical anthropology. To a great extent American physical anthropology *is* what Hooton put in his monumental textbook, *Up From the Ape*. His students comprise virtually all of the second generation of American physical anthropologists—most of Hooton's contemporaries had museum or anatomy department positions and had few or no anthropology students. Hooton's students founded physical anthropology programs all around the country, and most American physical anthropologists today can trace their academic lineage back to him.[65] In 1928, Carleton Coon was Hooton's second Harvard graduate. Of all his students, Coon had the fewest qualms about bridge building over the gap between dry bones and living people. Coon differed from Hooton in trying to extend the ancestry of the living races into the past, using evidence from the fossil record. But Coon's view of race was much like Hooton's even if its application to fossils was more like Keith and Gates.

COON'S POLYGENISM

Coon arrived at the University of Pennsylvania in 1948, after teaching at Harvard since 1934. He left Harvard just one year after being promoted to full professor, a position he received upon his return from his wartime activities in the OSS (the forerunner of the CIA). He had remained affiliated with Harvard since his graduation in 1928. Anthropology was very holistic in those days; much of what we consider physical anthropology was still practiced by anatomists and physicians, and it would be more accurate to describe Coon simply as an anthropologist. He was a generalist—an ethnographer (he studied the customs of living peoples), an archaeologist (he dug), and a physical anthropologist (he was interested in race and in skeletal material). He promoted the synthesis of the subdisciplines and was quite successful in some areas. For example, in a 1950 paper on culture and biology, he examined the interface between culture and genetic change. Examining ethnographic data from an evolutionary perspective, Coon analyzed cultural differences in sexual selection as expressed in mate choice, female fertility, and child survivorship in terms of how genetic change is affected. This was at a time when anthropologists were first turning to examine the details of how culture influences human evolution, a focus pioneered by some of Hooton's students,[66] and Coon's statement was the first detailed exposition of it. His pioneering efforts exemplified an idea developing throughout physical anthropology, a focus on the relationship between evolution and culture that became, and should have remained, the profession's unifying principle.[67]

Coon was from a well-to-do New England family, crusty, hard-boiled and independent, and he was attracted to anthropology at an early age by its promise of adventure. The field fit well with Coon's self-image as a rugged, worldly explorer-adventurer, laughing in the face of danger—the image of Indiana Jones. He had no particular interest in scientific philosophy, he was not a great theoretician, but he was fundamentally interested in people. Coon had a self-avowed facility with languages—and indeed virtually all subjects except plane geometry—and found adventure and excitement in his field work. He was attracted to the biological aspects of anthropology by Hooton when he was still an undergraduate at Harvard, and he maintained a biological focus throughout his career. This, by definition, meant a focus on race. Nevertheless, unlike most of Hooton's other students he was first and foremost an anthropologist, and he was truly a generalist, much of whose work would be labeled ethnographic today. He was really not an evolutionary biologist in the sense that physical anthropologists dealing with human evolution are nowadays; he had no principal interest in broad biological theories or paradigms but was mainly concerned about physiological mechanisms that allowed given adaptations to environmental stresses and were the prime movers for adaptive variation. He was interested in race, and although he had a typological concept of race, much of his work focused on the adaptive significance of racial features. He wanted to do more than simply pigeonhole people into racial types, and his work represented a significant break with the physical anthropology of his predecessors.[68]

Yet in his thinking about geographic variation (races) and temporal variation (stages of human evolution), Coon was a typologist. Although he believed he was bringing evolutionary theory to the question of race and race origins—and at the time he was widely perceived as having done so—his was a meld of the new *and* the old, not a substitution of the new *for* the old. Coon never showed the commitment to the importance of variation that is the foundation of the new synthesis—the importance of understanding the *pattern of variation* within as well as between populations. He used evolutionary theory to the extent that he was an adaptationist: he recognized that racial features evolved and was interested in the adaptive significance of *typical* racial forms. We do not mean to imply that he misunderstood selection; certainly, he recognized that differential fertility and mortality acted on variation to cause change in populations—this was an important part of his 1950 paper. But this somehow never seemed to become truly incorporated into his scientific world view and when considering race he thought about types, and how to describe them, not variation and how to describe it. His questions were basically typological. While he recognized that races were complicated and that "clinal zones" (regions of intergradation between two different things) existed between geographically varying parts of species (i.e., subspecies), he believed these subspecies represented distinct types, delineated by their differing adaptations and distinct boundaries:

> Over the border, which may be a natural barrier such as a range of mountains or a patch of desert, or even a critical isotherm, may be found another subspecies of the same species, equally well established in a state of equilibrium with its environment. As the two environments differ in certain details, so do the two genetic structures of its occupants. . . . In each territory natural selection keeps the gene structure of the local subspecies constant by also eliminating unfavorable genes that flow over the border.[69]

But there is more. At the geographic nucleus of subspecies one finds relatively pure types which have not been diluted by admixture. He wrote further, in *The Origin of Races:* "Each of the five subspecies recognized in this book was firmly and uniquely installed in its geographic center. Between the nuclei of these five centers lie intermediate regions . . ."[70] Minor variation came from mixing between these types. It was a hurdle to overcome or ignore, not to examine for its own meaning, and he spent much of his career trying to establish "true" divisions of humankind and further refining and defining racial categories metrically and morphologically.

The subspecies could represent temporal as well as geographic variation. In its temporal use, Coon described subspecies not as a division of a continuum, but similar to the geographic use as "steps in a single evolutionary line."[71] Thus Coon treated racial categories and evolutionary stages the same way, as concrete and meaningful, and in a sense he organized his thinking about humanity around these typologies. When he met people in virtually any context, he immediately classified them in terms of ethnic or racial origin. While he did not see this typologising as prejudice, and he was generally very good-natured about the racial or ethnic groups he recognized, he clearly had expectations of the kinds of behavior he could expect from members of different racial groups. There was more to this than the famous comparison of a Chinese scholar and a Aboriginal Indigenous Australian that caused him so much grief, but this episode was telling. In *The Origin*, in a set of pictures of individuals from different races and some fossils and their reconstructions, the last picture contrasted "the Alpha and Omega of *Homo sapiens:* An Australian Aboriginal woman with a cranial capacity of under 1,000 cc. (Topsy, a Tiwi); and a Chinese sage with a brain nearly twice that size (Dr. Li Chi, the renowned archaeologist and director of Academia Sinica.)"[72]

The "isms" in this are overwhelming, especially given his contention that the Mongoloid race crossed the "*sapiens* threshold" first and thereby evolved the furthest, while Aboriginal Indigenous Australians "come closest, of any living peoples, to the *erectus-sapiens* threshold."[73] Topsy was a small-brained Aboriginal Indigenous Australian woman (although one might wonder how the brain size of a living person is known), presumably because Aboriginal Indigenous Australians crossed the "*sapiens* threshold" much later, she was not even worthy of having her Tiwi language name reported ("Topsy" is an English nickname). As Montagu puts Coon's position: "Of course there are cultural differences, but the implication is no matter what cultural advantages Topsy [with her small brain] had been afforded, or those of her children, neither she nor they could have achieved what Dr. Chi [sic!] has achieved [with his large brain].[74]

Fundamentally, Coon could not shake (more important, never saw the importance of shaking) the typological way of dealing with human variation that most of us grew up with in one way or another. This did nothing to dilute the importance of most of his work on human variation and adaptation. He was a major contributor to the field, particularly in regard to physiological responses to stress in the living peoples he studied.

Coon was born in 1904 in Wakefield, Massachusetts, where both sides of his family had lived for several generations. His paternal great-grandfather settled in Wakefield, newly arrived from Cornwall, and his maternal great-grandfather moved there from Maine. His family was well connected and wealthy, and both his father and grandfather were interested in travel and foreign places. Although they were in Wakefield for only a few generations, his family was an integral part of the history of the town. Coon grew up in a place and time where all citizens of his little town were discussed in terms of their religious and ethnic identity as indicators of their social status. He lived there until being sent to school at Andover in 1919, having been expelled from Wakefield High School for destructive boyish pranks. After graduating from Andover, he went on to Harvard, where he pursued his ambition to be an explorer. Among other things, he took Egyptology (he had learned hieroglyphics at Andover) and was especially interested in learning Arabic for several reasons, but most of all because "I wanted to learn to pray flawlessly so that when I became an explorer I might pass as a proper Muslim, like my fabled heroes, Sir Richard Burton and Charles Doughty."[75]

As a sophomore he took Anthropology I with Hooton and was hooked. He changed his major from English to anthropology and in the next few years took all anthropology classes offered, also working in Hooton's laboratory. He graduated in 1925 and continued as a graduate student, receiving his Ph.D. in 1928 at the age of 24. Coon did his field work in Morocco on the tribes of the Rif; his dissertation was titled *Fundamental Racial and Cultural Characteristics of the Berbers of North Africa as Exemplified by the Riffians*, exemplifying both his grounding in four-field anthropology and his descriptive/classificatory focus. He worked, as did all Harvard students at the time, with the whole faculty, but he had a special relationship with Hooton, and considered himself his student and later his protégé. While he was attracted to the romantic and exotic aspects of the field, he always maintained a focus on furthering Hooton's studies on race, sometimes in the context of his travels and sometimes in more mundane analytical environments. He remained affiliated with Harvard and continued research in North Africa and Albania (where he was arrested by the Serbian authorities and accused of being a Croatian spy), and a little later in Saudi Arabia, returning between stints to live in Cambridge where he was mostly supported by his father. In 1934 Coon was hired in a real position when he became an instructor at Harvard. He was promoted to assistant professor in 1938, replacing Roland Dixon, who died in 1936, but soon left to joint the fledgling OSS in 1941, an ideal choice for someone of his disposition and training. He returned to Harvard academia in 1945, appointed as an associate professor and rose to full professor by the time he left in 1948. He taught and published on race: *The Races of Europe* came out in 1939; his first joint book, with Garn and Birdsell, *Races, a Study of the Problems of Race Formation in Man*, was published in 1950. He also coauthored a general textbook and reader during this time. During the war he brought his knowledge of the Arab world to his OSS position and recounted his many adventures in his autobiography and his book, *A North African Story*. Returning to the stodgy world of Harvard academics must have been difficult.

Coon left Harvard for Penn in 1948, lured by a faculty more sympathetic to his adventurous tastes and the promise of virtually unlimited time for field work and a curatorship at the University Museum. He had only to teach a half course one semester a year, if so doing did not interfere with his field work. That offer would bring anyone to Philadelphia, even those like Coon with a horror of having to live in a rowhouse in the city. But he didn't need to worry about that. Coon was soon at home on the Philadelphia Main Line, with a direct train route to both the university and his club on Rittenhouse Square, surrounded by neighbors, "Proper Philadelphians" who, like the Coons, maintained New England ties with summer houses in Maine or Gloucester.

The University Museum, located near the university's border with West Philadelphia, a slowly decaying neighborhood, stood in splendid contrast to its urban surroundings, housing a faculty of whom many were out of touch with the changing world around them.

This myopia was by no means characteristic of all anthropologists, but in a sense a clash of classes had entered anthropology (reflecting trends within many academic disciplines) where a growing number of nonelite intellectuals spawned a still growing dichotomy between different social and academic philosophies. Anthropology at a growing number of institutions was characterized as removed from the ivory tower and deeply involved in the surrounding sociopolitical realities. Many anthropologists were scientists who thought their work directly applied to the world in which they lived and who felt a responsibility to that world. Yet other academics, often attached to old, elite institutions, remained distanced from social reality and, although they were social scientists, felt no responsibility to society. They used their science, their positivistic belief that data speak for themselves, and their status as "objective observers," as buffers between themselves and the rather uncomfortable reality of the less fortunate world around them. Coon was one of these positivist scientists. To the end of his days he maintained the importance of his objectivity and blamed Franz Boas and his followers (who arguably both intellectualized anthropology and made it relevant to the real world) for destroying the field. Indeed they may have destroyed Coon's vision of an anthropology more adequately represented in an Indiana Jones movie than in any recent pages of the *American Anthropologist* or *Current Anthropology*.

One might think Coon's relationships with the subjects of his ethnographies is paradoxical in this regard. He became involved with and had true affection for the people he encountered; for example, he invited a Riffian friend back to Cambridge with him and said in a letter to his granddaughter shortly before he died that "nothing I have written ever since compares with these books [his early books on the Rif] *because I was a Riffian*" (Coon's italics).[76] However, one never gets the feeling that he thought like a Riffian, shared Rif concerns or world views, or truly thought that it was important to be able to do so. He saw his work and his relationships with informants as a series of adventures, and he enjoyed his contacts enormously. Yet, he was strikingly unaware of social inequities and the responsibilities of colonial regimes of his region. He was the epitome of the colonialist anthropologist. Coon was proud that Hooton favorably compared him to T. E. Lawrence (Lawrence of Arabia), reproducing that passage in his autobiography:[77] "[Coon is] a bit like Colonel Lawrence and a great deal like Sir Richard Burton, possibly a little erratic, and with more than a touch of genius."[78] Coon later met the Colonel's brother, William Lawrence, who told him that after T. E. Lawrence sold many of his books off, Coon's books on the Rif were still among his favorites. Coon was clearly pleased that the admiration he felt for Lawrence was mutual.

We think the comparison between Coon and Lawrence is an apt one. Lawrence epitomized the romantic image Coon had of himself and of the adventurer-explorer anthropologist he saw representing the discipline. In fact, arguments surrounding the nature of Lawrence's attitudes and relationships with the Arabs can be equally applied to Coon. Both men were flexible and adapted well to very different lifeways and indeed felt it was important to do so in order to know the people well. Both spoke Arabic and lived among their subjects, conforming outwardly to their customs for the sake of acceptance. Both developed a great deal of respect for the people they visited. Yet modern Arab analysts generally depict Lawrence as a representative of colonial interests and world views, with an intrinsic assumption of European superiority. The same appears to be true of Coon. While he relished individual relationships, he never questioned the validity of Western influence (domination) over the rest of the world and in a sense treats his positive relationships with informants as fulfilling the "white man's burden." Except for the occasional danger his presence conferred in some exciting situations, he believed his contacts with informants were in their interests. He was convinced absolutely of European superiority as evidenced by European cultural achievements, and he felt this was most likely a consequence of evolutionary history. This notion was incorporated in his 1962 book *The Origin of Races*.

Given his lack of involvement in the social disparities of the early 1960s, Coon was truly surprised at the reception *The Origin of Races* received from the anthropological and biological communities and the general public. Coon was suddenly thrust into the political limelight as his work became a center of controversy, adopted by the right and maligned by the left. Although he clearly loved to be the center of attention, this was not the kind he bargained for. *The Origin of Races* pushed American anthropology into an uncomfortable, face-to-face confrontation with its own racism, bringing to a head controversy that had been brewing for years. *The Origin* was at once an excellent and a terrible book. Its avowed purpose was to bring evolutionary principles to Franz Weidenreich's ideas of human evolution. Weidenreich, who died in 1948, was a German anatomist and paleoanthropologist who postulated that some regional features, such as the flat faces of East

Asians, were as old as the human occupations of different regions.[79] Coon set out to write an evolutionary history of the races, each of which he believed had "followed a pathway of its own through the labyrinth of time."[80] Each had been molded in a different fashion to meet the needs of different environments and each had reached its own level on the evolutionary scale. Coon set out to recount this history.

When *The Origin of Races* was published in 1962 it was the most comprehensive treatment of the human fossil record yet published. In it Coon also discusses virtually all of the topics that are necessary for an understanding of human prehistory—humans as social animals, the relationship between social and biological evolution, the order primates, human growth and development, human adaptations, the archaeological record, virtually everything that is relevant to modern biological anthropology. Coon covered evolutionary principles and most of the fossil record in detail, and like no other worker before him except Weidenreich, Coon paid attention to geographic variation as well as temporal change in his treatment of human evolution. *The Origin of Races* was one of the most valuable source books ever published, and most paleoanthropologists would have been quick to use it in their introductory classes, if not for one fatal flaw: Coon's interpretation of the fossil record and other data he presents.

Based largely on the measurements he collected or could find in earlier writings,[81] and the craniofacial features earlier emphasized by Weidenreich, Coon traced the modern races into the past, before the emergence of *Homo sapiens* to a prehuman species known as *Homo erectus*. Coon believed that humankind could be divided into five major geographic races, "human subspecies," able to interbreed but effectively isolated from each other. These were the subspecies of *Homo erectus*, and his thesis was that they independently became the subspecies of *Homo sapiens*. The subspecies were geographically defined, one centered on each of the continents (except Africa, which had two): the Caucasoids (Europe and Western Asia), Mongoloids (Central, South, and East Asia), Australoids (Greater Australia), and Congoids and Capoids (African). Where they met, they bred, so along the peripheries of the geographic centers for the different subspecies were "clinal zones" that were effectively zones of hybridization. Each subspecies had biologically based attributes that typified them and differentiated them from one another. Each subspecies had its own evolutionary history that transcended species. Each was older than *Homo sapiens*.

Coon argued each of these subspecies crossed a rubicon, a "*Homo sapiens* threshold," at some point during its evolutionary history, and this was the major transition to humanity. The threshold was not an arbitrary division between fossils imposed by human taxonomists, but rather reflected an important biological change, arguably the most important biological change in human evolution, because this change made us human. The core of Coon's claim is that the subspecies of *Homo erectus* evolved into *Homo sapiens* not once, but five independent times, and that these times were different. Human races maintained their integrity throughout the entire Pleistocene (for at least 1.75 million years). Subsequently, humanity was achieved independently for each of them, and for these fundamental reasons Coon's scheme was a polygenic one.

The idea that the races evolved at different rates, crossing a *sapiens* threshold at different times, is quite like Haeckel's (who thought the races were different species, but recognized there were no distinct boundaries between them and that they could interbreed). Haeckel believed each race had traveled a different distance from an apelike ancestral condition, stopping at different points along the evolutionary trajectory. This is not much different from Coon's contention that the races to cross the *sapiens* threshold first evolved the furthest. Coon, like Haeckel, does admit to genic exchange between the subspecies. His defenders[82] point to this fact as they contend he was not a true polygenist because his scheme was not one of independent evolution. But his belief that the races evolved at different rates and became human at different times surely shows that *independent* evolution was the overwhelming signal he read from the fossil record. After all, as the Russian-born émigré geneticist Theodosius Dobzhansky (1900–1975) noted in his critical review of *The Origin:*

> The specific unity of mankind was maintained throughout its history by gene flow due to migration. . . . Excepting through such gene flow, repeated origins of the same species are so improbable that this conjecture is not worthy of serious consideration; and given gene flow, it becomes fallacious to say that a species has originated repeatedly, and even more fallacious to contend that it has originated five times, or any other number above one.[83]

Gene flow is an important element in the evolutionary process. It is a potential cause of evolutionary change, one part of a more general force of evolution called genic exchange. As we use the terms today, genic exchange includes two mechanisms of change. Migration involves the movement of genes when it is caused by individuals moving, and includes individuals entering or leaving a population, introducing or removing genetic material. The second, gene flow, the movement of genes as entities, might occur when populations come in contact and mates are exchanged, when a member or members of one social unit join another, or when mate exchanges are formalized as in exogamy rules that forbid choosing a mate from one's own group. These can cause genic exchanges to extend over wide regions, just as a wave crest moves across a body of water without any particular water molecule moving very far. Dobzhansky uses gene flow differently, as a synonym for genic exchange. This process could not have played a significant role in Coon's model, as Dobzhansky realized, and thus Coon brought polygenism into the second half of the 20th century.

Coon believed that crossing the *sapiens* threshold was the result of mutations that caused endocrine and neurological changes that ended in the major differences between the species: namely, in "intelligence, self control, and the abilities to provide food efficiently and to get along well in groups."[84] The threshold could be recognized based on several morphological features, the most important of which was brain size, but also changes in curvature of the cranial bones and tooth size, particularly in tooth-size/brain-size ratios.[85] The rubicon was set at "about 1,250 to 1,300 cc," although Coon recognized that some *Homo erectus* specimens may be larger than that and some *Homo sapiens* are smaller.

Given the prominence of brain size in his evolutionary model, in *The Origin* Coon says surprisingly little about its importance in modern humans. However, he implies that it has great significance and meant to write a book on the brain and behavior.[86] He stressed the importance of brain size for language and culture (a word he avoids for its lack of agreed upon meaning, although it is quite important in his earlier writings). As the primary feature of the *sapiens* threshold, large brain size, along with the endocrine changes that Coon believed were concomitant with brain expansion, was the primary source of our humanity. Coon implied in *The Origin* that brain size is linked to intelligence and subsequently defended the proposition. In his 1981 autobiography[87] he disdains the oft-used argument against a brain size and intelligence link: the contention that the well-known author Anatole France's small brain (1,000 cc) proves that the two are unrelated. France's brain, Coon contends, is not actually as small as thought (this is in response to some critical comments made by Ashley Montagu in his review of *The Origin*[88]). Although Coon presents no data on brain size variation within living populations (surprisingly, in a book so laden with facts and figures about humans), he does point to small brains in some living Australians as evidence for their of lack of evolution much past the *sapiens* threshold. There is also the "Alpha and Omega" photograph, with comment.

Coon's book linking the brain and behavior was published posthumously in 1982 as *Racial Adaptations.* In it, he proposed the theory that brain size increased from a late *Homo erectus* mean value of some 1,200 cc to 1,600 cc in a single successful mutation that he believed doubled the cortical surface area (the cortex, or outer layer, of the brain is where most of its electrical activity is observed.[89] When this mutation spread throughout one of the subspecies, it had become human and had crossed the *sapiens* threshold. The mutation began somewhere in Eurasia, and reached Australasia and Africa last.

While Coon considered the threshold real, and believed that *Homo erectus* and *Homo sapiens* were truly different species, and that *Homo sapiens* had a different genetic makeup that allowed for more successful *sapiens* adaptations, he did not rule out genic exchanges between the species. The brain size mutation could spread between adjacent subspecies by genic exchanges. Therefore interbreeding between *Homo erectus* and *Homo sapiens* must have been possible and would help the humanization process.

In fact, all modern human populations do not average 1,600 cc; indeed, few do. The worldwide variation in brain size corresponds to a number of variables, including climate.[90] What accounts for modern brain size variation? Coon suggests:

> Our evidence indicates that wherever and whenever they crossed the threshold to *Homo sapiens*, all of the races acquired cranial capacities of that size, whether or not the environments in which they lived made them need such large and expansive brains. If they didn't, their brains grew smaller until they had reached their respective optima. . . . [Brain size] fell back to *Homo erectus* level or a little higher in the tropics.[91]

In Coon's polygenic model, the length of time a subspecies had been in the *sapiens* state was linked to cultural achievement. While he noted that even the most "advanced" races included culturally backward populations, particularly in conquered areas away from the ancestral homeland,[92] he also believed that

> it is a fair inference that fossil men now extinct were less gifted than their descendants who have larger brains, that the subspecies which crossed the evolutionary threshold into the category of *Homo sapiens* the earliest have evolved the most, and that the obvious correlation between the length of time a subspecies has been in the *sapiens* state and the levels of

civilization attained by some of its populations may be related phenomena.[93]

Even before its publication, *The Origin of Races* was panned.[94] It sparked enormous controversy in the anthropological community, and with an increasingly socially conscious academic body, it was generally poorly received. It is too bad, but perhaps unavoidable, that many of the very good aspects of the book were overlooked in the face of what to many was a polygenism-based blatant racism. Coon was attacked on scientific grounds, which, though valid, were socially motivated. It was criticism that stemmed not only from ideological differences about race and politics, understandings of evolution and species, but also from the sociological makeup of the field itself.

Coon's most vocal critics were Dobzhansky and Montagu, whose criticisms are perhaps best expressed in a famous *Current Anthropology* exchange,[95] and the debate over Coon and his work went on. Washburn, who was president of the *American Anthropological Association* at the time, was highly intolerant of the book. In his address to the 1962 meeting of the association he blasted *The Origin* and its author: "Sherry tore into Carl like you wouldn't believe."[96] Much of the audience stood up, cheering, at the end of the speech. It was widely perceived that *The Origin* was a racist book and this anthropological audience knew it; not from reading the enormous manuscript, but from the fact that is had been warmly received by the racist press.

Washburn, speaking for many with his "new physical anthropology," was not only against racism but against race itself as an organizing principle for research. In the published version of his speech (which eyewitnesses agree differed markedly from the speech he gave), Washburn wrote: "Race, then, is a useful concept only if one is concerned with the kind of anatomical, genetical, and structural differences which were in time past important in the origin of races. Race in human thinking is a very minor concept."[97]

These comments fell on more than receptive ears. This was a time when young Turks of the profession were claiming there was no such thing as race at all.[98] It was a quite different direction than Coon had traveled.

THE RACISM ISSUE

Coon pointedly thought about race as the object of his research. He was a selectionist and understood the adaptive significance of most racial features. In fact, most of his work on race is about just that. He understood that morphological differences were products of the environment, and therefore he had to work to establish links between culture and race. Like the polygenists before him, he established this on the basis of brain sizes, not only of living people, but of their fossil ancestors. But at the same time the sentiments in his *magnum opus* about race, *Origin of Races*, were racist and they were dangerous. Various right-wing individuals and groups were quick to quote more racist sections of the book, even prior to its publication, for all the racist reasons people embraced polygenism a century ago. It is true, as Coon complained in his autobiography, that his critics do misrepresent him to some extent by ignoring his mention of genic exchanges in the text. They criticize him for the absurdity of having the independent evolution of five subspecies to the *sapiens* state, but of course, he does attribute some importance to population contacts and mixture. The telling point, though, is that these subspecies became human *independently*, and it mattered *when* they did so. This is the kernel of validity in the criticisms.

Coon did little to defend himself except to say he was not a racist and was simply reporting objectively derived science. He wrote in response to Dobzhansky's later (1968) criticisms: "Were the evolution of fruit flies a prime social and political issue, Dobzhansky might easily find himself in the same situation which he and his followers have tried to place me."[99]

But as Marks quips, "We can marvel at Coon's naiveté"[100] in this statement, and in any event, in spite of protestations to the contrary, Coon was not a dispassionate observer. Nor was he unaware of the import his work would have. He recognized that while the mechanisms for crossing the *sapiens* threshold would not be of as much social concern, importance would be attributed to his postulated times for and order of the threshold crossings. Almost unbelievably, he thought this would be due to a snobbishness over order of arrival held by the general public. He did not agree there should be:

> The implication is that whoever came first is thereby best, a logical fallacy because in some environments climatic pressures cause some subspecies to become adapted more rapidly than they do in others. [Yet, unfortunately], it makes a difference in status whether your ancestors came

over on the *Mayflower* or on an immigrant ship in the early 1900s, all else being equal.[101]

Based on the misguided snobbery of the common man, Coon expected criticism from the left "not from American or other so called Blacks, but from their so-called white protagonists."[102] Coon believed that painting him as racist resulted from fallacious thinking, because in his view the races started out with equal capacities, and crossing the threshold first had nothing to do with who was best at the time. Again, we marvel at Coon's naiveté. His racism does not lie in his depiction of the relative equality of races *before* they crossed the *sapiens* threshold; his racism occurs in evolutionary arguments for *modern* inequality! Coon himself (sadly restating Haeckel) proposes that cultural achievements of each race are related to length of time in the sapient state, and so while no innate qualities of different races may have allowed them to cross the threshold at various times, the time of crossing was important for their future cultural development and their "level on the evolutionary scale." In truth, Coon does not need to be misquoted or misrepresented to find strong evidence of his views of racial inequality.

The point has been made in many places that racism in historical figures can only be judged in the context of their times. How do we interpret statements and phrases that would be clearly racist if uttered in the 1990s when they come from the pen of Thomas Jefferson? One way is to compare his writings with those of other commentators on the human condition of the same time period. Using this method it is more difficult to consider Jefferson a racist. He was inevitably constrained by his culture and prevailing climates of opinion.

Admittedly, the America into which Coon was born and raised was extremely racist. Schools and public facilities were segregated. African-Americans often couldn't vote and were relegated to a virtually invisible subservient class with little political access. The establishment, including much of the intellectual establishment, held entrenched values that predicated inequality. No doubt, Coon was raised in an environment where assumptions of nonwhite inferiority were unchallenged. In spite of this, however, it is difficult to escape the conclusion that Coon was a racist,[103] particularly as his own social context came to include the academic community.

By the time that Coon wrote *The Origin of Races*, challenges to racist assumptions and values were prevalent in the political and intellectual world around him. The civil rights movement, while still in its youth, was making gains and had powerful proponents that actively supported it. Freedom marches, sit-ins, and other forms of organized passive resistance were well under way and attracting considerable attention. Certainly any well-read American and—one would suppose, any academic American—would have thought about these challenges to his value system, however well entrenched.

Within anthropology there were strong antiracist sentiments that escalated following the Holocaust. The brutal lessons learned, and the unfortunate role anthropology played in the philosophy of the Third Reich, caused many within American anthropology to turn inward and examine American racism and the role or duty of anthropology to combat this evil. From its racist origins, anthropology was becoming an increasingly liberal discipline that actively supported civil rights. Two decades before Coon published *The Origin of Races*, Ashley Montagu had written *Man's Most Dangerous Myth*, and since then others continued to present arguments against the very existence of race. Anthropologists had drawn up the UNESCO statement on race that deemphasized biological differences and encouraged recognition of the importance of cultural diversity. The racism in *The Origin*, while probably a fair reflection of the intellectual attitudes of the country as a whole, countered at least one large channel in the mainstream of physical anthropology.

Therefore the racism inherent in *The Origin of Races* is not a consequence of historicism but of Coon's world view, tenaciously maintained in spite of (or perhaps because of) exposure to different world views and changing social tides within both anthropology and the society it is part of. Coon represented and embraced a conservative side of the field, one that was losing ground as anthropology came to hold more liberal views. But it was not simply a matter of Coon keeping old-fashioned ideas that were being swept away by a new liberal tradition. Twentieth-century American anthropology was continually pulled by two different factions epitomized by Harvard and Columbia. Each represented a very different social aspect of academia. Harvard epitomized the "old-boy" tradition in which anthropology was an elitist profession. Coon saw himself within it akin to the old English explorers of the Victorian era. Moneyed and socially connected, they would go to the colonies, have great adventures, shoot some lions, talk condescendingly but kindly to the natives who were carrying their bags (whom they often imagined were their friends), and come back to the club, have whiskey and tell of their exciting adventures to the other members. Coon identified with this society. He treated anthropology as a sort of explorers' club, with science thrown in to give it intellectual credibility. His science was positivist, and he thought of it as completely objective. This is how he saw his life and anthropology, as a statement about himself. He was not a racist in the sense that he wanted to discriminate actively against the underclasses, but there is no doubt that he had absolutely no sense of social responsibility. In fact, he felt this diluted the objectivity that was necessary in science.

On the other hand, the New York school, whose apical figure was Boas, was quite the opposite. Dominated by a central European intellectual tradition, these workers brought a different kind of science and theory to the discipline and emphasized a holism in anthropology that focused on culture and relativism.[104] Boas especially worked on race, society and evolution, and he had strong social commitments. He was a fiery intellectual, not an old boy.

Boas trained as a physicist in Germany, receiving his doctorate in 1881. He began as a philosophical materialist, reaching his maturity in the heyday of monism.[105] But much like Coon, he had a growing interest in exploration and took additional training in Berlin as a geographer (which in Germany at that time greatly overlapped with ethnography), learning what he needed to know to travel away from civilization. During this time he began to shift away from his earlier materialism, and he was intellectually prepared for what he would learn of other cultures when he mounted an expedition to Baffinland in 1883. Slowly, his interest shifted from questions about Eskimo migration routes to a desire to understand what determines human behavior. He was deeply impressed by how the customs and traditions of the Eskimos he lived with, expressed in their technology, society, and psychology, held together. What he had yet to learn was their arbitrariness and his inability to show the causal effect of geography, meaning the environment, on their forms. The relations were clearly much more complex than he expected, and he soon came to realize that they must be seen in their historic context, later clarified as their *cultural* context. These intellectual developments finally drew him to anthropology, a field that he first embraced and then broadened and finally reinvented in its modern form.

By the time Boas first came to Columbia, in 1896, he was deeply concerned with questions of race and racial features.[106] He had researched problems of variation and change in both Europe and America, and would soon do much more. It was then that he laid out his vision for the development of anthropology in America.[107] It was based on the theoretical orientation he was developing in response to the prevalent evolutionism that linked culture to biology and arranged races, with their cultures, on an evolutionary scale. Focusing on technology, he realized cultural elements were so widely diffused that no civilization "was the product of the genius of a single people."[108] The diffusion process was historical, and he came to understand that "historical events appear to have been much more potent in leading races to civilization than their faculty, and it follows that achievements of races do not warrant us to assume that one race is more highly gifted than the other."[109]

Thus, Boas was out to show there were no necessary links between biology and culture or race and behavior. Most of the 19th-century theories linked biological change and social change. This was one of the main justifications of ranking the races "biologically" in terms of their evolutionary advancement based on their cultural achievements. Boas brought the concept of cultural relativism to the issue of whether this linkage was valid and raised the question of whether one could be ranked relative to another, whether cultural change could be described as progress. Certainly some aspects of culture such as technology could be said to advance, but in general, changes could not be described as progressive. There is no sense in which one kinship system or language could be characterized as more advanced than another. The perspective of cultural relativism is that cultures must be viewed as a functioning whole, their elements evaluated only in the cultural context. Combining his understanding that biological change could not cause cultural change because it was much too slow, and his realization that significant biological differences were not inherited, Boas taught that cultural evolution was not a history of progress or of race. Cultures could not be lined up in a sequence of evolutionary stages. The biology of different human populations was not the cause of differences in their cultures.

These were times when the subdisciplines of anthropology were taught together, and there were no narrowly specialized cultural anthropologists, anthropological linguists, archaeologists, or physical anthropologists. Anthropologists were all four-field, but they had different emphases in New York and Cambridge. As the subdisciplines became differentiated, the Boasian school would spawn the new generation of ethnologists and Harvard the physical anthropologists, the field continuing on those lines drawn in the early part of the century. However, although it trained relatively few physical anthropologists, Columbia had a profound effect on the field. It may be no accident that Hooton's student Sherry Washburn, perhaps more responsible than anyone else for the new face of postwar physical anthropology, went first to Columbia with his Harvard degree.

Coon was left behind, but he was not merely an anachronism. He did straddle changing times in physical anthropology, the shift from a focus on race to an evolutionary core that was incompatible with such typological notions. He tried to bring both together and failed. But this is not where he got into trouble; it was in his clear insensitivity to social issues, born of his attitude toward anthropology that was engendered by his social class and training. Coon did not meet disaster because he was old-fashioned. He was distanced from his critics by a philosophical divide—he practiced a different, and not altogether older, kind of anthropology that became increasingly invalidated as the field professionalized, attracting members of society who were not from the typical source populations of the elite universities.[110] An increasingly liberal society, conscious of and worried about race, spoke to the more liberal elements of the field and increased their dominance within it.

Coon felt embittered for the remainder of his life by the dominance of the academic left. They attacked him as racist, made him feel ostracized from his profession, and reminded him that his field had, for all intents and purposes, changed from one dominated (and funded) by private men of wealth to an open discipline that largely exists on public funding. He attributed this "downfall" of anthropology to Boas, who he claimed treated him with disdain and always disliked his work on race. He bemoaned the fact that

Carleton Coon.[111]

> the snake of racial consciousness had raised its head out of the Central European bulrushes, largely through the cult leadership of Franz Boas. His devotees leaked introspection into our curricula, turning both physical and social anthropology into political forums . . . as . . . the Boasian doctrine spread, expeditions, research and publications gradually ceased to be funded by the elite and well-to-do. Public money had to be sought and the subjects had to meet the public's taste. . . . Anthropology is fragmented and in the public domain.[112]

The Coon episode reflects a number of things, perhaps most important that it was no longer possible to combine the modern understanding of evolution with any form of polygenism. The new evolutionary synthesis and separate origins just don't mix. Coon's problems came from more than a single person or school. Nor was he simply an anachronism, a man out of time. His most egregious errors lay in his misuse of science; he made inferences that the data could not support. As Marks recently put it: "Coon's mistakes were inferring race from fossils, using cultural criteria for ranking races, and ranking races on very poor evidence by inferring different times for becoming human."[113] The data did not speak for themselves; it was unwarranted to analyze human evolution as Coon did, even accepting the validity of the same observations that were important in his thinking. Many of his critical mistakes lay in his use of science—made from within his own field of expertise and caused by his self-avowed goal to mix race, racial typologies, and human evolution.

7

THE STRAW MAN

WHEN MILFORD WENT TO SCHOOL, fifth grade was the "science grade." The board of education deemed Chicago's children old enough at this stage to learn the wheres and whys of the world, a welcome break from trying to learn the state song. Science was presented in broad and exciting ways, at least at Jamieson elementary school on the city's north side. This was one of many of the board's experiments, and much more successful than the first one Milford encountered. His generation was taught to read by word recognition rather than by sounding out syllables, plaguing him with spelling problems that were never corrected until computer spell-checks were invented.

Milford's science year did not have a great start. There was a science fair in the fall, and as his project Milford brought some salamanders he had found in northern Illinois; he presented them in a wooden cage with a large water bowl to keep their skins wet. Unfortunately, the projects were kept over the weekend and the water dried up, as did the salamanders. Later that year there were student papers with presentations. Milford was assigned human evolution. At least the fossils were already dead.

Earnest Hooton's classic, unique text, *Up From the Ape*,[1] was a logical source of information, and the revised edition wasn't that old, but it was too difficult for a fifth grader. The more accessible books Milford used were *Meet Your Ancestors*[2] by Roy Chapman Andrews (1884–1960) and *Mankind So Far*[3] by William W. Howells, both first published in the mid-1940s. Andrews' book was more interesting for a fifth grader. He wrote not only of primitive prehistoric humans and their lives, but also of the exciting expeditions that sought their remains. Andrews worked at the American Museum of Natural History. As much a sportsman as a scholar, he led several American Museum exploratory expeditions, including a 1925 visit to central Asia in search of everything from dinosaurs in the Gobi Desert to the origin of *Homo sapiens.*

The views of human evolution these two books presented were quite different. Although now, 40 years later, he is much more familiar with the works of Howells, Milford remembers the stronger impact Andrews' book had on him as a boy, drawing him to the field. Howells' effect, however, extended well beyond elementary school. William W. Howells, a grandson of the famous author William Dean Howells, was Hooton's seventh Ph.D. student, a 1934 Harvard graduate.

Howells' views pervaded his popular book, but were most accurately expressed in a scientific paper he wrote at about the same time, 1942, "Fossil Man and the Origin of Races," which specifically addressed and thoroughly dismissed the polygenist approach to race origins. Howells was working at the University of Wisconsin[4] while his polygenic professor remained very active at Harvard. Yet, Hooton's polygenist position was not mentioned. It is as interesting to see who Howells *didn't* target for criticism as it is to see who he did in this important essay published in the widely read *American Anthropologist*, the journal of the American Anthropological Association.

Howells' paper ignores Hooton's contention that the races had independent origins in different primate species. In fact, he only mentions Hooton in a footnote, citing his (purely Dixonian) conjecture that one possible explanation for the features of the Nordic race is that it represents mixture between modern humans and Neandertals.

He does mention Hooton's friend Arthur Keith, citing several editions of *The Antiquity of Man* as discussing the fossils but with "no attempt to make a real system out of the Hominidae."[5] But he ignores Keith's theoretical framework, expressed clearly in his address to the British Speleological Association[6] as in several earlier papers,[7] embracing the notion of a parallel, independent evolution of races. Instead he turned his attention to Franz Weidenreich, the paleoanthropologist who had strongly influenced both Keith and the author of the competing textbook, Andrews.

Weidenreich provided an excellent target: a well-published, active scholar who, then 68, epitomized the "old school" for Howells in figurative and literal ways. Coming to the United States from China, where he directed the Cenozoic Research Laboratory of the National Geological Survey of China that excavated the human remains at the famous Zhoukoudian ("Peking Man") fossil site, he was the model of the central European gentleman of the last century, complete with mannerisms, style, and accent. Howells, in his 1981 essay on Weidenreich's life, tells this clearly favorite story of a 1940 interaction with the United States National Museum's Aleš Hrdlička, the Bohemian-born founder of the American Association of Physical Anthropologists (or, as it is abbreviated, the AAPA):

> Weidenreich was presenting to the AAPA meetings in New York his study of brain size and skull form in dogs. He preserved strong marks of central Europe in his accent and ear, and so did Hrdlička, who thought he was talking about *ducks.* Since both words came out something like 'dahks,' things were resolved only after an interval of much confusion.[8]

He presented Weidenreich's views of human evolution as polygenism, requiring "parallel evolution going through a series of stages in all regions of the Old World." Howells described Weidenreich's theory as one of "races derived from parallel phyla"[9]—in fact, that "the present races have arisen from different strains of sub-*sapiens* species of the genus *Homo*, by a general process of convergence."[10]

Howells' focus, in his 1942 essay, was on the distinctness and extreme antiquity that races seemed to have according to Weidenreich's scheme. In later works, when Howells talked about ideas of race being distinct and having great antiquity, he would mention Weidenreich and Hooton in the same breath: "Weidenreich believed the first racial division in man was very ancient, as you have seen. Hooton also seems to have felt that races were at one time *more distinct and well defined.*"[11] He quotes Hooton as saying that most variation today is not between "pure racial types" (though Hooton did think that pure types once existed) and that we are not in the "primary race making phase" in our evolution, nor have we been for at least 10,000 years. Howells goes on: "This of course does not hold a candle to Weidenreich's estimate of the whole Pleistocene since races first diverged. *But both views strongly suggest a day when races were pristine and pure.* It is an idea which needs reexamination."[12]

Howells believed that fossils were too fragmentary to be used in this reexamination. He devoted most of his very active career to an "objective" analysis of the much more complete and infinitely more numerous skeletal remains of living people. By "objective," like Hooton he meant a mathematical analysis, but one based on examining the relationships and variation of many measurements at the same time, a task made practical on the high-speed computers first available in the 1960s. He became a pioneer in these multivariate techniques.

Howells clearly considered Weidenreich's views to be an extreme version of Hooton's polygenism. In his very popular 1959 textbook, *Mankind in the Making*, he summarized Weidenreich's views in the form of a candelabra, which he described as a "modified (and exaggerated) representation." According to Howells, Weidenreich's "central idea," which he expressed in a diagram of parallel vertical lines, "is the one I have described, of parallel evolution."[13]

The candelabra continued to be Howells' interpretation of Weidenreich, extending through his scientific career to the most recent of his highly successful popular books on human evolution, the 1993 *Getting Here*. In it, Howells says that the "Candelabra Hypothesis . . . was first expounded in 1939 by Weidenreich." He goes on to describe the candelabra as "an exaggerated mental diagram, in which populations of *Homo erectus* spread out widely at the base, as in a candelabrum, with separate local ascents from that point to the modern races of *Homo sapiens*."[14]

The intellectual relationship between Weidenreich and Howells was always contentious. The year after Howells' 1942 paper Weidenreich wrote a response, "The Neanderthal man and the ancestors of *Homo sapiens*," also published in the *American Anthropologist*.[15] He focused on Howells' claim that Neandertals were a distinct species and expressed special irritation about Howells' statement that "everybody agreed" with his interpretation of Neandertal—Weidenreich most

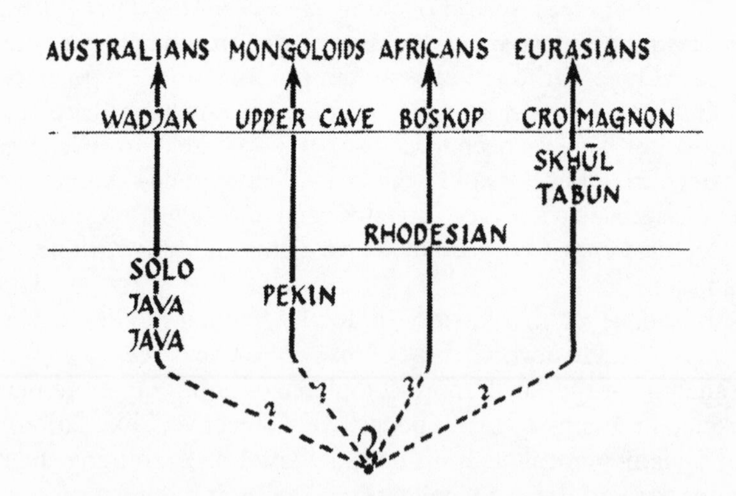

Weidenreich's theory as a candelabra. (From Howells, 1959.)

certainly did *not* Weidenreich argued that all Neandertals were not alike—he recognized three different groups—and that *some* Neandertals were ancestral to later humans. Elsewhere, still addressing Howells, he maintained that the later Neandertals of Europe, the "Classic Neandertals," were the one group which was *not* ancestral to moderns:

> The seeming anachronism that classic Neanderthal man lived later in Europe, in spite of being more primitive than the more advanced Ehringsdorf group, can be explained by the arrival of the Wiirm Glaciation. This drove the group into neighboring regions free from ice and brought with it a more arctic form possibly better adapted to the new environmental conditions.[16]

However, the other two Neandertal groups were ancestral to later Europeans, and the similarities of the three Neandertal groups showed they could not be linked as a distinct species "unless 'species'.. . stands merely for type and means no more than that this type differs from modern Man by certain features."[17] Howells, he contended, was both incorrect and self-contradictory in his claims.

Even inadvertently, not specifically addressing Howells, Weidenreich seems to have stood against much of what Howells championed. For instance, writing on metric analysis, one of Howells' major contributions to paleoanthropology, Weidenreich was quite critical:

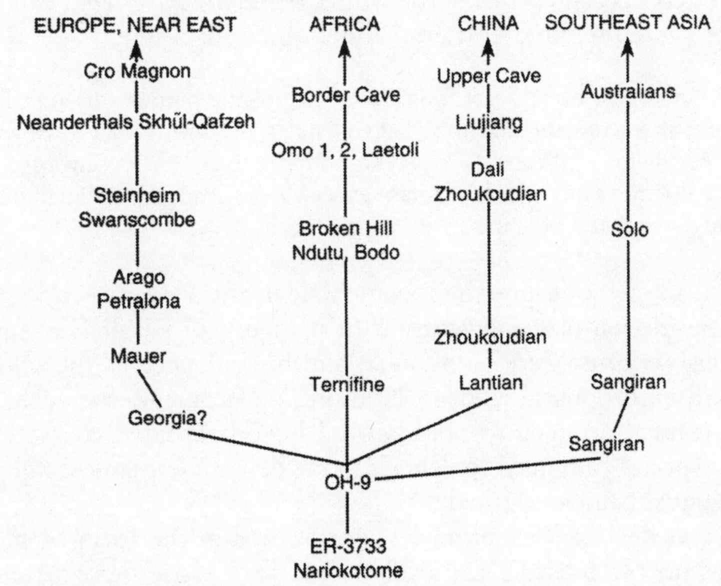

The candelabra. (From Howells, 1993.)

> There were anthropologists who spent all their time in finding new measurements, indices and angles not applied before, hoping they might make distinctions more secure. But only a few took pains to test whether the demonstrated differences were of any diagnostic value. Everyone who is familiar with this problem is aware of the lot of useless ballast which is carried through the literature and piles up further every day.[18]

Howells, in turn, did not much like Weidenreich's theory of human evolution or even his approach to evolutionary problems. In remarks he made as a symposium chairman in the 1950 Cold Spring Harbor Symposium on Quantitative Biology on the Origin and Evolution of Man, meeting just after Weidenreich died, he promoted the inductive approach in which "the data speak for themselves."

> Let us look at the fossils first as indicators of the general process and scheme of development, until we have far more of them, instead of feeling obliged to put them into a given line at whatever effort, or use them as links in a chain already visualized. This was Dr. Weidenreich's impatient attitude: he constructed a chart, which always reminded me of Dalton's table of atomic weights, not only giving all the known human fossils precise places, but providing other precise places, like a stamp album, for those not yet discovered.[19]

Howells characterized this periodic table of hominid evolution as a polygenic model, which he described as meaning: "Human evolution progressed in each corner of the world, *essentially apart* from what was happening in the other corners."[20] He protested:

> Weidenreich has at least four different evolving human varieties, living far apart, moving ahead by fits and starts, producing their own special peculiarities of form. . . . Yet these four careers at last converged to produce the same kind of man everywhere. And all, miraculously enough, breasted the tape at the same time.[21]

Many of our colleagues have come to identify Weidenreich this way, as the 20th-century polygenist with a theory of parallel evolution. Howells' writings were so popular, and his influence as the scion of Harvard anthropology, indeed Hooton's replacement, was such, that there seemed no need to read beyond his oft-repeated characterizations, especially no need to delve into the dense, Germanic-English of the scientist being described.

A good deal of Weidenreich's writing was in the form of monographs, on Neandertals and even earlier prehistoric human remains from Java and China. Why read these publications, now more than a half-century old? In American culture "old" is anything but "good," and it is widely observed among academics that students find virtually nothing more than five years old to be worth reading. It is not unusual for students (and professionals) to learn the ideas of the classic scholars in their field through secondary sources. In our time of increasing specialization and the information explosion this is becoming even more prevalent. There often just isn't time to keep up with everything, and secondary sources have become even more valuable. When paleoanthropologists or paleoanthropologists-in-training do read Weidenreich's monographs, it is most often for the anatomical details and comparisons (especially since many of the fossils described in these monographs have been lost or destroyed), not for the theories, broader views, or explanations that are also there.

The Weidenreich most scientists "know" is the result of his reputation in textbooks. General knowledge about Weidenreich's views comes from secondary sources, in books and papers meant for both students and the public written by educators and popularizers, which underscores the extent to which our field is a public science. Thus, to take some current examples: Francis Clark Howell, the recently retired Berkeley paleoanthropologist, wrote a paper summarizing recent developments in paleoanthropology, in which he cites, with approval, Howells' description of Weidenreich's ideas as a candelabra, and describes Weidenreich's theory as requiring "unparalleled parallelism"[22] (just as Howells did decades earlier[23]). The American Museum of Natural History paleontologist Ian Tattersall, in a popular book on Neandertals, depicts Weidenreich's "theory of independent descent of the major modern racial groups."[24] Authors as diverse as the Stanford geneticist L. Cavalli-Sforza and the Liverpool paleoanthropologist Bernard Wood refer to Weidenreich's "candelabra theory."[25] Roger Lewin, the widely read popular science writer, discusses Weidenreich in a recent book, *The Origin of Modern Humans.* He describes his theory as the "candelabra model" and provides a picture of a multicolored candelabra[26] illustrating a theory in which the evolution of different human races went through the same stages (or species, it is unclear which the colors represent) independently.

WEIDENREICH'S LIFE

Franz Weidenreich was a superb functional anatomist, perhaps the greatest ever to work in paleoanthropology, who during the latter part of his career described and analyzed many fossil hominid remains. He is most famous for his monographs on the "*Sinanthropus*" fossils from Zhoukoudian, the famous "Peking Man" material,[27] but the last 20 years of his life were devoted to the most detailed descriptions of human remains from several sites, on different continents, and the analysis of their significance in his developing scheme of human evolution. His publications continue to epitomize descriptive monographs for paleoanthropologists.

Weidenreich was born in 1873, in a village near Edenkoben in the Bavarian Palatinate of Germany.[28] He was educated in medicine and related sciences, receiving his M.D. from The University of Strasbourg in 1899. His interest and research were then in the field of hematology. He wrestled with the universal problem of how form is related to function, a theme woven through his many different studies over the course of his life, examining this relationship in the anatomy of the blood and lymph cells. For a long time thereafter he was known as "blood Weidenreich." For the next two years he retained positions at Strasbourg (assistant and then lecturer), where he worked under the famous German evolutionist Gustav Schwalbe and soon became his close personal friend and collaborator. After a brief hiatus when he worked with Paul Ehrlich at the University of Frankfurt, he returned again to the anatomy department at Strasbourg in 1902, where he stayed as professor of anatomy, from 1904 until 1918 at which time he was dismissed after the French took over Alsace-Lorraine. His dismissal was part of the general replacement of Germans by French in the region's infrastructure, but may well have also been induced by his political activities; he was president of the Democratic Party of Alsace-Lorraine and an active member of the municipal council of the city of Strasbourg. These were sensitive positions during the war years.

Weidenreich in Strasbourg. His family kindly allowed us to publish this private picture.

During this period, the prime years of his career if Weidenreich had been like most academics and had not (as the Chinese curse) "lived in interesting times," he greatly expanded his interests. From blood, he had reached out to other tissues of the body, among them the muscular and skeletal systems. Following 1904 he published repeatedly on the evolution of the human chin and its significance for speech. His interest in this and other aspects of functional anatomy as they pertain to human evolution were ignited during this period through his close relationship with Schwalbe. In 1906 Schwalbe invited Gorjanović-Kramberger to lecture in his department at Strasbourg. We can never be sure that Weidenreich was in the audience, but it seems likely. Perhaps they talked. Gorjanović also went on to publish on the evolution of the human chin, in 1908 and 1909, and Weidenreich's extraordinary 1937 monograph on the dentition of the Zhoukoudian (China) hominids was dedicated to Gorjanović.

In 1913 Weidenreich published on the pelvis and its transformation in human evolution as our ancestors became upright bipeds. Later he wrote extensively on the evolution of human feet, tracing their changes from grasping functions similar to gorilla feet to the human structure with a double-arched base and a powerful big toe. In those days his interest was focused on the elements of the human locomotor system and their evolution. Based on this anatomical system, one of the unique distinctions of humans, he reflected the views of Schwalbe and agreed with the earlier German scholars like Haeckel in deriving humans from an anthropoid stock. But Weidenreich's researches suggested it was an ancient divergence, not from any recent ape.

Looking for a good position after the war, he somewhat reluctantly settled for the professor of anatomy post at the University of Heidelberg in 1921, where he stayed until 1928. He began this period of his scientific studies with an analysis of the effects of muscular action on the form of the skeleton.[29] He became particularly interested in human evolutionary problems from the perspective of a functional morphologist, asking questions about how the body worked. Weidenreich first focused on the shift to bipedalism. He didn't particularly ask why humans became upright walkers, but was intent on detailing the morphological changes and biomechanical consequences this entailed. Throughout the 1920s he continued the thrust of his earlier work, publishing on the foot, on other aspects of the lower limb, on the spine and the skull, and later on the hand.[30] Weidenreich's interests continued to broaden and he was soon also publishing on the jaws and teeth, an important element in the studies of his later career. He did virtually no phylogenetic theorizing[31] at this stage of his life and was really interested in the development of anatomical systems; he was a comparative anatomist. However, it was during this period that he increasingly focused on the problem of how the functional adaptations he studied came to be and why variability persisted in the face of successful specializations.[32]

Weidenreich left Heidelberg to become the prestigious professor of anthropology at the University of Frankfurt am Main, where he remained until 1934. He founded the Institute for Physical Anthropology there, for the study of anatomical variation, and lectured in paleoanthropology at the nearby Senckenberg Museum (a natural history museum with a major focus on paleontology). In 1928 at the age of 55, he published a description of the Ehringsdorf cranial remains, his first major work in paleoanthropology, written while he was still at Heidelberg. The Ehringsdorf publication demonstrated his expertise with the European fossil record, developed during his close association with Schwalbe, who was, after all, the leading German paleoanthropologist of his generation. But his view extended far beyond Europe, even in this first serious foray into the details of paleoanthropology. As a good comparative anatomist Weidenreich published comparisons of the Ehringsdorf cranium, which he had reconstructed, with *all* of the then-known fossil specimens that related to Neandertal evolution. He discussed and pictured newly discovered specimens such as the Kabwe skull (also known as the Broken Hill skull) from Zambia, found in 1921, and the Zuttiyeh face from Israel, found in 1925.

Weidenreich saw human evolution as a problem of *anthropology* and not paleontology. He never worked on the fossil record of any other species except the human one; beginning with his initial work in hematology he was primarily a human-focused comparative anatomist, developing through the years a major emphasis on skeletal morphology. With his position now as an *anthropology* professor, he felt he could focus on human *evolution.* Yet, one oft-arising theme in the long after-dinner conversations he often had with his family was that the study of human evolution must begin with the details of human anatomy.

The Ehringsdorf work established Weidenreich as a theoretician in paleoanthropology as well. The same year, he published a paper, "The Evolution and Racial Types of *Homo primigenius*"[33] (Schwalbe's and most other central Europeans' term for Neandertals), in *Nature and Museum*, a popular publication of the Senckenberg Museum. This paper, his first theoretical one in paleoanthropology, advanced the theory that there was a "Neandertal phase" to human evolution, a theory much like that of his professor Gustav Schwalbe. This was just one year after Hrdlička's "Neanderthal Phase of Man"[34] was published in the *Journal of the Royal Anthropological Institute of Great Britain and Ireland.* The men were clearly in communication. Weidenreich was aware of Hrdlička's work and his opinion about the place of Neandertals in human evolution, recognizing it to be very similar to his own. Hrdlička, in turn, cited Weidenreich (as well as others such as Gorjanović-Kramberger) as believing that Neandertals "did not completely die out, but became gradually transformed into later human forms, from which in turn developed man of today."[35] He, quite properly, traced this belief to the "authoritative notions" of Schwalbe.

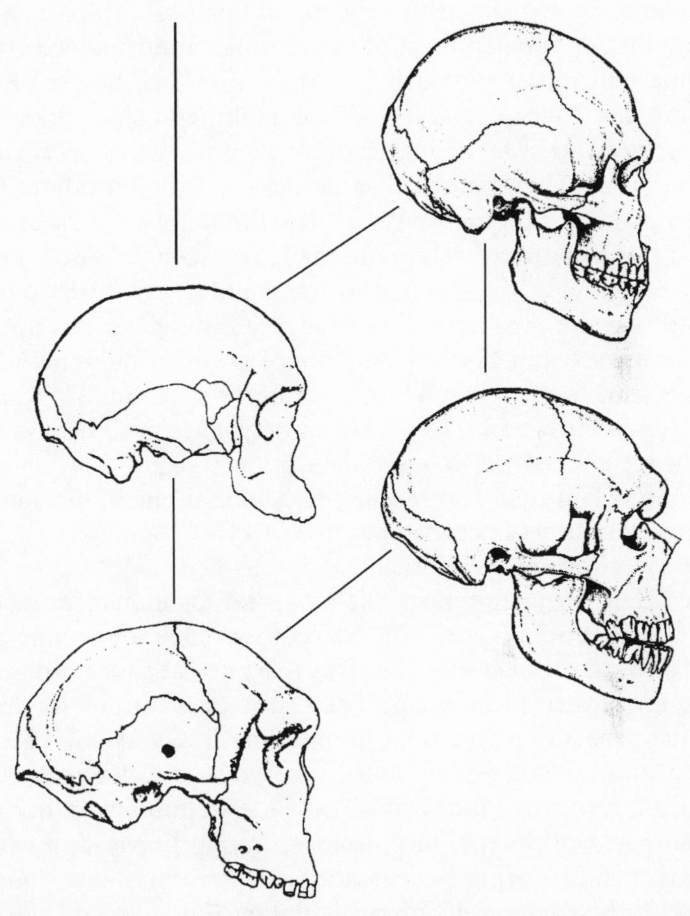

Weidenreich thought the Neandertals represented the same relation to later Europeans as Kabwe represented to later Africans. This figure[36] *illustrates his ideas, showing that in Europe the La Ferrassie Neandertal* (bottom right) *is ancestral to the later Upper Paleolithic Předmostí 3 male* (above), *just as in Africa Kabwe* (lower left) *is ancestral to the later Jebel Irhoud* (above). *But at the same time, he believed Kabwe was old enough to be ancestral to Neandertals such as La Ferrassie. Later archaic Africans such as Jebel Irhoud (unknown to him) were presumably potential ancestors for archaic post-Neandertal Europeans as well. La Ferrassie is from France, Předmostí from Moravia, Kabwe from Zambia, and Jebel Irhoud from Morocco.*

With his usual, and more often than not justified, disregard for the dates claimed for specimens, Weidenreich considered the Kabwe skull to be the earliest, most archaic *Homo primigenius*, ancestral to the further evolved European Neandertals. These two different Neandertal races, in his interpretation, were not just evolutionary stages. They also reflected geographic difference, and were in some measure ancestral to the correspondingly different African and European races of today.

Weidenreich was affected by the preoccupation with race and racial hygiene that dominated most of the German anthropological field at this time. His initial approach to variation in fossil humans incorporated the race concept, and in 1927 he published a short book on *Race and Body Form*. In it he examined two different ways of organizing human variation. The first, racial variation, was from his point of view a phenomenon of geography. Yet he saw the number of races as arbitrary. It was a matter of definition: the fewer the number of characters used for classifying, the smaller the number of races. Moreover, he recognized that the geographic basis for race differences was not stable, and that the races themselves were not of great antiquity because they were constantly changing. Thus, there never were "pure" races. As he wrote later: "Any search for stable archetypes . . . will be condemned to failure. . . . Crossing is not a late human acquisition which took place only when man had reached his modern phase, but must have been practiced ever since man began to evolve."[37]

The races were all hybrids, and always had been.[38]

The other organizing principle he found for human variation was that of constitutional type. This concept was common among anthropologists of the last century as well as this one. The idea is that people can be characterized by a correlated suite of features—for instance, muscular versus slim versus fat in one American version. These were not just meant to be descriptions of the body, but also related everything from features of the head to disease susceptibility to (in one later version) personality type. For instance, among Europeans, who were first classified according to constitutional type, the slender body type was said to be correlated with a tall and narrow nose and a relatively long head. Many workers of that time associated constitutional types with racial typologies. However, in his 1927 book Weidenreich showed that these constitutional types could be found in all races, albeit in different proportions, not only the general body form but (in his words) also "the related physiognomical traits." He pointed out that unlike races, which are *geographic* variations, the constitutional types are *individual* variations. However, like the races, the number of individuals who correspond to the "ideal" constitutional type is very small.[39]

Around the turn of the decade Weidenreich published several papers and articles on race, including one on the Jewish race. He was an active liberal and began to lecture publicly on the topic, and publish in the popular magazine of the Senckenberg Museum, as well as in more technical journals. Most of these dealt with the impermanent nature of races. In particular he noted that many so-called racial features, including those used by the increasingly important Nazis to demonstrate superiority, were actually related to the constitutional types that he had shown to crosscut all racial groups. Weidenreich had already provided ample grounds for Nazi ire. In *Race and Body Form*, he had noted that the elongated head form that Nazis believed characterized the "Nordic type" their racial purification laws were aiming for did *not* characterize the greatest Germans, including Beethoven, Goethe, Kant, Leibnitz, and Schiller. Now he argued that several so-called racial features used to characterize certain European populations were adaptive responses to climate and altitude and could easily change from generation to generation. This was not all well received by the anthropological academic community of the time, increasingly dominated by race-hygiene workers, and Weidenreich uncomfortably noted that his teachings "contradicted Nazi ideals."[40]

Although he didn't consider himself a Jew,[41] Weidenreich was of Jewish descent and his position was increasingly difficult; finally, he saw the handwriting on the wall. In June of 1933, six months after Hitler was named chancellor, he wrote to Franz Boas at Columbia University that he had been able to keep his job for so long because he was "not politically active."[42] For some time he believed he could continue working in the Nazi state, even suggesting to Boas a year later that they jointly establish an institute for the study of the biology of Jews. But at last Nazi biopolicy caught up with even apolitical, nonreligious Jews;[43] Weidenreich was obliged to resign from his professorship because he was considered Jewish (he was lucky to remain as long as he did—employment of Jews in academic positions was illegal after the civil service law of April 1933) . His physical anthropology institute was closed down, as the idea of an anatomically based physical anthropology fell into disrepute.[44]

Under the influence of Germany's leading physical anthropologist, the notorious[45] Eugen Fischer (1874–1964), the domain of physical anthropology broadened to include genetics, blood-type variation, and constitutional types, but its focus shifted to become racial studies (*Rassenkunde*). Concerns of the discipline moved away from morphology and systematics, as physical anthropologists became increasingly active in public science, framing eugenic policy, determining paternity, and establishing links between behavioral (psychological) and racial variables. The paleoanthropologist who described the famous Steinheim skull and posited a chimpanzeelike ancestor for humans, Hans Weinert, became an adviser for the Office of Racial Policy,[46] and others who had published on the human fossil record went to work for the SS.

Weidenreich left Germany in 1934, at the age of 61, and accepted a position at the University of Chicago as professor of anatomy. But he left for Beijing, China, the next year to replace the Canadian Davidson Black, who had just died unexpectedly. He was chosen by the China Medical Board of the Rockefeller Foundation, on the recommendation of Keith,[47] to be appointed the visiting professor of anatomy at Peking Union Medical College and honorary director of the Cenozoic Research Laboratory, which the Foundation funded. Thus began the most productive part of his career as a paleoanthropologist. Weidenreich spent the wartime years working in China, overseeing the Zhoukoudian excavations and, as detailed in a deluge of papers and monographs, analyzing the fossil human remains from that site. Following his training and earlier work in Europe, he became intimately acquainted with the fossil record of a second region, East Asia.

The most famous remains from the Lower Cave at Zhoukoudian were those of "Peking Man," a sample that many paleoanthropologists formally place in the species "*Homo erectus.*" Weidenreich called them "*Sinanthropus pekinensis*" for convenience, although in reality he believed they were *Homo sapiens.* His use of "*Sinanthropus pekinensis*" was a convenience

> . . . without any "generic" or "specific" meaning or, in other words, as a "latinization" of Peking Man. . . . it would not be correct to call our fossil "*Homo pekinensis*" or "*Homo erectus pekinensis*"; it would be best to call it "*Homo sapiens erectus pekinensis.*" Otherwise it would appear as a proper "species," different from "*Homo sapiens,*" which remains doubtful, to say the least.[48]

The "*Sinanthropus*" remains were of a type of human living in northern China 400,000 to 600,000 years ago. The Zhoukoudian sample of "*Sinanthropus*" is a large and diverse accumulation. Specimens were discovered in excavations continuing to the present, leading to one of the largest human collections from a single site, more than 45 individuals. The Zhoukoudian site consists of a number of caves clustered in a range of low hills bordering the town of Zhoukoudian, located just 50 km to the southwest of Beijing.

The Zhoukoudian cave overlooks a broad plain, an ecotone boundary between evergreen forest and lower, more open grasslands. Humans lived in the rock shelter portion of the large Lower Cave, at or near its opening. The site represents the most northern temperate extension of the human range at that time, and did until the last 200,000 years. The climate for most of the period of occupation was not much different from today's, which is similar to that in the northern United States in terms of seasonal extremes and temperature averages. During the span of the human occupation, the climate oscillated between colder and warmer periods several times.

The Zhoukoudian skulls were among the most archaic human fossils then known.[49] They were long and low, with thick projecting browridges and a strong angle at the cranial rear. Yet, their individual cranial capacities were within the normal human range, although their average was

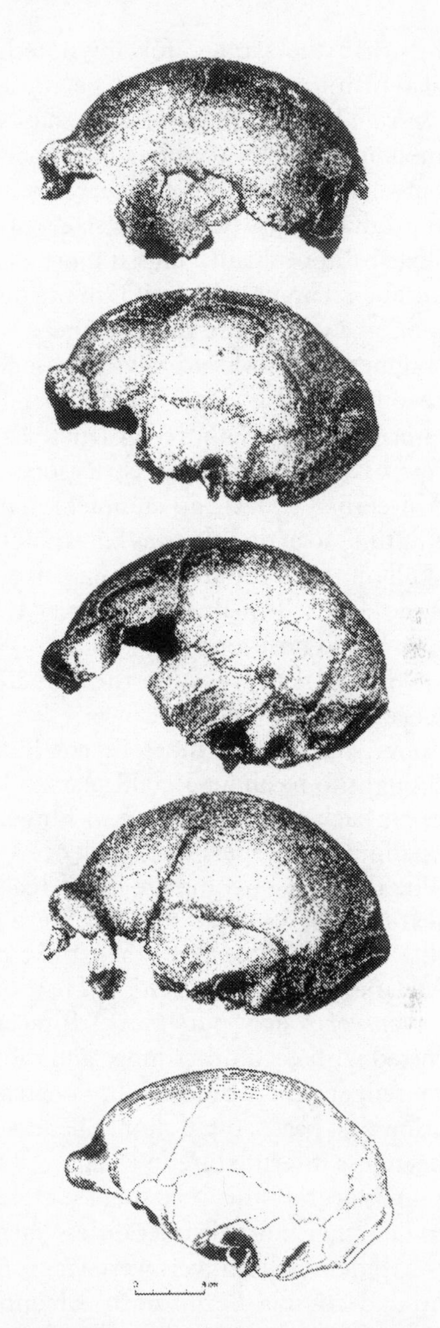

The five crania from Zhoukoudian known to Weidenreich.[50] *The bottom E1 cranium is the oldest. The three locus L crania* (the three above E1) *include the female that was reconstructed* (L2, highest of the three), *and the two males L1 and L3* (lowest). *D1* (at the top) *is the least complete. All but the bottom drawing are Weidenreich's.*

less than the average for any population today. The mandibles were also primitive; they were large and robust, and had strongly receding chins. They differed in size much more than the crania. It is normal for men and women to differ in human populations. When a difference between the sexes involves features that are not directly associated with reproduction, such as body size or the shape of the nose, it is termed sexual dimorphism. This dimorphism is much greater in the Zhoukoudian mandibles than it is in the mandibles of any living human population. Among the remains there were no complete faces, just some fragments of several individuals mixed together. Weidenreich finally reconstructed a single female cranium from one set of these; much later Carleton Coon reconstructed a male specimen, using a cast, but few have seen this reconstruction which is at the Harvard Peabody Museum.[51] In fact, no complete male cranium of an East Asian from this time span or older was known until the discovery of the Sangiran 17 skull on the island of Java, some 30 years later. The Zhoukoudian faces were large (the male much larger), with flat cheeks facing forward, and low, broad noses.

"*Sinanthropus*" was not the only human sample from Zhoukoudian. Located just above the Lower Cave, at the Upper Cave or Shandingdong, some much more recent human specimens were discovered, thought to be about 30,000 years old. Their analysis brought Weidenreich back to the question of human races, as he found in the three Shandingdong specimens evidence against the concept of "pure races." The three most complete individuals, he felt, "show certain common features" of an Asian sort, but were different enough to typify "three different racial elements, best to be classified as primitive Mongoloid, Melanesoid and Eskimoid."[52] He took this variation as proof that there never were once "pure races." If races had been "pure" in the past, and mixed with each other more and more over time until achieving their present state in which no pure races are left, we would expect that variation of a past sample should be less than today's, as there would have been less intermixture in the past. The marked variation Weidenreich found, even in this very small sample, showed that races did not seem to be either more distinct or less variable in the past.

Weidenreich's travels were far from over. In 1937 he left to visit the up-and-coming Berlin-born paleontologist G.H.R. von Koenigswald (1902–1982) in Java. This was where the famous "Java Man" skullcap was found near the village of Trinil by a young Dutch doctor who was one of Haeckel's disciples.[53] Eugène Dubois (1858–1941) found the Trinil skullcap in 1891, and it was described in great detail by Weidenreich's professor, Schwalbe, when Weidenreich worked in his laboratory. Von Koenigswald went to Java to find better specimens of this enigmatic hominid, "*Pithecanthropus erectus*," with the hope that more complete remains would resolve the controversy that continued to swirl around the Trinil specimen.[54] The issue was whether the Trinil specimen was an ape or a human. Trinil was only the top of a cranial vault, broken well above the ears, so it was impossible to tell how tall the skull was. A tall cranium would be a hominid, but a low one could be an ape. The controversy was fueled by the pronouncements of Dubois himself. Assuming that a femur found in the same excavation belonged to the same specimen as the skullcap, he likened it to a giant gibbon. It's not that he thought it was apelike; quite the contrary, Dubois thought that Trinil was a real "missing link" between humans and their apelike ancestors. As a missing link, the hominid should be about halfway between ape and human. But the body weight estimated from the femur made the Trinil skull seem to have a large, humanlike brain size, relative to body weight. Dubois may then have come to liken Trinil to a giant gibbon because, if it had a gibbon's proportions, the estimated body weight for Trinil would be much larger, and therefore its *relative* brain size would be smaller, midway between human and ape, as he expected the missing link to be.[55] And so the debate raged on.

Von Koenigswald's discoveries clarified this and many other issues. Funded by the Carnegie Institute of Washington, and later by the Rockefeller Foundation through Weidenreich's Cenozoic Research Laboratory, von Koenigswald was wildly successful, and in 1938 Weidenreich visited him to study the important specimens he was recovering from the area around the village of Sangiran. Von Koenigswald went to Beijing with new material the following year, and they made comparisons.

In a joint paper,[56] they reviewed the new discoveries and suggested that the Javan and Chinese remains were two races of the same species—what Weidenreich called geographic specialization or horizontal differentiation. They attributed the differences to race because they thought the two samples were about the same age—we now know that the Javan remains are approximately twice as old. The Sangiran crania were smaller than their Zhoukoudian counterparts, and at that time no remains of faces were known (faces were discovered later). The Javan crania had flatter and more receding foreheads, and their browridges were straighter across and more evenly aligned with the forehead (Zhoukoudian skulls had a distinct angle between the top of the browridge and the more rounded, higher forehead).

Vertical differentiation, or different evolutionary phases, presented

CRANIAL CAPACITIES
(in cubic centimeters)[57]

	Female average	Male average
Indonesian Sangiran	875 (*n*=5)	1032 (*n*=2)
Chinese Zhoukoudian	965 (*n*=3)	1078 (*n*=4)
Indonesian Ngandong	1093 (*n*=2)	1177 (*n*=4)
Aboriginal Indigenous Australian	1119 (*n*=22)	1239 (*n*=51)

in the Javan fossils. Human remains from Ngandong, on the Solo River not far from Sangiran, were much younger, perhaps from Neandertal times,[58] but were clearly, in Weidenreich's mind, descendants of the earlier Javan specimens.[59] The Ngandong remains, all skullcaps but for two tibias (shin bones), were a different evolutionary phase for several reasons, perhaps the most important being that they are larger—the average brain size when considering the men and women separately is similar to some living Australian populations. The Aboriginal Indigenous Australians, descendants of the Ngandong folk, have heads that are, on the average, about 4 percent larger.

The Ngandong people were generally more robust (thicker cranial bone, and stronger muscle attachments) than the earlier Sangiran folk, and had cranial proportions much more like modern humans. Yet they shared many cranial features with the earlier Javans, and Weidenreich considered them a link between the Sangiran folk and Aboriginal Indigenous Australians, one of the modern evolutionary phases of the region. This combination of horizontal and vertical differentiation is not unlike the way Weidenreich had analyzed the Neandertal races a decade earlier.

Weidenreich's ideas about the pattern of human evolution were jelling. In 1938 he attended the International Congress on Anthropological and Ethnological Sciences in Copenhagen, reading a paper on hominid classification that he published the next year in the *Bulletin of the Geological Society of China* as one of six wide-ranging lectures.[60] The

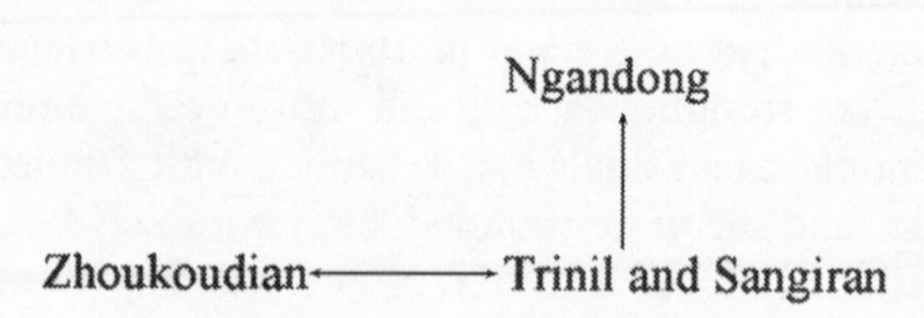

next year he came to Chicago to attend the American Anthropological Association meetings. The paper he read there was published the next year in the *American Anthropologist.* In these publications he laid out the conclusions he had drawn from his Far Eastern experience, his contention that human evolution was not a series of speciations and replacements,[61] with change resulting as each more successful species took the place of the previous one. He wrote: "We have no other choice but to abandon the generally accepted contention of there having been only one center of evolution from which, from time to time, new types would have sprung, and spread over the whole world, dislodging the older ones."[62]

Instead, in the *Polycentric* model he developed, there were four centers of human evolution: Western Asia, Northeast Asia, Australasia, and Africa. As he described it:

> More and more I am coming to the impression that, just as mankind of today represents a morphologic and genetic unity in spite of being divided into manifold races, so has it been during the entire time of evolution. While man was passing through different phases, each one of which was characterized by certain features common to all individuals of the same stage, there existed . . . different types deviating from each other with regard to secondary features. These secondary divergences have to be races as regional differentiations, and, therefore, as correspondent to the racial dissimilarities of present man.[63]

Weidenreich continued comparative analyses of the Zhoukoudian material, but the war stretched on. Excavations at Zhoukoudian ceased in July 1937 because of the war, and the area became dangerous—two workers left to guard the important site were killed by the Japanese. The commander of the city's occupation forces denied Weidenreich access and, instead, tried to excavate further himself but failed without creating too much damage. In the spring of 1941 Weidenreich's American visa had expired forcing him to return for a year to renew it. He left for the American Museum of Natural History in New York with his notes and photographs and a complete set of casts and drawings of the specimens. Asked to carry the Zhoukoudian remains with him, he refused. By contract they were to remain in China, and as a stateless person he feared if he were stopped, they would be confiscated. Meanwhile in Java, von Koenigswald was imprisoned by the Japanese in a POW camp. He managed to save almost all of the human fossils by hiding them with friends or in some cases burying them in milk cartons, but the Zhoukoudian fossils, including the bones from Shandingdong, were lost.[64] By the end of 1941 it was clear that Beijing was soon to be captured by the Japanese army. The human fossils were packed for shipment to the United States, and actually left Beijing in foot lockers to be shipped to New York with the American Navy, but were never seen again. It is widely assumed that the fossils were destroyed, but their fate remains unclear and the story has provided fodder for several books and much exciting speculation. The disaster was in some sense mitigated by the extraordinary detail of Weidenreich's monographs, and the fact that he made excellent casts[65] of virtually all the specimens (the only materials not preserved this way were the worn isolated teeth). Although casts don't replace the originals, they do provide an important record. A number of original casts were widely distributed before the original specimens disappeared. Nevertheless, Weidenreich was devastated by the loss of the Zhoukoudian material.

At the American Museum of Natural History, Weidenreich finished his descriptions of the Zhoukoudian crania from the materials he had brought with him. He then went on to describe von Koenigswald's newer specimens[66] from the casts he had sent; the war was still on, and Weidenreich was unsure of either von Koenigswald's or his fossils' fates. But all were safe. When Weidenreich received word of this, he approached Paul Fejos, research director of the Viking Fund (later the Wenner-Gren Foundation) with a request for funds to bring von Koenigswald and his fossils to America,[67] and their collaborations continued at the American Museum after the war. He also wrote papers on the Australian specimen from Keilor (about 15,000 years old), the Uzbekistani Neandertal from Teshik Tash, and even a short note on the early Aboriginal Indigenous American from Tepexpan.[68] When Weidenreich died, in July of 1948, he was in the middle of another major monograph describing the Ngandong specimens. Von Koenigswald edited it for posthumous publication and wrote its introduction.[69]

Many things were unusual about Franz Weidenreich. Much of what he accomplished was groundbreaking in paleoanthropology, but perhaps the most important "first" in his career was his hands-on experience with the human fossil record of three main regions of the world:

1. Europe, where he was trained and did his early work, especially on Ehringsdorf and its implications
2. North Asia, as the result of the Zhoukoudian discoveries from both the half million or more years old Lower Cave ("Peking Man") and the Upper Cave (Shandingdong) where the human remains were only some 30,000 years in age
3. Australasia, both the Javan and the Australian fossil remains

It must be kept in mind just how unusual this was. This kind of experience is virtually unheard of in human evolutionary studies, where, even today, most workers are specialized, tied to a single region and a limited time frame. Since for most of this century the majority of the students of human evolution were Europeans, Europe was extensively studied, and there developed an implicit belief that the European fossil record represented the pattern of worldwide evolution. Workers like Weidenreich's early mentor, Schwalbe, saw human evolution as a linear sequence of evolutionary phases, leading to modern man, one of which was a Neandertal stage. Weidenreich believed this, too, but was different in that he saw geographic variation as also having significance in the evolution of humans as a whole. There was a "paleoanthropine" stage in his scheme; perhaps you could call it the "Neandertal" stage, but he didn't; the Neandertals were a Eurasian race represented by the "Tabun group" within this evolutionary stage. The Eurocentrism of the day also left many human paleontologists with the idea that extra-European affairs didn't really matter: Europe was indicative of what was happening the world over, and if not, it was the pinnacle of evolutionary innovation and the source of all important changes. Earlier this century, paleoanthropologists who denied any possibility of a Neandertal ancestry for Europeans were more than willing to admit an archaic ancestry for other races. Take the case of Marcellin Boule (1861–1942) and Henri Vallois (1889–1977). These two prominent French paleoanthropologists are as responsible as anybody for the precept that Neandertals were a distinct human sideline, leaving no descendants. But they, like others, really meant only the *European Neandertals.* It was perfectly acceptable for *other* races to have a Neandertal ancestry. In their widely read textbook *Fossil Men* they wrote:

> We know that Neandertal man existed in Asia as he did in Europe, and no doubt he lived there at the same period as our own continent. But there he no longer appears in isolation; some of his forms are linked with the Prehominians, others with the Men of the Upper Paleolithic. Thus, whereas the Neandertal man of Europe occupies the position of a type apart, of whose origin we are ignorant and which seems, according to all the evidence, to have vanished without issue, *the little we know about this type in Asia shows it as included within a regular evolutionary sequence.*[70]

The relatively few paleoanthropologists who worked outside of Europe did so because they believed their region of choice was particularly important for understanding human *origins* and would bring them to the cradle of humanity, and not because they were interested in geographic variation. Louis Leakey was certainly attracted to East Africa for that reason, while workers like von Koenigswald and Dubois before him believed, following Haeckel, that Java was the most likely place to look for human origins. Weidenreich was interested in origins as well—human origins were the focus of his most popular book, *Apes, Giants, and Man*—but he was exceptional in having detailed *multi*regional experience with a focus on the pattern of evolution.

Writers Erik Trinkaus and Pat Shipman describe Weidenreich, at the end of his life, as seeming to "project the wrong image: he was a European scientist of the prewar era. . . . Weidenreich was in his mid-70's; he was old, he was tired, and his ideas seemed out of date."[71] Yet we can hear his pithy remarks and sense of humor reverberating across the decades. In his last publication (published posthumously), written at the age of 73, with a not-unusual sarcasm he quipped:

> It took 40 years before Neanderthal Man was recognized as a special human form and not pushed aside as a pathological variant of modern man, and it took 40 more years before Dubois's *Pithecanthropus erectus*, originally described as a giant ape, was acknowledged as a normal sized hominid. . . . Therefore there is at least a remote chance that as a consequence of the evolution of anthropological science and scientists, *Meganthropus* and *Gigantopithecus* may be admitted to the hominid family at about 2000 A.D.[72]

Here is a scientist who first embarked on a career in paleoanthropology when he was older than either of us, old enough, in fact, for early retirement at many universities. At the age of 62 he was forced to make a new life for himself away from his native land, and in doing so he began the most successful and prolific part of his career that continued until his death. At his age, we hope to be as old and tired as he was.

POLYCENTRIC EVOLUTION

Is Weidenreich's Polycentric evolution actually polygenic evolution as Howells described it? Weidenreich didn't think so. Polycentrism, as Weidenreich was beginning to envisage it in 1939, was a description of what the Javan remains increasingly showed him: there were different ways of becoming human, different pathways and even different rates of change. Kabwe, he had realized a decade earlier, was vertically differentiated from the Neandertals by being more archaic, but it was also horizontally differentiated by being more African. Zhoukoudian folk had distinctive Chinese features, some unique to the Chinese, and in aggregate the features made them more like North Asians than like any other living population. Early on he recognized four such linking features[73] (later there were more, and others have refined and expanded the list):

- A coal-shovel-like form for the inner surfaces of the upper incisors (shovel-shaped incisors)
- An internal thickening on the inner surface of the mandibular body (*torus mandibularis*)
- An extra bone where the cranial sutures meet at the back of the head (Inca bone)
- A thickening of the bone, from front to back, along the top of the head (sagittal keel)

Yet the variation at the much more recent Shandingdong also reinforced his earlier realization that the earlier populations were *not* Chinese in a racial sense. The races were constantly changing, not because of hybridization from once pure forms but because of the shifting requirements of the environment. Were the races once pure and unmixed, the Upper Cave crania would be far more homogeneous because they were older, Weidenreich felt. Then came the Javan discoveries and the links he found in them, from Sangiran to Ngandong to the Late Pleistocene Australians.

Thus Weidenreich had discarded two models of human evolution. He did not think that from ancient times there were once pure races that had come to grade evenly into each other because of hybridization, but he also did not believe there was just one place where human characters evolved and then spread:

> There must have been, not one, but several, centers where man developed. But we should be completely at a loss if someone should ask on which special spot the decisive step was made that led from a simian creature into man. There was not just one evolutionary step. Evolution went on wherever man may have lived, and each place may have been a center of both general development and special racial strains.[74]

In 1939 he had clearly come to the critical realization of Polycentrism—past stages of human evolution (as he called them) had populations divided by race, different races in different regions, just as the present stage is divided. As his ideas developed, he first explained this pattern by the isolation of these races from each other and characterized the same changes in different regions as parallelism. These were as different as could be from the elements of the evolutionary model he would soon develop, but even then there was one thing he was definite about: his ideas *were not polygenism.* He addressed this specifically and pointed out that he had never implied that there was "an origin from different primate or anthropoid branches gradually transforming into a uniform type by convergence."[75] The differences, from early on, were of races and not of species. The features that differed were really minor compared with the major changes in human evolution he had studied, such as bipedalism.

By 1940, in the essay that prompted Howells' first comments on the issue, Weidenreich summarized the meaning of new fossil discoveries with the comment that "it can be taken for granted that none of the new specimens reveals any particularity which could be interpreted as a plain indication of separate development," but he still considered the regional evolutions to be rather independent of each other. Yet he realized that as races of the same species they were closely related to each other and suggested subspecies names be given to the geographic variants of each evolutionary stage—for instance, instead of "*Pithecanthropus erectus*," "*Homo erectus javanensis*," instead of "*Sinanthropus pekinensis*," "*Homo erectus pekinensis.*" Three years later, as mentioned above, he dropped even these and suggested all fossils be admitted to *Homo sapiens* in a formal taxonomic sense, citing the agreement of the Russian-born, highly influential geneticist Theodosius Dobzhansky who formally proposed this in 1944. However, Weidenreich was unwilling to put his money where his mouth was, continuing to use genus names like *Sinanthropus* to refer to what he regarded as subspecies. His reasoning was that the old names were embedded in earlier writings and to change them would be confusing. In the case of the Zhoukoudian remains, he was reluctant to change the name from "*Sinanthropus*" to "*Pithecanthropus*" (the name of the Javan specimens he admitted "*Sinanthropus*" was extremely similar to) as he was being urged to do, because of "respect for Davidson Black and his work. . . . It would, indeed, have been rather tactless had I begun my task as Black's successor by eliminating the name he had coined."[76] Moreover, he noted that he was only proposing to go with the flow: "In paleontology it always was and still is the custom to give generic or specific names to each new type without much concern for the type or relationship with other types known."[77]

His nomenclatural changes, even if they were never put into practice, reflect the way he perceived humanity. He regarded all human forms, living and fossil, as a single species for two reasons:

1. Even the most distinct geographic races were not distinct types but graded into each other with numerous intermediate forms.
2. The whole of human variation is less than that of domesticated species, for most characters "overwhelmingly less."

A single species, of course, has a single, monogenic origin. And so his perception of the human species became a second line of argument against polygenism—that old precept which, in Germany, had evolved into the biopolicy of the Nazi regime, dislodged him from the land of his birth, placed two of his daughters and no doubt untold friends and relatives in concentration camps, including his mother-in-law who was murdered, and committed endless horrors in the name of bioscience. He never addressed this directly, and over his later career made only the sparsest remarks about his own life circumstances. But it is difficult to believe that it was ever far from his mind.

Weidenreich contrasted his ideas with the theories of Count Gobineau, who like many others in the middle of the 19th century considered the different races to be different species, and all inferior to the white one. He singled out Gobineau's work, in particular, because it was heavily promoted by the Nazi propaganda machine. Dipping into pre-Darwinian polygenism, Gobineau had questioned whether Negroes are included when the Bible refers to "man." For Weidenreich this presented a way of addressing more current theorizing, and he remarked, "It is strange to note that Gobineau's ideas about Negroes are not fundamentally discordant with those of certain modern taxonomists and geneticists."[78] The geneticist he had in mind was the British plant geneticist R. Ruggles Gates.

In a 1946 essay, "Generic, Specific, and Subspecific Characters in Human Evolution," Weidenreich addressed a resurgence of polygenism, most blatantly in the work of Gates but more widely as Gates had the backing of both Hooton and Coon.[79] Gates, in an essay on "Phylogeny and Classification of Hominids and Anthropoids," attacked Weidenreich for his denial that races were different species and that their evolutionary histories were separate. He called Weidenreich biased on religious grounds, accused him of naive and muddled thinking and of misrepresentation and misinterpretation of the material because Weidenreich did not believe the one thing he is (ironically) usually associated with—parallel evolution of the human races! It must be remembered that even during the war, polygenism was not such a socially reprehensible theory. Gates was supported by several scientists, most importantly Hooton, as noted above the preeminent physical anthropologist at Harvard, probably the most powerful physical anthropologist in the country.

Gates, in fact, found a source for his ideas in Weidenreich's work, just as Keith did a few years earlier.

> It appears to me that we are justified in regarding Mongoloid, Australoid, Caucasoid, and Negroid types of man as representing separate species, each with various geographic races clearly defined. . . . The justification for this view will depend upon how long these four specific types have been evolving in separate geographic areas, more or less isolated from each other.[80]

It was the length of time of separation and not the amount of anatomical distinction that meant they were species and made Weidenreich's work crucial to Gates's argument. No wonder Gates was so critical of Weidenreich, who neither cited him nor backed him. Instead, in a thinly veiled reference to Gates's work, Weidenreich quipped that "raising the differences between racial groups to the rank of specific differences by giving these groups specific names is nothing but an attempt to exaggerate dissimilarities by the application of a taxonomic trick."[81]

It is therefore surprising to find Weidenreich linked to Gates, but he was! Right after Weidenreich's death, Howells wrote that

> the literature contains various illicit references to parallelism, orthogenesis, species, and so on. I refer especially to Gates and Weidenreich, who have not, I think, seen what they are implying by some of their deductions. Weidenreich's chart of human lines developing during the Pleistocene calls for an unparalleled parallelism, in fact for an incredible convergence.[82]

But in life Weidenreich continued to speak out pointedly *against* the parallelism and isolation of polygenism. In the preface of his popular 1946 book *Apes, Giants, and Man*, he discussed the arbitrary nature of racial divisions. He expressed his certainty that all people are cross-fertile and that they can and do interbreed. In the view he developed, human variations are minor, and specifically contradicting Haeckel, he argued that because differences among groups of humans were much smaller than among domesticated animal species, these animal breeds make a valid basis for modeling human variation because, for all intents and purposes, humans *are* domesticated, albeit *self*-domesticated (an idea he had written about much earlier).[83]

Finally, Weidenreich completed his development of the Polycentric model we understand today. Denying any notion of independent evolution or parallelism, he contended that human evolution was best understood as a network of interconnected populations that retained regional continuity for at least some features. Each of the four major evolutionary centers he perceived retained differences at a racial level,[84] and their features could be directly related to the features of races of today.

> Human forms preceding those of modern man were distributed all over the entire Old World and differed typically from each other, just as is true of any present geographical variation. In addition, the ancient Javanese forms, *Pithecanthropus* and *Homo soloensis*, agree in typical but minor details with certain fossil and recent Australian tribes of today so perfectly that they give evidence of a continuous line of evolution leading from the mysterious Java forms to the modern Australian Bushman. The same holds good for the African branch. Rhodesian man can be linked through the intermediate stages to the Florisbad man and the Boskop man, both recovered from southern African soil.[85]

These past races were not the present ones. Weidenreich took the persistence of certain traits to be evidence of genetic continuity, not racial identity—for instance, continuity between the Chinese "*Sinanthropus*" remains that he studied in so much detail and Mongoloid populations of today such as East Asians and Aboriginal Indigenous Americans. Now it is certainly true that the humans of a half-million years ago are ancestral to *Homo sapiens*. But as the genetic anthropologist Bob Williams explains, the fact that some traits that were common in East Asia a half-million years ago are still more common in East Asia than elsewhere, does not make the ancient East Asians "primitive" Mongoloids.[86] For this to be correct, it would have to mean that the Mongoloid race persisted for a half-million years (if not more), and would imply that the other characteristics of living East Asians were present in these remote ancestors. It would attribute an underlying, unchanging reality to the Mongoloid classification, a valid manifestation of essentialism or typology. Combinations of features were seen to persist for long periods of time, but the continuity was of features, not of populations. "Races are not more constant than species; like the latter, they continuously change their character in the course of time."[87] According to Weidenreich, it was manifestly *not*, as Coon was to later say, "*Sinanthropus* and the Mongoloids."[88]

Weidenreich's model involved more than simply the vertical lines of continuity; at the same time, there was significant gene flow between the evolutionary centers, and the human species retained its unity as a whole. He believed the general pattern was that most past populations had several ancestors, and those that did not become extinct almost certainly had several descendants. For instance, in summarizing his understanding of Australasian evolution, he wrote:

> . . . at least one line leads from *Pithecanthropus* and *Homo soloensis* to the Australian aborigines of today. This does not mean, of course, that I believe all the Australians of today can be traced back to *Pithecanthropus* or that they are the sole descendants of the *Pithecanthropus-Homo soloensis* line.[89]

Weidenreich thereby believed that a significant aspect of the skeletal morphology of Aboriginal Indigenous Australians was to be found in their regional antecedents, the Javan Sangiran people. However, he never regarded this as a unique or isolated ancestral-descendant relation. He posited that some of the ancestry of the Australians is to be found elsewhere, and that not all of the descendants of the Pleistocene Javanese live in Australia.

Weidenreich illustrated his evolutionary network as something like a trellis, a network with vertical lines passing through the main stages of evolution, horizontally separated into the main centers of evolution (distribution and specialization), and diagonal connections between them reflecting the patterns of genetic interchange. He considered it an actual representation of how populations exchanged genes, "a graphic presentation of the conception of the hominid group as one species."[90] It was this trellis that Howells and others rendered as a candelabra by omitting the interconnections.

This graphic representation illustrated his conception of how past and present human groups are related, and does not emphasize the vertical connections over the diagonal or horizontal ones, although he is presented as doing so by virtually everybody who has abstracted his model as a candelabra. His published writings broadly emphasized (what he called) the continued hybridization of human populations, and he was quite specific about this pattern of population relations in an unpublished book on human evolution.[91] In it he divided the human fossil record into ten broad phases, just as he did in the trellis shown above: the first phase was the oldest, and the tenth phase was modern humanity. He wrote:

> If the Hominidae are one species in the genetical sense and an exchange of genes was possible in phases I to IX [the prehistoric phases] as it is possible in phase X (modern man), the commonly used form to represent their lineage gives an entirely wrong idea. The tree with a com-

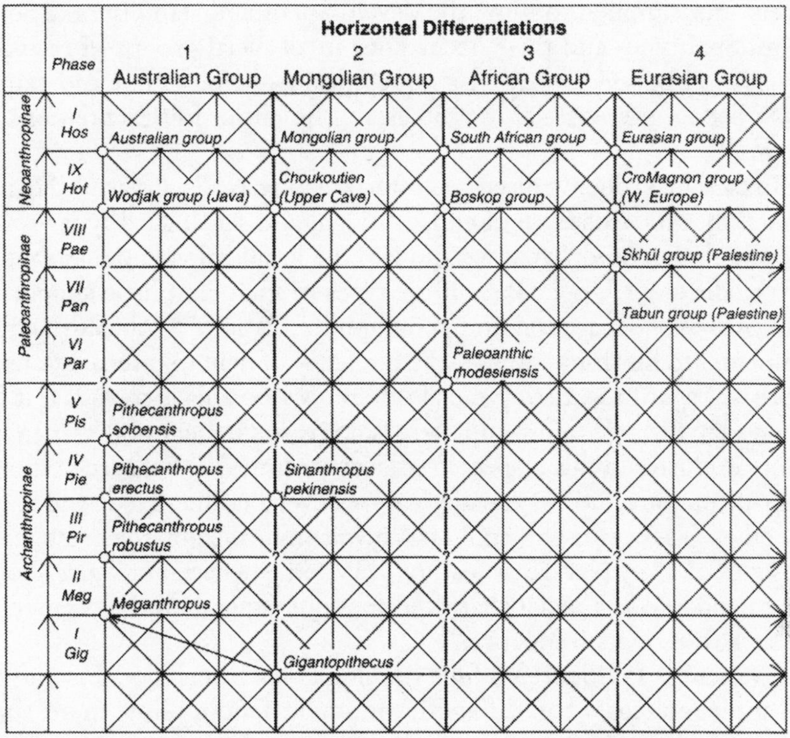

Weidenreich's trellis:[92] *a network of populations connected by gene exchanges.*

mon stem and more or less abundant ramifications leaves no possibility to indicate graphically an exchange of genes. The branches and sub-branches appear to evolve completely independent of each other once they have deviated. In reality, there must have been intercommunications between the branches. The graph which best fits this perception is a network. Its interconnections indicate the lines along which the exchange of genes could be effectuated.

This was a break with Haeckel, and even his mentor Schwalbe. Weidenreich would not arrange human fossils on a genealogical tree. Insofar as it might apply to human evolution, he did not accept the premise of the "ladder *versus* bush" debate.

Weidenreich's much earlier interest in constitutional types had become important in his thinking once again. He was able to resolve the two seemingly contradictory organizing principles for human variation, race and constitutional type. The significance of a worldwide distribution for the constitutional types, the fact that they were found in every race, implied to him "the development and stability of . . . constitutional types and their occurrence in all racial groups of mankind today . . . give evidence that geographical isolation is not and cannot have been a prerequisite for the establishment of [differentiations] in man."[93]

In fact, he explicitly discussed interbreeding as all-pervasive. He reasoned that the existence of constitutional and serological types across all populations and "what we know of the localization and distribution of the different Upper Paleolithic types of man lend little support to the theory of separation and geographic isolation."[94] Absolutely fundamental to his thinking was "the tendency of man to interbreed without any regard to existing racial differences . . . this is so today; it has been so in historic times, and there is no reason to believe that man was more exclusive in this respect in still earlier times."[95] Because of the extent of mixture, no diagnostic features ever became fixed, and races, whether past or present, could not be distinguished by the presence or absence of a key feature or features. "Whether a certain feature should be considered as a racial character does not depend on the occurrence as such, but on its frequency."[96]

Even after Weidenreich fully developed his trellis model, some paleoanthropologists had difficulties in understanding how there could be common evolutionary changes across the human species without parallelism; some even had difficulties in understanding the model. In *The Fossil Trail*, the present American Museum paleontologist, Ian Tattersall, described it as of "mind-boggling complexity," in which the diagonal lines were "meaningless in the original, as far as I can tell."[97] Not surprisingly, and like other popularizers discussed earlier in this chapter, Tattersall believed Weidenreich's evolutionary centers were meant to be "parallel lineages,"[98] and wondered how they could all evolve the same way.

But explaining parallel evolution was not Weidenreich's problem. His was quite the opposite. His trellis model started with the premise that the regions were strongly interconnected. From his understanding of trait distributions and the evolutionary forces that change them, it is clear that the notions of universal hybridization and persistent genic exchanges were most important in his thinking. But if true, this would make it difficult, perhaps impossible, for regional features ever to become established or be maintained. This regional continuity—and not common evolutionary trends—was his problem to account for. How could he explain why there was any regional continuity at all; why wasn't any distinct regional evolution simply swamped out by this process?

In fact, he never adequately dealt with this problem. The origin and establishment of regional features were not issues he ever seems to have considered, but he did address a final problem, created when the pattern of transforming morphology did not fit the chronology. According to the dates of fossil remains considered accurate at the time, modern humans appeared even before the earliest archaic remains he knew, the Zhoukoudian and Sangiran specimens. This was all out of the order he expected, and at first he questioned the early, seemingly modern specimens like Kanam and Piltdown with the presumption that their dates were incorrect. Weidenreich had a lifelong suspicion of dates, that in the event was more often shown to be right than wrong. He long held that if the fossils in hand could be laid out on a table in order of their anatomical evolution, it would most likely be their temporal order as well, and any dates to the contrary should be considered dubious.

> If all hominid types and their variations, regardless of time and space, are taken into consideration, their arrangement in a continuous evolutionary line, leading from the most primitive state to the most advanced, does not meet with any difficulty. Neither gaps nor deviations are recognizable.[99]

As for the specimens whose chronological position does not match this anatomical sequence, he continued, "the geological determination of some of the specimens enumerated . . . is not above all doubt." In fact, in the case of Kanam, a small fragment of mandible, it turned out to have been both misdated and misanalyzed, as the piece was first thought to have a chin, but later it was shown to be a pathological bone growth.[100] In the case of Piltdown, it was a fraud, but that is another story.[101]

Weidenreich, in fact, knew Piltdown was a fraud; he was one of the few paleoanthropologists aware of it and willing to say so (Gorjanović-Kramberger, as we noted in Chapter 5, suspected the same but would only publish his misgivings in Croatian).[102] The Piltdown fossils were discovered between 1912 and 1915 in a gravel quarry near the River Thames in Kent County, England. They were associated with animal bones and crude flint tools. The deposits were very ancient, making this find the earliest human fossil then known. The find consisted of cranial fragments that were very modern in appearance when reconstructed, associated with an apelike mandible and some molar teeth. The diagnostically important canine tooth was missing. Arthur Smith-Woodward (1864–1944), who was reconstructing the specimen, carved one out of wood to fit his reconstruction, and just several months later a canine was found at the site that closely matched his fabrication. Still, nobody in the group of scientists studying Piltdown was suspicious. The combination of features in the specimen reconstructed from the skull, jaw, and teeth suggested that modern cranial traits, and implicitly human mental characteristics, evolved earlier than other human features such as the dentition. This important milestone in human evolution was a European (English) phenomenon. But unbeknownst to anyone but the perpetrator, until the chemical testing of 1953,[103] Piltdown was a carefully constructed fraud, composed of fragments of a recent human skull, the mandible of an orangutan, an orangutan canine filed down to flatten its pointed crown, and a mix of some ape and human teeth. All were stained to look like the real fossil animal bones from the site. This strange combination was widely, almost universally, accepted as a valid fossil, certainly by the five different scientists who reconstructed it, including Arthur Keith who tried twice.

One of the reconstructors was a German scholar, Heinz Friedrichs. Weidenreich, in one of his earliest forays into paleoanthropology, wrote the introduction for Friedrichs' description of Piltdown.[104] In it, Weidenreich simply expressed his disbelief that the pieces used in the Piltdown reconstruction were from a single specimen, or even a single species. He (quite correctly) asserted, from his study of the anatomy, that the skull was that of a modern human and the jaw was of an orangutan. He didn't know what to make of the canine and did not seem willing to entertain the possibility that it was simply manufactured to look as it did, as part of a fraud. But he was virtually alone in his opinion that the fragments didn't go together, and more than a decade later his conclusions were not yet accepted.

> I am only wondering why, if a human vault, a simian mandible, and an anonymous "canine" were combined into a new form, the other animal bones and teeth found in the same spot were not added to the . . . combination? . . . I do not believe in miracles . . . the sooner the chimaera . . . is erased from the list of human fossils, the better for science.[105]

Even without Kanam and Piltdown, there were more specimens that seemed out of place, such as the modern-appearing cranial rear from Swanscombe, thought to be the same age as the Zhoukoudian remains. In the light of today's knowledge Swanscombe is neither so old nor so modern as Weidenreich thought then, but he felt compelled to account for it. He did so using his understanding of how natural selection worked (this is discussed in the next chapter). Weidenreich believed Darwinian selection could only prune out features that already existed in a population, when they became disadvantageous because of changing conditions. This is why he distinguished the races, past and present, by the number of anthropoid characteristics they had lost, their differential retention of the "simian stigmata." We must assume, he concluded, "that development was not going on simultaneously everywhere but was accelerated in one place and retarded in another . . . [so] all the discrepancies between morphologic and chronologic sequence of the known types of fossil man could be understood."[106] It is tempting to say he should have had the courage of his convictions and realized this and other "problem" specimens were out of sequence because their dates were wrong, but these comments in reality reflect "Germanic stigmata." In the long tradition of European anthropological science, he thought the Caucasoid race was the most evolved. But unlike his predecessors, he posed this in careful terms because he did *not* mean it in the sense of being more advanced, but rather in the sense of having lost more anthropoid features. The living races, for him, were equally advanced (they were in the same evolutionary stage), and he always regarded the differences between races, in any event, as being minor.

RACE AND HUMAN EVOLUTION

What was Weidenreich's view on race? Pat Shipman, author of *The Evolution of Racism*, writes:

> Weidenreich's theory had the interesting effect of placing the origin of races and racial differences far back in time. These were not recent evolutionary developments, his work implied, but deeply rooted, long-standing distinctions. But his work was too typological in orientation; he had also underestimated the variability within each regional group.[107]

Race, it seems, was intertwined with Weidenreich's evolutionary precepts, and was critical to them. But did he believe that races were ancient, long-lasting "types" of humanity? He began his last long essay on race (Chapter 4 from *Apes, Giants, and Man*, "The human races: principles of their classification and organization") with a quote from one of Hooton's mentors, Roland Dixon: "Race is not a permanent entity, something static . . . it is dynamic and is slowly developing and changing."[108] One of the seemingly paradoxical aspects of Weidenreich's writings is the inconsistency of this precept with his contention that there is horizontal differentiation in ancient as well as modern humans and that some racial features appear long ago.

In fact, Weidenreich's writings on race seem fraught with contradictions. In dismissing the theory of a single center of human origins Weidenreich had given human racial variation great antiquity. But this did not imply antiquity for the *living* races. He considered the human races to be real entities and, following Dobzhansky, he proposed that they should be classified morphologically based on features that have four criteria: heredity; stability for long periods of time; limited influence of external factors on variability; and expressions independent of age and sex. Thus on the one hand he seemed to see racial classification as useful, yet on the other he recognized races as arbitrary and certainly realized races are not static and the characteristics one associates with races of today may not apply to races in the past. He believed evolutionary rates differed between the centers of evolution, at least as far as the loss of anthropoid features is concerned, but then recognized that what is "primitive" is relative and will differ from one center to another. He spoke strongly against racism, especially the assignment of different mental attributes and abilities to races, and realized the limitations of (what he called) usual anthropological methods in defining or recognizing races. Yet he often wrote typologically, as though races had an essential reality and the difficulties lay with anthropologists' attempts to describe them.

In part, the problem is that, on the issue of race, Weidenreich was a populational thinker with a typologist's vocabulary, burdened by an essentialist upbringing and an often imprecise use of English. He understood modern humans as a polytypic species (a species with geographic subdivisions, i.e., distinct races), but also understood that the races were always changing and were not typological units. He was absolutely unique in extending this idea of human variation into the past. The populational bent to his work is evident in his investigation of the distribution of constitutional types, mentioned above, and also in much of his other work on modern humans. Weidenreich frequently spoke out *against* typologizing human variation. One example he liked to use was taken from the works of the Swiss anthropologist D. Schaafhausen, who examined a very limited sample of 250 draftees from three German-speaking cantons for evidence of racial "purity." He used only four characters, broadly divided, such as small, medium, and tall stature. These gave 108 possible combinations, although only 40 were actually realized in the limited sample. What was surprising is that the combination that racial thinkers considered typical for the Nordic race, the race believed to be most common among the Germans—tall, long-headed, light eyes, blond hair—*was not found in a single individual.* The combinations for three other races that might be found among the Germans were only recognized in 9.6 percent of the sample. The other 90.4 percent were not "racially pure." Most of these Germans from a very limited area could be considered hybrids, although as Weidenreich points out:

> The interpretation of individuals presenting combinations other than the conventional ones as "hybrids" is based on the presumption that the features considered as characteristic of a race of today were even more pronounced in the past . . . in other words that the races were once purer than they are today and that their character has changed by interbreeding.[109]

This, of course, was not Weidenreich's view, but it does reflect a contradiction he needed to resolve. If there were never once pure races, why could modern racial features be found in past populations?

The issue is whether today's races extended into the past. Weidenreich did not think they did. He realized "the existence of racial types in the past identifiable with those of today would demonstrate that their differentiation and fixation must have taken place long before Upper Paleolithic times."[110] This is a possibility that Weidenreich was not prepared to believe. He was sure there never were barriers.

But what is it, then, that extended into the past? What was regional continuity based on? The key question Weidenreich asks is whether and to what extent the races of today are traceable races of earlier phases of man's development? The answer he gives is that "typical features of Australian aborigines occurred in *Pithecanthropus* and *Homo soloensis*,"[111] or that "some of the characteristic features of *Sinanthropus* reappear in certain Mongoloid groups of today."[112] *But* "this statement does not mean that modern Mongols derived exclusively from *Sinanthropus* nor that *Sinanthropus* did not give rise to other races."[113] The evidence for these relationships was found in the study of features, because "morphological characters are the only ones available for palaeoanthropologists."[114] "Features" is the key word here, as he wrote "all the facts available indicate that racial characteristics make their appearance as individual variations."[115] But the features demonstrate relationships among past and living races; they do not reflect their identity.

These past races, then, are not the same as the modern ones for at least two reasons. First, while there is significant local, or regional, ancestry, it is not *unique* ancestry, and a past race has descendants in all living races, just as a living race has ancestors in all past ones. Second, Weidenreich regarded races as transitory and not permanent because of what he called hybridity, or continued gene flow, as evidenced by the fact that the same constitutional types were found in all races. Fossil discoveries such as Shandingdong confirmed his view that there never were pure races in the past. *Therefore, regional continuity is not the continuity of populations or races, but the continuity of features.*

If there were never pure races, how could he go on to discuss the hybridity of most living people? What would they be hybrids between? His solution to the contradiction of hybridity and pure races, finally, is found in questioning what "pure race" can mean in the context of his Polycentric evolution:

> If the overwhelming majority of present mankind consists of "hybrids," for the reason that very few individuals correspond to the demands of a detailed racial scheme regarded as indispensable, one may ask: Where are the "pure" individuals who produce the hybrids anew each day? Every dog show exhibits "pure" breeds of the Doberman pinscher and makes us acquainted, at the same time, with the "impurities" of this breed. Although the Doberman itself produces marked mongrels when crossed with other races, the history of its race tells us that it was first bred in 1865, by crossing Manchester terriers, Great Danes, sheep dogs and setters. Nevertheless, today the notoriously hybrid Doberman pinscher is generally acknowledged as a "pure" race. What, then, makes the difference between hybrids and "pure" types, regardless of whether we are dealing with dogs or human beings?[116]

But there never was a solution to the other contradiction, of how regional features could be established without isolation:

> Man has changed not only his general but also his racial characteristics continuously, as a consequence of partly a continuous acquisition of new properties and partly of crossing. If so, how could distinct races ever have developed and persisted? . . . The usual answer to this question is that the human groups have been kept isolated by insurmountable geographical barriers after they acquired their special qualities. [But] the development and stability of constitutional and serological types, and their occurrence in all racial groups of mankind today . . . give evidence that geographical isolation is not and cannot have been a prerequisite. . . . What we know of the localization and distribution of the different Upper Paleolithic types of modern man lends little support to the theory of separation and geographical isolation.[117]

The Polycentric theory is the forerunner of the tenet of Multiregional evolution that humans evolved everywhere, because everywhere was always part of the whole. It was for the Multiregionalists of some four decades later to finally resolve the contradiction between regional continuities and continued, ubiquitous crossing.

THE INTERPRETERS

Weidenreich had no doctoral students after he became a paleoanthropologist. During the final years he worked at Heidelberg and while he was professor of anthropology in Frankfurt, he granted no advanced degrees in anthropology, and after that he never worked at an institution where paleoanthropologists could get degrees. This had important implications for his influence in paleoanthropology. Students are an important conduit of ideas from one generation to the next.[118] Doctoral students are often active proselytizers of their professor's theories. Even when they choose the opposite track and attack their professor's work, this can help bring it to prominence. But a different, more subtle, but more important way that students can strongly influence the acceptance of ideas is when they embrace their professor's theory as the framework for their research. They can articulate a theory and use it as the basis for establishing further understandings that can be drawn from it, and in doing so help transform the theory into a part of the body of accepted knowledge. This does not, of course, prove a theory correct, but it does act to expand its influence. Weidenreich's influence was limited—he had no students to proselytize, attack, or further articulate his Polycentric model. And, he was up against a phalanx of graduates of the most prolific physical anthropology professor of his times, or if not always up against them, at least left to their tender mercies.

Of course, he was not totally without influence; he had an impact on a great many people, although, ironically, most of them were polygenists. Keith's 1936 paper mentions him because Weidenreich's writings contributed to his shift to polygenism through his demonstrating regional evolution in China. His interpretation of Weidenreich was similar to Howells', writing in his autobiography "quite independently of me Dr. Franz Weidenreich formulated in 1943 a theory of the origin of modern races *from separate early Pleistocene ancestors* [italics ours].[119] He also influenced Gates, as noted above. However, perhaps the most prominent crediting of Weidenreich is in the introduction of Coon's *The Origin of Races.*

Coon dedicated his book to Weidenreich, and then, in the introduction, he wrote:

> While I was writing *The Races of Europe* in Cambridge, Massachusetts, he was busy in New York studying the Sinanthropus remains. At that time he concluded that the peculiarities that made Sinanthropus distinct from other fossil men were of two kinds, evolutionary and racial. From the evolutionary point of view Sinanthropus was more primitive than any known living population. *Racially he was Mongoloid.*[120]

Quite unlike Weidenreich, Coon "tried to see how many racial lines, including the Mongoloid, could be traced back to the first instance that any kind of man had appeared on the earth." Ironically their very different precepts of the evolutionary pattern became inexorably confused with each other, until finally Coon came to be described as an amplification of Weidenreich;[121] they were thought to be essentially the same,[122] and Coon was described as providing an explanation for Weidenreich's ideas in evolutionary theory.[123] Because Coon developed a fundamentally polygenic model, and because of the reception given to his claims that the races became human at different times and Australians and Africans became human last, this association managed to totally discredit Weidenreich. But apart from their recognition of regional continuity, Coon and Weidenreich were not at all the same. Coon had provided an evolutionary mechanism that was an answer to the wrong question, starting from a presumption of isolation and parallelism that was contrary to Weidenreich's views and alien to his beliefs. Just the opposite from Coon, Weidenreich started with the presumption of ubiquitous, constant gene flow. He needed to explain how the regional continuity he observed could possibly occur. Coon never addressed this, or ever understood it as Weidenreich's real problem. Indeed, Coon never saw it as a problem at all since he accepted the virtual isolation of the races.

Coon, nevertheless, was much more accurate than Howells in imparting Weidenreich's views; he acknowledges the importance of gene flow in polycentric evolution. What he fails to represent is that interbreeding was critical and fundamental to Weidenreich's ideas. For Coon, gene flow was almost an afterthought, something that needed to be acknowledged as a necessary part of his explanation for human similarity (so that he wouldn't be tarred with the convergence brush), but he thought it occurred minimally. Here there was a contradiction. On the one hand, he would explain important evolutionary similarities, such as increasing brain size, as the consequences of genes that could spread across racial boundaries, but then he envisaged the races evolving at different rates. Races could be evolving almost side-by-side, but so independently that they became "sapient" at quite different times. This is a contradiction he never resolved.

So, in fact, what Coon actually accomplished was a continued distortion of Weidenreich's ideas; they were represented as something that was actually a cross between Weidenreich's real ideas and Hooton's, which is how Coon himself viewed human evolution. He was far more influenced by Hooton than by Weidenreich, which is apparent in his creation of polygenic consequences out of polycentric evolution. We can now see he never convincingly applied evolutionary theory to the problem; his attempts to do so flew in the face of evolutionary thinking, and as we noted in the last chapter Coon's book was criticized as antievolutionary by some of the foremost evolutionists and geneticists of the time.

Weidenreich often seems to be better known among anthropologists by what others wrote about him than from his own prolific record. The major secondary Weidenreich sources are Coon and Howells; therefore many people see Weidenreich through eyes heavily influenced by Hooton. Coon and Howells treated Weidenreich's thinking similarly, in many respects. Both attributed a far more typological outlook to Weidenreich than his writings reveal. They attributed virtual or real parallelism to his polycentric theory and contended his scheme demonstrated the great antiquity of races. Howells' view of race was progressive in his middle and later career. He strongly criticizes the idea of classifying race on the basis of physical features. We think he fundamentally agreed with Weidenreich in terms of race, except for the way the two thought of features; Weidenreich could separate the idea of features from race, while for Howells features are what led to racial typologies. But his interpretation of Weidenreich as a polygenist was not benign. Particularly damaging to an accurate understanding of Weidenreich's views was the candelabra in Howells' widely read books. Even now, writings on the intellectual history of the time describe Howells' change as an "exaggeration" or "simplification" of the trellis.[124] But there is no sense in which the candelabra either exaggerates or simplifies Weidenreich's thinking—it simply misrepresents it.

So Coon and Howells both misrepresented Weidenreich's ideas, but to different ends. They held opposing views of whether Weidenreich, as each saw him, was essentially correct. Coon supported Weidenreich's theory as he interpreted it and venerated Weidenreich. But Howells condemned Weidenreich, representing him as a polygenist who theorized evolution by parallelism. Ironically, both Coon and Howells may have been ascribing Hooton's polygenic ideas to Weidenreich.

Weidenreich considered all living humans and human fossils[125] members of the same species and his ideas on the nature of race were the underlying foundation of his polycentric scheme. Weidenreich never thought there were such things as pure races, and he certainly didn't consider the races of today to be the same as those of the Early Pleistocene. He simply thought regional differentiation was old and that some of these features were maintained. One really needs to understand different people's views on what races are to understand how they fit in an evolutionary perspective. The typological notions of race attributed to him have made it difficult to understand Weidenreich's model, which suggested the regional continuity of traits representing "links" to the past, *not* the continuity of races themselves. For Weidenreich races were evolving entities that were constantly changing. While he mentioned adaptation, especially with regard to how races attained their final forms of today, as far as he was concerned the primary source of this change was interbreeding. It is true that Weidenreich had his own contradiction to contend with—he needed an evolutionary mechanism to reconcile this idea of evolution with the continuity of traits in regions. After rejecting an isolation explanation, he had failed to find a rationale for how regional continuity could occur. But Coon didn't provide it either. Instead, Coon tried to furnish explanations for worldwide change in the face of virtual isolation. He couldn't furnish an evolutionary explanation for Weidenreich's theory because he did not share Weidenreich's view of race and therefore had the problem backward.

8

FUNCTIONAL MORPHOLOGY, ORTHOGENESIS, AND THE DUBOIS SYNDROME

WHEN THEY WERE FIRST PUBLISHED, the Mitochondrial Eve results[1] were clearly incongruous with Multiregional evolution, and we wondered how the two could be reconciled. Milford drafted a quick response to the first paper that laid out the Eve theory, a jointly authored work[2] in the prestigious journal *Nature* on mtDNA and human evolution (based on Rebecca Cann's dissertation written for Allan Wilson). The Cann thesis assumed mitochondrial DNA (mtDNA) in the cytoplasm evolved by accumulating random mutations. The more mutational differences, the more time separated mitochondrial lines. Since the rate of mutations could be determined by comparing the *amount of difference* between two species to the *time since they diverged* as determined from the fossil record, mitochondrial dissimilarity made a clock that could time divergences—that "holy grail" of Eve theory genetics, the "molecular clock." The incongruities with Multiregional evolution came from the clock, specifically the divergence times that were calculated for human mtDNA. Cann contended that her work showed all extant human mtDNA, and therefore they reasoned all modern humans, had a common, unique, recent origin in Africa. All other human lines from this time of origin became extinct, without offspring. This meant that the vast majority of human fossils represented populations that left no descendants. Because the new mitochondrial work emerged from an older genetics tradition of maternal inheritance, Milford sent the draft of his reply off to our Uncle Ernst for comment.

Ernst Caspari (1909–1988) was a biologist, a geneticist with the strong interest in history that sometimes comes in later life with the realization that one has lived through (as he would say) "exceeeedingly" interesting times. Ernst witnessed biology's struggle to come to grips with Darwinism, to provide explanations for evolutionary and developmental transformations and for the diversification of species, when only a very limited understanding of genetics and its importance to these problems was realized. He understood the attraction of the non-Darwinian evolutionary theories that were so popular in the earlier decades of this century and the appeal of the hereditary mechanisms they proposed. Ernst told wonderful stories of genetics as an emerging discipline and of its principal actors. And most important he had the comprehensive breadth of view needed to place the motivations and ideas of others, even the non-Darwinians who did not share his perspectives, into an understandable referential framework. Listening to Ernst, the ideas of some Lamarckians and orthogeneticists, who we used to dismiss as unscientific, made sense in the context of their times, reflecting attempts to provide evolutionary explanations for problems that are still debated, although now within a paradigm in which an important role for natural selection (differential reproductive success) is assumed. Such a role was not obvious earlier this century, especially in some parts of the German genetics community.

Ernst was born in Berlin into a family of Jewish descent. He was raised there and subsequently took a Ph.D. in 1933 with Alfred Kühn (1885–1968), the famous developmental geneticist at Göttingen, who was the central figure of a so-called comprehensive school, one of the styles of thought characteristic of the German genetics community as it existed in the first 40 years of this century.[3] This school was defined by theoretically oriented geneticists, with broad interests not only within science but also in philosophy and the arts. Ernst's dissertation[4] work was on the moth *Ephestia*. He performed successful gonad transplants, from moths who had pigmented eyes, that subsequently caused eye color to change in moths with unpigmented eyes. What Ernst had demonstrated was that the gene product that controlled pigmentation was a diffusible substance which he called a hormone (later identified as kynurenine). Pigmentation could be achieved by inheriting the appropriate gene that controlled hormone production or by a transplanted gonad that produced the hormone. This work was repeated two years later in a laboratory overtly competing with Kühn's. George Beadle's research was on eye color in the fruit fly, *Drosophila*. Later, working with Edward Tatum on the much faster developing bread mold, *Neurospora*, Beadle won the Nobel Prize in 1958 for the "one gene–one enzyme" hypothesis that the earlier work inspired.

After finishing his degree at Göttingen, Ernst stayed on as an assistant in Kühn's lab. Kühn was part of a vibrant academic community—he was friends with the nuclear physicists James Franck and Max Born and a number of other left-wing faculty members. Kühn's political views were decidedly liberal. After the Nazis came to power, many in his intellectual circle of friends emigrated, and Kühn kept Ernst as an assistant, even after the civil service law of April 1933. Kühn was nervous about this, writing to a colleague, Otto Koehler, that he needed to keep a low profile to avoid drawing attention to the Jewish members of his lab that he never reported.[5] Ernst left Germany in 1935, taking a position as an assistant in microbiology at the University of Istanbul where he stayed for three years. Although he felt academically isolated, he thoroughly enjoyed Istanbul, reveling in the beauty of the city and its historic importance as the crossroads of Eastern and Western empires. He immigrated to the United States in 1938 to take a fellowship at Lafayette College in Pennsylvania (arranged by the mouse geneticist L. C. Dunn). He continued his genetic research there and taught comparative anatomy. After several academic positions he was offered the chairmanship at the University of Rochester in 1960, where he remained until his death.

Ernst Caspari in 1938.

CYTOPLASMIC INHERITANCE AND LAMARCKISM

We soon received Ernst's comments, longer than the paper Milford drafted as his first response. He was delighted that attention was being paid to the genetics of mitochondria, as it was an issue that he dealt with in earlier phases of his career. He had been interested in cytoplasmic inheritance[6] for decades, writing a major review of it in 1948, and was absolutely convinced that this maternal aspect of inheritance was also under Darwinian selection. He was helpful in developing our understanding of the implications of the differences between nuclear genetics, which we were used to, and the mtDNA cloning process. Long discussions ensued over the role cytoplasmic inheritance played in German genetics and the strong commitment in interwar Germany to non-Darwinian evolutionary theories based on it.

Ernst's early research was in developmental genetics, a particularly popular focus in interwar Germany, when most American (and many other) geneticists were more interested in transmission genetics.[7] Ernst's adviser, Kühn, was involved in work on cytoplasmic inheritance and supported the "plasmon theory" of his friend and colleague Fritz von Wettstein (1895–1945), a botanist from the laboratory of Carl Correns (1864–1933),[8] who was the most eminent of the *plasmon* theorists. *Plasmon* referred to the genetic elements in the cyto*plasm*, as opposed to the *genom*, the genetic elements in the nucleus.[9] In fact, the German focus on cytoplasmic inheritance during the interwar years was a consequence of their work in developmental genetics.[10] Interest in cytoplasmic inheritance originated with late 19th-century pre-Mendelian theories of cell differentiation and division and developed during the flurry of theorizing and experimentation that accompanied the rediscovery of Mendelian genetics. At that time, cytoplasmic inheritance took on meanings beyond what we, today, attribute to the DNA carried by bodies in the cytoplasm such as mitochondria (in animals) or chloroplasts (in plants). It also included the expectation that there was a transmission of organisms in symbiosis with the cell, a transmission of self-perpetuating patterns (liver cells begetting liver cells), and the perpetuation of cellular organization.

This was an exciting period, when the concept of the cell was changing radically and connections were being drawn between cell studies and evolutionary theory. German scientists were at the forefront of this work, with August Weismann (1834–1914) positing early on that there was a direct relationship between the newly discovered cell nucleus, heredity, and evolution.[11] The question was how the same hereditary message in each cell's nucleus could oversee different functions in different cells. Weismann proposed that the hereditary material in the nucleus differentiated at each cell division, with different parts going to different daughter cells, accounting for how the cells within a developing organism specialized and differentiated. This would imply that cell differentiation and specialization were due to unequal distributions of nuclear material to daughter cells during cell division. He hypothesized the zygote, or newly fertilized egg, begins with one library in the nucleus, and as cells divide, different cells and their offspring get different parts of it. But in 1892 Hans Driesch performed experiments showing that as sea urchin cells divided, the nuclear material in each daughter cell was the same. This seems obvious now, with the sure knowledge that all cells have the same nuclear DNA, but it was not so, then. Furthermore, Driesch's discovery implied an important cytoplasmic role. Since *all* of the nuclear material went to each of the daughter cells, it seemed as though cell *differentiation* must be the result of processes acting in the cytoplasm. Each cell nucleus gets the same library, but different parts are read. This fit in well with ideas akin to preformation that were very popular in Germany. According to the preformation concept, there was an agent intrinsic to the zygotes that governed the course of an individual's subsequent development. Many experiments seemed to support the idea that after fertilization, the cytoplasm was home to something that governed at least some developmental change, if not also evolutionary change.

Perhaps the strongest support for cytoplasmic inheritance came from the discovery, by Carl Correns[12] and Erwin Baur (1876–1933), of nonreciprocal crosses in plants, strongly indicating a nonnuclear, maternally linked aspect to heredity. Occasionally when two plant strains were crossed, certain characteristics of the offspring always mirrored that found in the strain that contributed the egg. It was reasoned that if inheritance was purely nuclear, it should not matter which of the strains the egg-contributor came from. There should be no systematic difference between hybrids pollinated by one or the other of the strains because the male and female hereditary material should be the same. Correns assumed pollen from the male contributes only nuclear material, while cytoplasmic materials as well come with the egg that is fertilized. If the cytoplasm contributes to the inheritance, it would matter which strain contributed the egg. But there was no uniform opinion on this, and some, like Weismann,[13] denied it completely. Several of his colleagues in the German-speaking world believed that the hereditary material was in the chloroplasts or other plastids; some thought that these could be introduced by males through the pollen tube (Erwin Baur); others (Kühn's friend Fritz von Wettstein, for instance) hypothesized that there was nonparticulate hereditary material in the cytoplasm, thereby naming and describing the *plasmon* and distinguishing it from the *genom*.

Although the majority of the geneticists working on these problems held evolutionary views sympathetic with some form of Darwinism,[14] many took evidence for cytoplasmic inheritance as support for dualist theories of evolution (meaning Darwinism plus something else). Dualist theories were very prevalent in interwar Germany and held that while some forms of variation were inherited, other more complex kinds of features changed when acted on directly by the environment through the cytoplasm. This aspect of evolution reflected a then-accepted interpretation of Darwinism based on Darwin's belief in the inheritance of acquired characters. There was no necessary contradiction perceived between Darwin's natural selection and the ability to inherit characteristics acquired during life.[15] Selection was the mechanism that led to adaptation, while acquired characters were part of the variation that selection acted on (selection needs inherited variations to act upon, in order to create change).

Most biologists in interwar Germany continued to believe that there were multiple evolutionary mechanisms.[16] No one doubted that selection occurred, but many considered that it could not account for the major differences found between species, let alone between higher taxa. They saw its role mainly as pruning existing morphological variation. Selection could not create something new; it could not account for differences in archetype, or *Bauplan*. The problem was, what did? Some geneticists found the explanation in a form of cytoplasmic transmission, while many paleontologists and biologists turned to use inheritance, part of the neo-Lamarckian theory that developed after Darwin, to explain major changes. Neo-Lamarckism discards Lamarck's contention that evolution tends toward perfection, a form of finalism.[17] Instead, it changed the theory into one of use inheritance, as Darwin considered it.

BAUPLAN

The idea of an archetype, or *Bauplan*, comes from the major discoveries of the 19th-century biologists. They learned that animals in nature did not exist along a continuum of undivided forms, but were discrete entities (Mayr, in his development of the modern species concept,[18] calls this the discovery of multidimensional species). *Bauplan* was the way these discrete entities were described, and was often taken to mean "type," in the essentialist sense of having a distinct underlying nature. In this meaning, animal types are constant and unvarying, separated by unbridgeable gaps. But this is not the historic, and generally current, biological meaning.[19] *Bauplan*, in the German literature, refers to an abstraction of either the ancestral form of a group or the basic physiological/structural framework that it shares. In today's taxonomy, groups are defined by descent, uniquely sharing a common ancestor. But in considering the *Bauplan*, descent groups are further described by their structural similarities derived from their common ancestor. If "primates" are a valid taxon, there would have to be a generalized primate *Bauplan*.

Bauplan refers to underlying structural principles that characterize higher taxonomic categories such as species. The continuous variation within species is fluid and has a number of causes, but *Baupläne* are very conservative, and remain stable, in fact becoming more and more inflexible over the course of evolution.[20] Today, "morphotype" has taken on this meaning, in modern comparative anatomy, and even strong Darwinians recognize the idea that a species may be characterized by a common form defined by anatomical features. Explanations for the stability of *Baupläne* vary, but most recognize that a conservative developmental system is built into the genome, providing restraints during development that limit the amount of variation possible.

While many aspects of Darwinism were acknowledged by the international scientific community quite readily, natural selection as the major evolutionary agent took longer to be accepted. It seemed as though there were two kinds of evolutionary change—major and minor, as it were—and selection caused only minor changes. This understanding came from the way variation was divided, reminiscent of, or perhaps an extension of, older nonevolutionary essentialist thinking about variation. One kind of variation seemed to involve changeable elements that differed throughout a population; the other involved static elements that represented the differences between species, or higher taxa, often those reflecting *Baupläne*. This distinction, between the mechanisms of microevolution and macroevolution, continues to be debated today;[21] even now, there are evolutionists who contend that the variation between higher taxa remains to be satisfactorily accounted for. And there were other impediments to accepting a more general, and more important, role for selection in the process of evolution. The populational premise of selection was incompatible with the Platonic, typological world views that underlay much of Western thought and science. Successful selection depends on variation—without it there is nothing different to select—but variability is what the type concept is *not* about. How could variations in one type of animal (read: species) possibly lead to the appearance of another type?

Natural selection also was incompatible with finalism, the idea that there is an ultimate "purpose" to evolution. Moreover, the probabilistic nature of selection was difficult to understand and, at first, there was no probability theory in mathematics that could be used for its underpinnings. Instead, there was an underlying expectation that something as complex and wonderful as higher organisms and major adaptive complexes could not have developed by "mere chance." This sort of sentiment derives from a misunderstanding of the way selection works.[22] A widely cited example is found in the analogy of a chimpanzee at a typewriter. The minuscule chances of the animal composing all the works of Shakespeare by randomly hitting keys were compared with the equally minimal chances of natural selection fashioning a human being. Instead, selection was envisaged merely as a culling force for unfit individuals; the creative aspects of new combinations of genes could not be appreciated without probability theory because the interactive effects of genetic systems with many genes could not be abstractly grasped.[23]

It still remains difficult for people to come to grips with the idea of totally nondirected or nonprogressive evolutionary changes, when patterning is so evident in their consequences; so earlier this century selection, as it was understood, could not be seen as a way to make major changes in structural types. How, then, did new *Baupläne* appear?

NEO-LAMARCKISM

For these reasons until the modern synthesis, many ideas about evolution proliferated, some distinctly non-Darwinian. Most evolutionary scientists accepted the notion that natural selection was the agent of some change, but could not see how selection could account for the major differences one finds between species or higher taxa. They believed something else was needed to account for different *Baupläne*. Evolutionary mechanisms were postulated and described at several morphological levels. Broadly, these can be considered orthogenetic and neo-Lamarckian. In the former the evolution of the organism is thought to be directed by internal factors; there is a developmental program, often a trend toward increasing specialization, that the organism is programmed to carry out. This is a very Newtonian idea;[24] just as the initial conditions of the universe controlled its present state, so the initial conditions for life control its present state. Orthogeneticists view evolution as a process of unfolding this already-encoded program, controlled by inner directions.

Neo-Lamarckian evolution is shaped by the environment; both the agents of change and the new variations successful change requires have sources that are external to the organism itself. It lacks the progressionist interpretation of evolution implicit in orthogenesis (and in Lamarck's formulation), and thereby is more compatible with other aspects of Darwinism. In the early part of this century both neo-Lamarckian and orthogenetic views were held by most German paleontologists. While Darwinism was very popular in German science in the years following the *Origin of Species* largely because of Haeckel, as the fossil record became better known, selection was found to be more and more inadequate as an explanation.[25] Scientists did not yet understand how selection could cause the large changes that were apparent in the fossil record. Fossils, it will be remembered, were not of great interest to Haeckel, and he remained uninfluenced by these developments. There was a series of more or less successive schools in German paleontology, and finally Lamarckism, after a long reign, was replaced as the dominant evolutionary model by orthogenesis.[26] But neo-Lamarckism, particularly the aspects of the older Lamarckian ideas such as inheritance of acquired characteristics, remained popular until the modern synthesis, when the biological disciplines, from developmental and population genetics to morphology and paleontology, were finally able to account for macroevolution as well as microevolution within a Darwinian framework.

Neo-Lamarckism was extremely popular for a number of reasons, social and scientific. The idea of direct environmental modification, without the elements of individual competition that Darwinian thinking required, translated nicely to the social factors in Victorian England that were affecting a newly emergent middle class.[27] The idea of transformation due to environmental influence spoke to the proponents of a new professionalism, wherein individual achievement was replacing the hereditary class system as a main criterion for entrance and success in some (notably academic and medical) professions. In Germany during the same time period, the time was ripe to embrace Darwinian ideas, albeit with their own German twist.[28] There was no entrenched class system in the newly unified Germany and the professions were more or less open. German senses of competition found the Haeckelian interpretation of Darwinism, human evolution driven by competition between the races, extremely attractive, so Darwinian thinking had an early stronghold.

Neo-Lamarckism was scientifically appealing as well. Certainly in the pre-Mendelian days it is easy to see that Lamarckism was as believable as Darwinian selection in accounting for aspects of evolution. The mechanisms producing the variation that selection acts on were as unknown as the supposed mechanisms governing the transmission of acquired characteristics. In fact, a variety of different mechanisms were proposed to account for the transmission of acquired characteristics, just as a number of mechanisms were suggested to account for heritable variation. Later in this century, even when the principles of mutation and recombination as sources of variation became better understood, neo-Lamarckian paleontologists had trouble accepting that these could account for really large changes, for instance, changes in *Bauplan*. Paleontologists referred to the mutations that geneticists were inducing in *Drosophila* as *kinker* (meaning "trifles").[29] They created changes that caused new problems like malformed wings or eyelessness, not new species, to appear. Thus, the neo-Lamarckian paleontologists of the 1920s did not feel that Mendelian genetics, at least as it was understood at that time, provided the answer to macroevolutionary problems.[30]

CYTOPLASMIC INHERITANCE

German paleontologists and geneticists split over how to explain major evolutionary changes. Darwin's theory of natural selection was not sufficient, and in fact had fallen out of favor with many biological scientists. Interwar German paleontologists and field biologists, who dealt with phenotypic variation and macroevolution daily, believed some Lamarckian mechanism must account for the obvious adaptations of form to the environment.[31] But, unlike the paleontologists, most geneticists (Mendelian and cytoplasmic) rejected Lamarckism. Geneticists accepted the idea that natural selection played a role in the evolutionary process. But what role? Many Mendelian geneticists had rejected Darwinism as an overarching explanation for all evolution. Early in the century mutations were promoted as the explanation for major evolutionary changes. But after 1910 the experimental basis for favorable macromutations seemed increasingly questionable, and some turned away from examining the role of genetics in the evolutionary process altogether.[32] Population genetics theory, one possible key for reconciliation, was strongly developed in Russia, but it did not spread widely and was soon extinguished there.[33] Those geneticists who retained a Darwinian perspective and continued to focus on evolutionary problems were selectionists, even those who were not strictly Mendelian. Mendelians argued that both kinds of variation, continuous and archetypal, were controlled by genes in the nucleus. But other geneticists, while also selectionists, continued to believe there was a dual evolutionary process. From their understanding of developmental genetics, they envisaged two levels of evolutionary change, based on two types of inheritance. There were both nuclear genes and another site of genetic control, in the cytoplasm, governing a different class of traits.

Eventually, the concept of cytoplasmic inheritance spread to the morphologists. Although many geneticists didn't agree with it,[34] evidence for cytoplasmic inheritance was used to support a "*Grundstock*" hypothesis[35] that was accepted by many biologists and paleontologists. The *Grundstock* hypothesis was that evolutionarily important traits, such as those that distinguish higher taxa as part of their different *Baupläne*, were governed by extrachromosomal sources that could be modified by environmental forces more powerful than selection. The *Grundstock* itself had not been clearly located, but it was generally thought to be in the cytoplasm. It was known empirically that chromosomal genes remained stable in the face of environmental change, and it was hypothesized that this additional form of heredity was more responsive to external conditions. Minor evolutionary changes could occur through mutation and selection, just as the geneticists claimed, but paleontologists thought major changes required direct environmental action and believed these major changes involved structures that were so complex, often composed of many separate, interrelated features, that their evolution could not be the culmination of mutations (the explanation they thought would be required to account for their evolution by Darwinian means).

The concept of "dauermodifications"[36] was developed, encapsulating additional evidence supporting the case of cytoplasmic inheritance. Dauermodifications were thought to be long-lasting changes induced by the environment. Because they were persistent, they were thought to have become hereditary. These would continue to be passed on for many generations, but unlike other hereditary traits they were not permanent. Early experiments suggested dauermodifications were transmitted maternally from one generation to the next—therefore, through the cytoplasm and not through the nucleus. A number of genetic experiments supported this idea. Emil Fischer seemed to demonstrate that cold-induced color changes in butterflies were passed on, and the most famous of the dauermodification supporters, the selectionist Victor Jollos (1887–1941), showed arsenic tolerance in successive generations of *Paramecium* could be increased, and that this could last for many generations, although not forever without continued stress.

Because natural selection pruning modifications controlled by the nucleus did not seem sufficient as an explanation for evolution, paleontologists turned to the inheritance of acquired characteristics to help explain their observations, even as many geneticists had found evidence suggesting there was modifiable cytoplasmic inheritance. In dauermodifications there seemed to be a link between them. The important additional element was provided by the dauermodifications because they were held up as examples of neo-Lamarckian change. Jollos was by no means a neo-Lamarckian, but the neo-Lamarckian biologists and paleontologists saw his results as promising, indicating the first steps toward the discovery of a mechanism able to account for the transmission of acquired characteristics. The neo-Lamarckians thought perhaps the permanent changes they saw in the evolutionary record represented dauermodifications that had become immutable through constant exposure to the source of the modification.

Yet the *plasmon* theorists were ambivalent about dauermodifications because they recognized their problem: the modifications did not seem to last more than a few generations, and while the mechanism leading to permanence proposed by the neo-Lamarckians was possible, there was no evidence to show it was correct. *Plasmon* geneticists were torn between embracing the support extranuclear inheritance provided for their ideas or, alternatively, distancing themselves from the ephemeral nature of such transformations. However, in spite of these reservations, the *Grundstock* idea was influential, so much so that in 1937 the population geneticist Theodosius Dobzhansky found it necessary not only to emphasize its refutation, but also to undermine the stronger evidence for cytoplasmic inheritance.[37] But the *Grundstock* hypothesis continued to permeate the thinking of at least some geneticists. Erwin Baur and N. Timofeeff-Resovsky, both very strong selectionists of the time, allowed that there may be other sources of inheritance outside of the nucleus. Although they were pretty sure selection would eventually be shown to account for macroevolution, they didn't rule out the possibility of other mechanisms.[38] Later, in 1945, the great evolutionary geneticist Sewall Wright, whose mathematics of population genetics proved to be perhaps *the* fundamental basis for the new synthesis, did not feel he could discount the evidence for dauermodifications and sought to explain them.

In sum, the idea of cytoplasmic inheritance appealed greatly to morphologists because they needed another form of inheritance to explain *Baupläne*. It was also appealing to some orthogeneticists who could see the cytoplasm as the home of a template for development and who felt the probabilistic nature of selection, combined with the complexity of functional systems, could not account for long-term trends. It was popular with neo-Lamarckians who saw in it a heredity system acted on directly by the environment, especially when cytoplasmic inheritance was taken in conjunction with the dauermodification research. Many of the neo-Lamarckians were paleontologists who recognized trends in the fossil record and doubted selection could account for their long-lasting evolutionary direction. So for many good reasons non-Darwinian evolutionary mechanisms were embraced, particularly by those biologists and paleontologists who were broadly trained, as many German scientists were.

Perhaps the time had come to try for an evolutionary synthesis of the macroevolutionary changes studied by paleontologists and the microevolutionary observations of the geneticists. The problem such a synthesis faced was to reconcile in the same way a need of both paleontologists and geneticists: a need to understand what mechanisms beyond natural selection can account for major evolutionary change, the diversity of *Baupläne*. Many workers from both disciplines already seemed to agree natural selection wasn't enough.

THE 1929 MEETING

In the fall of 1929 a famous scientific meeting was held in Tübingen, a joint congress of the German Paleontological and Genetic Societies. Its purpose was wistfully synthetic—to discuss, and hopefully bridge, the gap between the largely (but not completely) selectionist geneticists and the largely (but not completely) neo-Lamarckian paleontologists. The selectionists needed to show convincingly how macroevolutionary change could take place through mutation and selection, and they did not have the analytical tools or the knowledge to do it. They still could only envision selection as a sieve or a pruner and could not adequately explain its role as a creative force. Continued experimentation clouded the role of mutations as a possible cause of major change because important ones were never observed. The neo-Lamarckians also needed to show a mechanism that accounted for major change and also for the evolutionary trends that fit organisms to their environments. They further needed to understand how environmentally altered morphological systems could be transmitted, and of course they could not do it. Each of these groups had only part of the picture, and part of what they had was wrong. Each could refute the tenets of the other, but they seemed to have developed differences so irreconcilable that they no longer spoke to each other, but past each other. Later, it was widely perceived that "in various confrontations between the two camps there was no evidence of a willingness to compromise; all the argument was directed toward trying to prove that the other camp was wrong."[39] For us, engaged in the modern human origins debates, how familiar it sounds!

Into this fray stepped none other than the new professor of anthropology at the University of Frankfurt, the functional anatomist and paleoanthropologist Franz Weidenreich. He was chosen to bring the factions together, as the keynote speaker for the first joint session of the societies.

Weidenreich opened the Tübingen conference with an attempt at rapprochement.[40] He argued that genetics and neo-Lamarckism could be reconciled and the understanding of the mechanics of inheritance at that time was just too weak to rule out the possibility that mechanisms that were acted on directly by the environment could exist. Weidenreich dismissed mutations as he described the features that some macroevolution theorists proudly claimed as their consequences (for instance, eye color or number of bristles on the fruit flies they favored breeding in their genetic experiments) as *Kinkerlitzchen*, or itsy-bitsies.[41] He also pointed out two places where natural selection seemed insufficient to be solely responsible for evolutionary change. First was the paleontological observation that anatomical alterations paralleled environmental change, fitting the species to the changing environment. To this he added the implications of his own studies of functional anatomy; these alterations affected entire structural systems made up of interrelated parts, not just single features. The problem he proposed was that the structural systems functioned in concert and it was unclear how they could respond all together. How could mutations and selection alone change all these things—everything involved in a well-adapted *Bauplan?*

Weidenreich, like many biologists who studied the past, was considered a neo-Lamarckian. Later he believed he was considered neo-Lamarckian because he was a functional morphologist, concerned with the relation of skeletal form and function,[42] and there is much truth in this observation. Interested in paleoanthropology since the time of his close association with Schwalbe, his research had increasingly focused on the interconnected functional changes that accompanied human evolution. As bipedal locomotion evolved in humans, Weidenreich showed that changes were not just in the locomotor system, the pelvis and foot, but also in the spine and cranium.[43]

Functional morphologists of the 1920s often turned to neo-Lamarckism because before the modern synthesis, Darwinian selection could not explain adaptations and adaptive trends, whereas neo-Lamarckism could. Furthermore, functional morphology and adaptation often *sound* like use inheritance, even though the mechanism that transforms functional adaptations into inherited variation is not neo-Lamarckian. The results of functional changes, adaptations, are often described in teleological terms, and the focus is really on the means to an adaptive end. Functional morphologists dealing with bones commonly examine an evolutionary problem by analyzing the physical stresses that come from skeletal functions and seek to understand how bones respond to them. They generally do not look at the evolutionary mechanisms for achieving adaptations, but, instead, consider them caused directly by changes in bone tissue. While the evolutionary mechanisms are populational and the process of anatomical evolution is based on the variation in a population, functional change isn't usually studied that way. Most morphologists do not think populationally because they are focused on the mechanics of a biological system that is usually within a single individual. So although they may examine both developmental and evolutionary links to adaptation, they do not really think about evolutionary process. Functional morphologists often have a different focus, not nonevolutionary but one centered on development and the relationships of biological systems within an organism and how these are adapted to the environment. Deeming functional studies in the 1920s as neo-Lamarckian was prevalent and not especially odious.[44]

So in his Tübingen speech, Weidenreich noted that while morphologists could accept that genes were stable, a point the Mendelians emphasized, this did not necessarily mean that mutations and selection were the only evolutionary mechanisms. Agreeing with many geneticists, he further noted that mutations were not a sufficient cause for major change; most mutations that geneticists studied were rare and trivial and not of evolutionary importance, implying to him that more important genes might have a different source and be less stable in the face of environmental fluctuation than the nuclear genes the geneticists were familiar with. He also questioned whether mutations were truly random and independent of the environment, arguing that H. J. Muller's recent work on x-rays as mutagens showed that mutations could be induced by the environment. These mutations could not be studied for a long enough period (that is, geologic time) to tell whether they had the potential to direct significant structural change.

For these reasons, Weidenreich cautiously turned to cytoplasmic inheritance as a potential genetic explanation for neo-Lamarckism[45] and as a potential way of linking genetics to the problem of macroevolutionary change. Now, the opposite was not the case. Acceptance of cytoplasmic inheritance did not necessarily mean the acceptance of neo-Lamarckism; many of its German proponents explicitly rejected this possibility.[46] But for those who were neo-Lamarckians, this mechanism was no less credible than the mechanisms being proposed to explain Mendelian inheritance. Weidenreich emphasized that the research on dauermodifications could indicate how the environment might directly affect change on whole systems, and the system he had in mind was the skeletal one. It was already realized by the German bone biologists that living bone is a malleable tissue that changes during life. Julius Wolff[47] showed that bone adapts functionally during life, changing its shape to best meet the forces acting on it, and Weidenreich wrote on the effects of muscle action on the skeleton.[48] Furthermore, developmental adaptations can be very similar to genetic (evolutionary) ones, and it made sense to explore links between developmental and genetic change. The responses of bone are the right modifications for a successful adaptive change, and if they could eventually become an inherited change, the species could evolve an altered structural system. Weidenreich suggested, following the dauermodification supporters, that the cytoplasmic "genes" (as we would say today) controlling the dauermodifications might eventually become as stable as the nuclear genes, if they were exposed to the environmental stimuli for a long enough period of time. Changes such as a specific response of bones to a particular stress might become permanent this way, if the stress continued for some number of generations.

Trying to reflect the interests and beliefs of the genetic community and combining this with his perspective from functional anatomy and paleontology, Weidenreich thought he had proposed a compromise. He had not. There is varying opinion over the value of the meeting. Of course, the perspective that different observers carried away from an attempt such as this is personal and depends on one's sympathies and biases. The well-known paleontologist Bernhard Rensch wrote that "mutual understanding of the different arguments was made possible, and this conference was perhaps the first step toward the later [modern] synthesis."[49] Others were less positive, and the general feeling seems to have been that it was a lost opportunity, even a disaster.[50]

Ernst Mayr's indignation reverberated more than a half-century later: "The paleontologists adopted the worst possible strategy. . . . They concentrated on trying to prove the existence of an inheritance of acquired characteristics, a subject which they were in no manner whatsoever qualified to discuss."[51]

Mayr suggested the meeting would have been more fruitful if the synthetic attempts had focused on solving the evolutionary problems the Mendelians were unable to explain, rather than on trying to account for neo-Lamarckian beliefs. But we believe the Mendelian geneticists were just as much to blame for its failure as the paleontologists. They did not accept the validity of the paleontologists' concerns, and the Mendelians do not appear to have been seeking answers to the macroevolutionary problems that concerned field biologists and paleontologists.[52] They clearly sent the wrong message to Weidenreich and other paleontologists. As the historian of science Jonathan Harwood put it, for the Mendelians

> the key to evolution lay not with paleontology's speculations but with genetic experiments. Since genetics had so far been unsuccessful in accounting for the emergence of complex adaptive traits, it was understandable why paleontology had resorted to the inheritance of acquired characteristics. But paleontologists would have to face up to hard genetic facts . . .[53]

The problem was that the "hard genetic facts" were hardly that. Gene structure and function were far from understood and it remained unclear how Mendelism could explain macroevolutionary change. This problem was recognized by much of the German genetic community[54] with its tradition of studying cytoplasmic inheritance, its great interest in dauermodifications, and its active examination of the possibility of directed mutations. More than in other regions, many geneticists of the German-speaking world were more or less sympathetic to neo-Lamarckian views. This may have been due in part to an emphasis on breadth of training; they tried to account for data from a large number of sources and did not yet realize that many of the soft-inheritance experiments were flawed.

But there was another reason for this sympathy, and for the intensity the debate engendered. It came from the nature-nurture argument as it was played out in the German political world and a confusion between the questions of whether genetic expressions could be modified by the environment and whether acquired characters could be inherited.[55] The neo-Lamarckian view was taken to downgrade the importance of inheritance in the expression of racial features. This was a position thought to be associated with Jews and Communists in Germany, and it undermined the Nazi and other racial hygienists whose eugenic programs were based on the validity of the Mendelian view: if racial traits were inherited they could be bred for or eliminated. The killing camps were racial hygiene in practice for the Nazis, they were the next step following the murdering of mentally and physically incapacitated citizens already underway when the camps were set up, because the Nazis treated social pathologies as biology and therefore inheritable, and believed these pathologies showed that Jews and Gypsies were degenerate races.

Interest in cytoplasmic inheritance continued in parts of the German genetic community but drew little attention elsewhere.[56] The victory of nuclear transmission genetics was almost complete, and the tie of cytoplasmic inheritance to use inheritance, especially during the time of the Lysenko affair, served to discredit it. There was a neo-Lamarckian taint from the dauermodification studies, despite Wright's attempts to explain them in a Darwinian framework.[57] In developmental genetics, the new understanding of gene action was quite simple and involved no input from the cytoplasm. Most of the focus was on transmission genetics, in any event, and this meant the detailed study of nuclear inheritance. When Uncle Ernst came to America in 1938, he wrote back to his adviser and mentor with some surprise that the German work on cytoplasmic inheritance was "almost unknown here . . . cytoplasmic inheritance has been thoroughly rejected by geneticists here."[58]

In fact, Ernst was partially responsible for a revival of interest in cytoplasmic inheritance, publishing a major review paper in English right after the war.[59] Ironically, this was at the time when new approaches were bringing a Darwinian credibility to use inheritance. C. H. Waddington pointed out[60] that there was the first Darwinian mechanism that could account for the inheritance of acquired characters, but this was inheritance in the nucleus, not the cytoplasm. This was *canalization.* Waddington argued that when a species changes its behavior or adapts in a way that altered ontogeny to produce a different adult form—for instance, the larger heart size that appears when human groups move to high altitudes (to make better use of the limited oxygen supply)[61]—there are two ways that the evolutionary process may respond. The first is to maintain plasticity in development so the structure continues to grow in response to the conditions it encounters (in this case, lower altitude, smaller heart; higher altitude, larger heart). An alternative response is to select for a changed developmental pathway. As the population evolves to respond optimally, it may be advantageous to buffer the response from variations in the cause, in this case so large hearts develop regardless of differences in the exact oxygen mix encountered by a growing individual. The response has become canalized, and the advantage of the developmental pathway changing this way is that the optimum adaptation will come to appear commonly, even when specific circumstances do not warrant it. As long as there is variation in developmental pathways, selection can establish a new one. The disadvantage of this change is the loss of developmental plasticity, and this means that canalization is not inevitable. But it is *possible*, and the point is that canalization provides a Darwinian mechanism that, under some circumstances, allows new characteristics that appear because of altered developmental conditions to become genetically encoded through a change in the ontogenetic process. Waddington believes that this is an acceptable explanation for the inheritance of some acquired characteristics and an explanation for the establishment of certain adaptations in a species.

Neo-Lamarckism in any form was now totally disassociated from work on cytoplasmic inheritance, and there is no discussion of it in Ernst's review article. But recognition of cytoplasmic inheritance never died, and is now experiencing a resurgence. The exploding field of mitochondrial genetics shows cytoplasmic inheritance to be an incontrovertible fact.[62]

ORTHOGENESIS AND BAUPLAN

Weidenreich was labeled both a neo-Lamarckian and an orthogeneticist, and to our modern ears this is the equivalent of calling someone a witch doctor. For us, these words have vitalist or mystical connotations and seem decidedly unscientific. But we need to examine what these concepts meant to *him*. The issues that made him consider rectilinear, or straight line, evolution as the result of more than natural selection are still issues today, although now their discussion can be incorporated into a Darwinian paradigm.

Largely as a result of the Tübingen meeting and papers ensuing from it, Weidenreich became widely known as a neo-Lamarckian. He has been, and continues to be, criticized for his non-Darwinian notions of evolutionary change. Indeed, he believed that this was why his early work on bipedalism as a functional system was ignored by his American colleagues in later years:

> Most of my contributions to the problem of correlation and adaptation have been ignored, even by authors dealing with the same question, perhaps because they were written in German, but certainly also because the time had not yet come for an approach to the problem of evolution from the functional side. Considerations that were suspected as "Lamarckian" were banned.[63]

It seems incredible now that a scientist so committed to functional explanations of variation disregarded natural selection as its cause. But his theoretical position must be seen the context of the times and of his training and professional life as a German anatomist and paleoanthropologist.

Weidenreich did think selection played some kind of a role in the continuous variation of populations, but that role was only as a "pruner"; he could not envision how selection could create major changes or be responsible for adaptive trends. However, while he first relied on a neo-Lamarckian explanation, he later gave it up: "The search of morphologists for environmental factors directing evolution lost much ground when the geneticists claimed that experiments on living organisms proved incontestably that acquired characters are not inheritable."[64] For Weidenreich as for the German paleontology community as a whole,[65] orthogenesis came to replace neo-Lamarckism as an explanation for evolutionary change beyond the ability of selection to create.

Orthogenic explanations have been common throughout the history of interpreting human evolution. Darwinian polygenists needed orthogenesis, or something like it, to explain how independently evolving races, really different species many of them thought, could all evolve the same way. Others needed teleological explanations. But Weidenreich was not a polygenist; he didn't need orthogenesis to explain parallel evolution because he didn't think there *was* parallel evolution. Nor was he particularly teleological in the vitalist sense—he didn't think evolution was directed by God. Moreover, he had a real understanding of continuous variation *within* species and the independent assortment of different traits, so essentialism was not critical to his thinking. So why did he embrace orthogenism?

At the root of Weidenreich's evolutionary ideas was his commitment to *Bauplan*, and this commitment can explain both his orthogenetic and neo-Lamarckian views. Like virtually all German morphologists and paleontologists and many geneticists, Weidenreich recognized two kinds of variation: continuous variation characterizing individuals and populations, and discontinuous variation dividing major categories—the archetypes or *Baupläne*. His continued reliance on non-

Darwinian mechanisms of evolutionary change stems from his belief that natural selection could not account for the origins and evolution of *Baupläne.* He could not understand how selection could change a *Bauplan,* since so many interdependent structures would have to change together.

Yet true evolution to Weidenreich involved the origin and adaptive evolution of major functional systems, with interrelated elements that must evolve together. Change in one part necessitates concomitant changes in the others. For *Baupläne* to change, the whole functional system has to change, and Weidenreich was convinced that while part of this change was adaptive, another part of it resulted from structural constraints on form, dictated by the system itself. These constraints accounted for orthogenetic evolution as he envisaged it.

For Weidenreich, bipedal locomotion was one of the most important unique human attributes and a major human adaptive complex. The human *Bauplan* was predicated on bipedalism. He saw it as an extremely complicated functional system that one way or another affected virtually all of human anatomy. He spent a large part of his career studying the details, and viewed numerous trends in human evolution, such as changes in the pelvis, lower limbs, foot, spine, and head, as interconnected consequences of bipedalism. Between 1913 and 1931 he published a series of detailed papers addressing the fundamental issue of "how far form and features of individual bones answer to the special requirements of erect position." The results of this work showed that the form of any element is dependent on the form of other elements—they are all correlated. "Any organization whatsoever can work only if all parts of which it is composed are adapted to each other and cooperate harmoniously."[66]

Weidenreich accepted the notion that there was progress in evolution, in the sense of increasing specialization along evolutionary pathways that were in a single direction "without any indication of deviation." This was orthogenesis, at least he called it that, although it is not completely clear that he used the term as others did. Mayr, for instance, defines orthogenesis as "a rectilinear trend in evolution (evolution in a straight line) due to a built-in finalistic principle,"[67] but Weidenreich said, "Should there be any aversion to the term orthogenesis because it may imply predestination, I have no objection to calling it rectolinear [*sic*] evolution."[68]

His main model for the evolutionary process was human evolution. He understood its details best, and we need to examine its details to see how functional constraints, rectilinear evolution, and *Bauplan* are related. From Weidenreich's studies of functional anatomy he reasoned that the direction of evolution was constrained by the intercorrelations between elements in a functional structure—in other words, by the *Bauplan.* Human bipedalism caused all sorts of concomitant changes that continued to be enacted in the gradual process of making the *Bauplan* more efficient. Weidenreich's notion of progress was just that, increasing specialization, and he believed this very definitely characterized human evolution. "If modes of evolution exist in which no specialization takes place, man is not a part of that category."[69] The postcranial remains of all fossil humans, like modern humans themselves, showed complete adaptation in this functional system, and therefore "it can be taken for granted that man achieved upright position as soon as he became morphologically discernible as man."[70]

The patterns of change in the cranium are more complex because there are several different trends involved and because they took place over a long time.[71] In addition to upright posture, humans also are unique in having a large, quite round brain, positioned atop the head and a face which is tucked under the braincase. This description of modern humans was one that Weidenreich saw evolving in the fossil record, contrasting with the more static bipedalism. If full bipedalism had been attained by the time of the 0.75-million-year-old Indonesian fossils, the same was not true of modern human cranial size and shape. Now, Weidenreich thought the evolution of brain size and shape was an ultimate consequence of bipedalism; since humans became upright, the skull has become better functionally adapted to the constraints of its new orientation atop, rather than in front of, the spine. The braincase "rolls up," using his terminology, bending around a transverse axis which runs approximately through the mandibular joints from side to side. In this way the front and back ends of the brain move closer together, resulting in the reduction of its length, and expansion of its height. The base of the cranium is shortened and deflected, resulting in an angled cranial base. Therefore he saw the shape of the skull as a consequence of specializations to the human bipedal *Bauplan.* The face also shrinks, becomes less prognathic, and positioned more under the braincase. These later changes are associated with general trends toward gracilization that effect the dentition and the robustness of the skeleton.

Brain size expansion itself, however, was a separate trend, independent of the orientation and shape of the brain, which were consequences of bipedalism. The trend in human evolution toward increasing brain size and specialization was thus only partially correlated with the primary bipedal specialization which Weidenreich saw as unique. The secondary specialization of increasing brain size, although "surpassing [the first] in significance," was not unique to humans. A phylogenetic trend for brain size increase was common in many mammalian taxa and also in reptiles and birds. However, Weidenreich argued that in humans the brain specialization went beyond increasing size: "The latest phase of specialization of the human brain must take place in the cortex and be independent of the actual size of the brain and, consequently, also that of the braincase."[72]

With these different but complexly related aspects to their evolution, all humans and human fossils demonstrated a unity of type as possessors of the same *Bauplan.* The significant changes in the fossil record were cranial changes, all thereby constrained within the same structural limits and the same intercorrelated functional system.

Ironically, Weidenreich's views on evolutionary mechanisms were used to explain how he could "advocate" polygenism, even though he never did. His polycentric model described major regional groups, interconnected by lines of gene flow, evolving in the same direction. Weidenreich had a populational, not a typological, view of races, but combined with his other assumptions this created questions he felt he needed to address about their evolution *within* a species: what kept the regional groups evolving the same way, and what kept them diverse? While his true problem lay with the second question, he focused on the first. For him the first question needed answering for much the same reason he found all evolutionary change problematic—Darwinian mechanisms seemed insufficient to explain it. While we realize he already had an answer—human groups evolved in the same way as the consequence of the continued genic exchanges of advantagous features he envisioned as part of his view on race—he didn't. He felt he needed to provide an explanation for the change itself and its widespread nature.

Weidenreich described humanity as made up of past and present populations that share not only common ancestry but also the same *Bauplan.* The evolution of these geographically dispersed populations was guided by the same functional limitations, constrained by the same intercorrelated matrix of structural elements, and orthogenesis explained their common evolutionary change. The problem with the orthogenesis explanation is that it could also be used to explain the evolution of isolated populations, and thereby Weidenreich's commitment to *Bauplan* as an "evolutionary force" inadvertently helped sustain the notion that Weidenreich believed there was parallel evolution of the races. Indeed, it is most ironic that had he believed that the races evolved in isolation, the two problems would have been solved; his orthogenic ideas could explain how such parallelism could occur, and the isolation would explain how regional features could be maintained. But he didn't believe this. His populational view of races as continuous, interconnected, and ephemeral entities did not allow the easy solution.

WHY A BIG BRAIN? THE DUBOIS SYNDROME

The expansion of the brain played a major causal role in Weidenreich's understanding of human evolution.[73] The correlations of the coevolving features were caused by function; for instance, the changing positions and leverages of the muscles that attach on the cranium, like the neck muscles, are the consequence of a larger cranial vault. He did not believe there was any other overriding relation that correlated these changes and contrasted his theory of structural correlations with Louis Bolk's theory of fetalization. Bolk (1866–1930), a Dutch anatomist, believed that evolution was characterized by a stoppage of the growth process, so the fetal characteristics of ancestral species became the adult characteristics of the descendants. This is why, he proposed, infant apes resemble human adults. This was quite the opposite of Haeckel's recapitulation theory (Chapter 5), and replaced it[74] as a heterochronic explanation of evolution by the turn of the century (heterochrony is change in the onset, timing, or rate of development). Bolk stood evolution on its head, arguing that the most fetalized races were the most evolved because they had developed farthest from the ancestral condition along the fetalization track.[75] Echoing the common scientific racism of the time, he found Europeans to be the most fetalized. Curiously, a similar idea was proposed in the first "Out-of-Africa theory" of this century. Julien Kollmann (1834–1918) coined the word "neoteny," meaning the preservation of ancestral juvenile stages in adult descendants. Applied phylogenetically, he reasoned the opposite of Bolk, that the human races evolved from African Pygmies, who were the most *primitive* of the races because they retained a juvenile ape form. The other races, he thought, had evolved farther from this African ancestor.[76]

Weidenreich's mentor, Schwalbe, rejected Kollmann's idea,[77] and Weidenreich rejected Bolk's. Weidenreich argued that features that were all said to demonstrate juvenile retentions, especially the large brain size relative to body size in adults, the rolling up of the cranial base, and so on, happened for a quite different reason. Humans retained juvenile ape cranial features because they were all linked to the mechanical consequences of evolving a relatively large brain. Baby apes have this relatively large brain because the brain achieves most of its adult size early in ontogeny, when the body is still small. The similarities between human adults and ape babies did not reflect an evolutionary process.

In fact, there is a different reason why Pygmies have a relative larger brain size than other, larger populations that both Schwalbe and his student Weidenreich missed—allometry, or relative growth. Size increase creates differences in scaling, and larger sizes often demand different proportions. This can be for mechanical reasons; for instance, elephants need legs that are relatively much thicker than mice legs because weight increases by roughly the cube of height, so that larger forms are relatively heavier (this is a very Weidenreichian argument, although one he never actually proposed). Scaling for brain size is less than proportional to body size increases.[78] This means that a population with half the body weight of another will not have half the brain weight, but a much larger brain; generally within humanity smaller populations have absolutely smaller but relatively larger heads. The greater the body size difference, the more distinct the difference in head proportion is.

Another Dutch scientist, Eugène Dubois, spent much of his career looking for evidence of a general law of brain allometry. He believed there was a universal expansion exponent for brain size of roughly ⅔ that applied all across mammalian evolution:

$$\text{brain weight} = \text{body weight}^{2/3}$$

Dubois used this relation to examine the relative brain size of his Trinil discoveries (discussed in Chapter 7). From the length of the femur found near the Trinil cranium (we now know this association was invalid, as the femur is much younger), he estimated the body size using a human model and then calculated Trinil's expected brain size from his formula. The Trinil relative brain size was much smaller than his expectations from the calculation—in fact, about the relative size of an ape's brain. But Dubois thought that his fossil was intermediate between apes and humans and, since he was unwilling to believe his formula was wrong, that the problem must lie elsewhere.[79] If instead a gibbon model was used to estimate the expected brain size, the Trinil estimate was greater, and using *this* estimate, the greater relative brain size would be intermediate between apes and humans, just where Dubois expected it to be.

Dubois appeared to change his mind in seemingly irrational ways; one day his fossils were human ancestors and the next they were the remains of a giant gibbon.[80] But there was a good reason why Dubois likened Trinil to a giant gibbon, not because he had given up his evolutionary interpretations about the place of Trinil in human evolution, but because he wanted to support them, and was unwilling to give up his belief in allometry as an overriding force pushing brain evolution.

What is historically interesting here is that the story of how Dubois changed his mind about Trinil and came to believe it was a giant gibbon became part of the legacy of paleoanthropology's history;[81] everybody knows it. But the *reasons* for it, and therefore its significance, have been lost from this mythic tale, even though they represented a much more important aspect of Dubois's life work. Dubois became known for the result (gibbon interpretation of one fossil), not its cause (his general theory of allometry), and in this way his work suffered a historic fate much like Weidenreich's—he became known for one aspect of his career that was misinterpreted because it was out of the context of the rest. We call this the Dubois syndrome.

Weidenreich's neo-Lamarckism and later his orthogenesis were almost unwilling interpretations of a functional anatomist turned paleoanthropologist who felt obligated to explain what his work showed and how he understood it. He could have provided a Darwinian explanation for both polycentrism and the evolution of structurally correlated systems. As Dobzhansky points out, addressing the second question, the one Weidenreich couldn't answer:

> [Weidenreich] states "the hominids have formed—and still form—one family, or in a strictly taxonomic sense, one species, and are all more or less related to each other in spite of manifold regional variations." In genetic terms this would mean that only geographic segregation, and none of the several known forms of reproductive isolation, kept apart from each other the different branches of the human species. A mutation or a genotype arisen in any one place is potentially able to reach all other human populations. . . . The question now arises why several hominid races, living in different parts of the world with quite different climates and environments, have all gone through similar series of changes? . . . Weidenreich attempts to answer this by an ingenuous theory that correlates the changes in several "characters" with a single fundamental trend, a gradual increase in brain size. This, however, becomes a theory of orthogenesis. . . . The basic fact which leads Weidenreich to assume parallel evolution of races is the presence of morphological traits binding the major racial subdivisions of living *Homo sapiens* to the primitive hominids which resided in the corresponding geographic regions during the Pleistocene. This interesting fact, however, may be accounted for in different ways. . . . hybridization of members of invading and indigenous races

> may permit the retention in a population . . . of some characteristics derived from remote indigenous forebears.[82]

This was not all of the answer, but it was a good start, albeit one that went nowhere because it was not noticed. Dobzhansky and Weidenreich increasingly paid attention to each other's work during the last five years of Weidenreich's life, as though they each realized that the other held a key to understanding the process of human evolution. While their mutual influence grew, in this case one can sense that Weidenreich didn't answer the question of how different human populations evolved the same way, or seem to notice Dobzhansky's answer, because his understanding of Darwinism was the one he gained as a *German* paleoanthropologist paying careful attention to the *German* genetics community. He was part of this intellectual community and internalized what his colleagues said and wrote. But they gave him the wrong understandings, and he needed to rely on neo-Lamarckism, and later his rectilinear orthogenesis. Finally, in the evolutionary community he became known for these and not for the polycentrism and the evolution of structurally correlated systems that were the real focus of his career.[83]

VALUE OF THE COMPREHENSIVE APPROACH

It is truly ironic that Weidenreich was not a Darwinian, for when dealing with intraspecific continuous variation he had a true appreciation of variability. This should have made him an excellent selectionist, had he the understanding of selection that came from the modern synthesis. This is clear from his writings on race, on constitutional types, on the composition and description of populations, and on his polycentric theory that described a network of human evolution. In the preceding chapter we hope we've made the case that he was not a typologist and had a populational approach to human variation. However, he was unable to apply the same understanding of populations to macroevolutionary problems which would have made him a Darwinian. He was too morphologically oriented. As a functional anatomist, he was not primary interested in populations; individuals, or more accurately parts of individuals, were his concern, as the interrelations of those parts explicated the human *Bauplan*. As a paleontologist, he focused on evolving structurally correlated systems made up of those parts. The same forces that explained modern human variation just didn't explain the *Bauplan.* The nature of the variation was not the same. Weidenreich was wrong not to embrace selection, but in many ways he was wrong for the right reasons. His eclipse of selection was in part a consequence of the very things that made him such a good and insightful scientist on other levels—a consequence of his breadth of vision.

Before Uncle Ernst finished his doctorate at the University of Göttingen, he moved back to his parents' home, then in Frankfurt, and attended the university for a while, also working at the nearby Senckenberg Museum of Natural History. He took courses from Weidenreich at that time, and family lore has it that he returned to Frankfurt just to learn from him. Weidenreich was not just one of his anatomy teachers; he also taught Ernst anthropology and did a lot to spark Ernst's interest in human evolution. Ernst retained Weidenreich's enthusiasm for anthropology, and Weidenreich contributed to Ernst's breadth of knowledge. Ernst was a geneticist and a selectionist. Yet he found nothing foolish or incomprehensible about Weidenreich's evolutionary views. It all made sense to him, as it fit into his understanding of history and broad scientific perspective. This kept him interested in anthropology as well; in later years, Ernst taught human evolution and published a few papers on general aspects of paleoanthropology. It seems that in his later years, as his genetic fields became more and more specialized and "pragmatic," he found in anthropology a way of expressing his more synthetic ideas. Anthropology's greatest strength lies here. In a world of increasing specialization, it remains a field at least theoretically committed to holism.

This is now threatened. As time goes on, we seem to be losing our breadth of vision. Just as the comprehensive school in German genetics gave way to more pragmatic approaches in genetics,[84] anthropology is becoming more specialized. Few scientists in any field have the breadth of knowledge that some of their colleagues valued so much only one or two generations ago. With this progressive narrowing of scope, we lose understanding, and it makes it easy to reduce the views of others to absurdity when we lose sight of their contexts. The Dubois syndrome is becoming more and more prevalent, we fear. We are fast reaching a time when we, even within a single discipline, are operating within many different frames of reference. There are no incentives to even recognize this fact and scientists continue to shake their heads at the "unbelievable stupidity" of their colleagues in closely related disciplines. We are losing our sense of the importance of history and, with it, much of our common ground.

9

CENTER AND EDGE

MILFORD FIRST VISITED ZAGREB IN 1976, as part of his research on the evolution of the masticatory apparatus. He went there to study the human fossils from the Croatian site of Krapina, the largest Neandertal collection in Europe. The site was especially important for understanding masticatory evolution as there were over 270 teeth and numerous upper and lower jaws preserved in the collection. Zagreb was a wonderful city, quite different from many of the other communist capitals then, with a large farmers' market, numerous consumer goods, and a westward orientation (in more ways than one—untold western movies were shot on the Croatian plains, where expenses could be kept low!). The food was marvelous, a mixture of traditional Croatian dishes, with special emphasis on fish and Dalmatian roasted lamb, Viennese finery (Croatia was part of the Austro-Hungarian empire), Turkish grills, kebobs and sweets because of the numerous invasions (every Croatian town, especially along the Adriatic coast, has a monument to some brave villager who held—or tried to hold—the town against the Turks), and newly introduced, hotly spiced Serbian dishes.

Even in a foreign, fundamentally interesting new environment, studying human fossil remains is, in the end, boring. At first there is the real thrill of seeing remains only read about before, or never known about at all. New fossil experiences, whether they are new specimens or new observations, are always exciting, but the work is long and mind numbing, overlain with the fear of making a mistake that would lie undiscovered until returning home (when it was too late to correct). Worse yet, the error might not be found at all. Some of the work is exciting, involving reconstruction or efforts to understand the warping or distortion of a bone or how it was broken and otherwise treated during the process of becoming a fossil. But during this phase of research the interesting scientific inquiry is often masked by the tedium of data collection. Most data collection involves repetitive observations that must be taken the same way, according to the same criteria, from specimen to specimen, site to site, and year to year, so data can be compared. It is boring, yet requires considerable concentration, and it is this very mind-numbing aspect of paleontological work that has the potential to lead to mistakes—that, and to desperation.

When you get home and begin the really interesting part of research, the analysis, a different mind-set begins. In the field all is care and accuracy. At home all is desperation, because the sample size is always too small. There are never enough measurements and observations for a specimen, never enough of a single observation to compare samples with as much accuracy as desired. So the search begins through the notes. A thought jotted down about a measurement that could not be taken because the bone was not preserved: "the height might be greater than 120 mm," becomes "height > 120 mm" in a data table. Later, typed into a computer the height is recorded as 120 mm (there is no ">" entry in a spreadsheet), and *voilà!*, a measurement is created where there was none on the specimen. It is too easy for this to occur, so care must be taken from the outset to avoid it.

Milford was working on Krapina fossils in Zagreb one afternoon, trying to fit pieces together because he was tired and feared he lacked the concentration necessary to collect systematic data accurately. Checking the edges of fragments for fit requires care and is repetitive, but it is tinged with the anticipation of real discovery and is a welcome break from the concentration required during data collection. Krapina specimens are very fragmentary, quite possibly because the human remains were trampled by the cave bears who were living in the cave after the Neandertals died there, and then more bones were broken after they were fossilized. The excavators working for Gorjanović had carefully unearthed the fragmentary materials, now all in the museum, each with an individual number and an entry in the main catalog. While several pieces had been joined together over the years, nobody had tried fitting the pieces together systematically before, and it was proving wildly successful.

Jakov Radovčić had been excavating fragmentary Neandertals from the nearby site of Vindija and since his training was in fish paleontology he was curious to learn more about human evolution. He had heard about Milford's visit and Crni's change of policy regarding reconstruction of the Krapina materials and wondered what the newcomer was doing with the specimens. Milford was in the midst of reconstruction when Jakov arrived to meet him. As their introductory conversation turned to the personal, Milford and Jakov soon learned they had things in common. Jakov, studying the opening of the Tethys Sea (later, this became part of the Mediterranean) and its effects on the evolution of bony fish, was able to go to the Chicago campus of the University of Illinois for his master's degree as the first Zagreb student to do so who was not related to someone in authority. With degrees from the same school, albeit different campuses, Milford and Jakov shared a link, however tenuous. Jakov and Milford found they shared a love of Chicago and of Neandertals.

Jakov was then working at the Institute for Quaternary Geology of the Yugoslav Academy of Sciences, under the directorship of Mirko Malez. He was assigned this job because of his paleontology experience, even though Jakov's expertise was fossil fish, and Malez worked on the terrestrial deposits found in Croatian caves of the last 200,000 years. Just a few years before, in 1974, Malez had begun excavation on the most important site of his career, the Vindija Cave, located about 80 km north of Zagreb. In it were thick deposits with animal remains (mostly cave bears). There were Middle Paleolithic (Mousterian) artifacts (the kind typically associated with Neandertals in Europe) and above them the earliest of the Upper Paleolithic industries (typically associated with "modern humans") and then even later archaeological debris and human bones. In all there were about 8 meters of rich cave materials. The potential Vindija held for understanding human evolution in Croatia came to overshadow fish for Jakov, though like Gorjanović before him, who was also a fish paleontologist, fish never completely lost their interest.

Malez and a group of young scientists and students from the institute were digging at the site. There was a real seesaw of intents and procedures, the young workers trying to excavate with the utmost precision and care, and Malez coming out in the late afternoon hours and weekends, trying to push them to work faster, faster, faster. "I will swallow that cave" was not only his favorite phrase, it was his war cry. But the visits were short and most of the excavations were done carefully, recording the all-important provenience for the majority of the human and archaeological remains.

Jakov's and Milford's friendship developed that year, growing stronger over conversations about excavation techniques and procedures, and the information that a good excavation should develop. Their discussions turned to the human materials in Malez's laboratory. As Milford and Jakov reviewed the remains together, Milford soon saw that some specimens weren't human at all! One toothless lower jaw fragment from the Šandalja Cave that Malez thought possibly to be an australopithecine, or certainly the earliest European, was actually a cave bear; and another "early human," a broken lower incisor, was that of an antelope. Who would tell Malez, a man often given to fits of temper? His employees at the institute balked; crossing Malez on these identifications would be foolhardy! Milford became the bearer of the bad tidings and thus began a productive professional, but roller-coaster personal relationship with Malez. "The mercurial Mirko Malez," Milford often thought to himself, but never dared say aloud.

The Vindija excavations were yielding many fragmentary human specimens, and Milford and Jakov set out to describe them accurately and compare them with the earlier Krapina remains. Here was the real importance of the fragments—not as just another late Neandertal site, but as one that could be compared with earlier Neandertals from a cave that was just a day's walk away. Later joined by several other Croatians working at the site, Milford's student Fred Smith, and Malez himself, they prepared an article for the publication *AJPA*. Vindija was late for a Neandertal sample, and the remains addressed one of the most interesting and controversial aspects of the Neandertal folk: what happened to them?

The research was done in 1978, and the Vindija paper submitted in 1980,[1] these dates sandwiching the 1979 research season that was so important to Milford's theoretical perspectives. That season Milford visited London, Zagreb, Beijing, Jokjakarta, and finally Canberra where he and Alan Thorne sketched out the framework for Multiregional evolution. The Vindija paper thus reflects Milford's frame of mind just before Multiregional evolution was developed. Two things are quite evident. First, he saw strong evidence of evolutionary continuity in the comparisons of these Middle Paleolithic central European samples. Continuity was much on his mind at that time, having recently been exposed to Alan. Still, Milford's thoughts remained focused on *worldwide* stages; he had not yet seen the Chinese fossils or reconstructed Sangiran 17. Consequently, he interpreted much of what he observed as changes that were not unique to central Europe; they were taking place everywhere. He did, however, think Neandertals were specifically ancestral to later Europeans. Milford had written earlier about Neandertals, in one of his first publications,[2] but he was coming to appreciate that the central European sample was quite special.

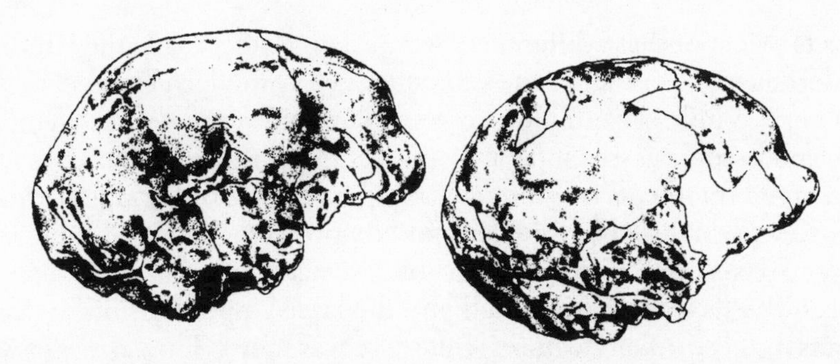

Perhaps Jelínek's best case for continuity in Moravia was based on the specimens from the Mladeč Cave. This figure[3] *shows skullcaps of males: Mladeč 5 and the Spy 2 Neandertal* (right). *Cranial size and shape, and many anatomical details, are extremely similar. In fact, some of the "Neandertal" features are better expressed in Mladeč than in Spy.*

The Moravian paleoanthropologist Jan Jelínek had been addressing this region for several decades,[4] emphasizing the transitional nature of the late Moravian Neandertals, and the numerous Neandertal characteristics found on the earliest "modern" humans of that region. The Moravian specimens outlined the similarities of Neandertals and the populations following them. Now the same thing seemed to be repeating itself just to the south, in Croatia.

At Vindija there was evidence of temporal change—toward modernization—*within* the Neandertals. Like the Moravians, these central Europeans shared characteristics that linked them to earlier populations and to the early modern Europeans living there somewhat later.[5] The Croatian Mousterian samples were many times larger than the samples found in Moravia and clearly demonstrated that Vindija was quite different from the earlier Krapina collection in many re-

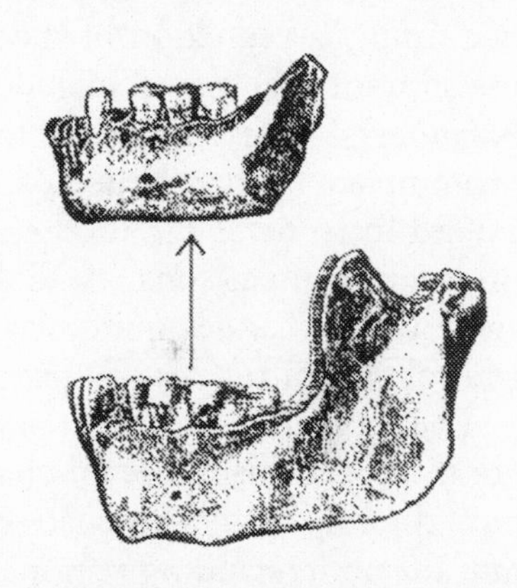

An example of the evolutionary pattern in Croatia is in this comparison of the Krapina J mandible (below) *and Vindija 231.*[6] *The later Vindija Neandertal is smaller and has a much better developed chin, evolutionary changes in the direction of post-Neandertal Europeans. Because of the direction of change, this was taken as an example of regional continuity, even though the changes were not unique to central Europe.*

spects. Most of these differences were changes that made the later Neandertals more similar to the subsequent early modern Europeans. For instance, Vindija mandibles were smaller, had more vertical symphyses, and in some cases moderately projecting chins. The numerous well-preserved fragments of Vindija frontal bones with browridges revealed the supraorbitals to be smaller and less projecting than in the earlier Neandertals. The Vindija supraorbital torus fragments always fell between the Krapina Neandertals and the later Upper Paleolithic Europeans in these salient features. But there was more. The *pattern* of ways in which the Vindija tori were smaller was particularly like that of later Europeans—the outer portion and especially the torus over the middle of the orbit thinned, but the central portion over the nose got thicker.

The later, Upper Paleolithic Europeans showed a continuation of this pattern. Their torus was thickest over the nose, often completely reduced over the middle of the orbit, while over the outer part of the orbit it was retained. This divided the brows into a central superciliary arch and an outside portion, the lateral torus. This superciliary arch was usually even thicker than the equivalent portion of the Neandertal supraorbitals, so the change from Neandertal to later populations was not simply a reduction of the browridge structure. Other changes in the supraorbital region include its flattening (the middle of the earlier Neandertal faces projected in a beaklike way that emphasized the prominence of their large noses) and a reduction in how far the brows extend in front of the forehead. Even though both samples are fragmentary, comparisons of Vindija to the earlier Krapina specimens show these changes quite clearly, and these comparisons formed the central part of the Vindija paper.

The second issue evident in the Vindija paper is that Milford did not yet interpret this temporal change as a specifically *regional* phenomenon; the role of genic exchanges in the evolutionary pattern, so important for understanding regional features, was still considered minimal. He thought of genic exchanges as migration. As far as the origins of Europeans were concerned, Milford and his coauthors argued that if

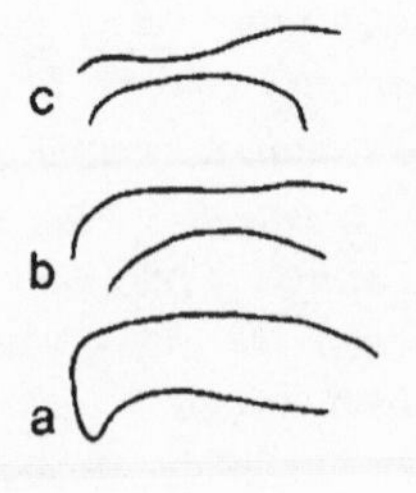

The pattern of reduction in the right side supraorbital torus.[7] *This schematic shows the intermediate position of the Vi 202 (b) supraorbital, compared with the earlier Krapina 28 Neandertal (a) and the later Upper Paleolithic Velika Pećina (c). Taken as an evolutionary sequence, the supraorbital torus is seen to thin out, first over the middle of the orbit and then on the outside, while the central portion over the nose becomes thicker.*

migration from a single source population of modern humans accounted for their appearance in Europe, it would have to account for the appearance of modern people everywhere (except, of course, where the moderns evolved). But there was no evidence suggesting a worldwide pattern of population replacement, let alone evidence of an influx of populations into central Europe. They concluded that there was no reason why the continued influence of selection, based on Mousterian technology, could not have caused the genetic changes that gave rise to the early modern European anatomy without significant genetic input from other populations. Milford was still constrained by his belief in phases or stages of human evolution. He was seeking worldwide trends. Evidence of evolutionary continuity was not evidence of regionalism (Milford still felt that regional features would be unrecognizable), but evidence of *in situ* change toward modernity.

As Multiregionalism developed just after the Vindija paper, Milford's thinking about Neandertals and human evolution in general changed in subtle but important ways. He no longer felt that stages adequately represented the evolutionary process; he now realized that while *in situ* change could occur regionally, genic exchanges were central to the overall evolutionary pattern. This was the greatest difference Multiregional evolution brought to the understanding of evolutionary change, and the questions asked of the fossil data. It was in the perception of how *in situ* evolutionary change took place through the key insight that genic exchanges were not the opposite of differentiation, they were its cause. Far-reaching gradations of anatomical differences were not disrupted by genic exchanges, *they depended on them*, and it was along these gradients that populations toward the extremes could differentiate and remain distinct.

Multiregional evolution led to the expectation that there was a significant role for genic exchanges whenever there was *in situ* evolutionary change. At the peripheries, especially in Europe where so much was known, the question of the specific part played by new genetic material in modern European origins became a central issue.

LACKING LINKS

Multiregional evolution views most human evolutionary changes as resulting from shifting balances between genic exchanges and selection: local conditions might change, advantageous genes might spread, or population movements might occur. This idea clearly did not develop directly out of Milford's earlier thinking, even his thinking just before the 1979 trip. Multiregional evolution preserved many, perhaps most, elements of the insights his research had provided, but it was a real paradigm shift. Afterward, he saw the world of paleoanthropology quite differently; really, he saw a changed world.

Why wasn't he more interested in Weidenreich's thinking before? It was certainly not for lack of experience with his papers and monographs; Weidenreich's discussions of the fossil record in Asia, his most detailed anatomical descriptions, were required reading for any serious student. What better training could there be than to read Weidenreich with a cast of the relevant specimen in full view, finding each feature, tracing each groove, until the unusual, unique anatomy of the long dead humans was fully understood and its identification became second nature? But his theories were something else. Weidenreich's Polycentric evolution did not seem to be a viable hypothesis, especially as simplified in the form of a candelabra, and until Multiregional evolution it had little influence on Milford's thinking. It was much more influential for Alan Thorne, precisely because his experience was with regional continuity outside of Europe, in a region where it was much less contentious, and one that Weidenreich had written about. In this Alan was exceptional.

The general lack of a Weidenreich school or influence in part was a historical accident or just plain bad luck; he had no paleoanthropology students to articulate and further his ideas, many transmission geneticists held antagonism toward him, and the extent to which he was misrepresented in secondary sources was a problem. But even seen in the most favorable light, the way Weidenreich meant it, two problems in the Polycentric evolution theory were never solved: the origin of geographic differences and the explanation of their continuity over time.

First, he never addressed the origin of Polycentric variation. In fact, the hominid origins theory he developed in *Apes, Giants, and Man*[8] was that the human line descended from a gigantic Asian ape and became much smaller through a series of evolutionary transformations, even as brain size increased. He never accepted the australopithecines of the Pliocene as hominid ancestors. But even as he was writing, the explosion of australopithecine discoveries from South Africa (and, later East Africa) had begun, and soon after his death the place of australopithecines in human evolution was quickly assured. His theories about origins were so off the mark that they were soon forgotten and in any event did not address how regional differences could have originated according to his model.

In addition, Weidenreich was never able to resolve the contradiction between the isolation that seemed to be required to establish regional distinctions, and the continued hybridization and absence of any "pure races" that his observations revealed. Weidenreich modeled these two contradictory elements in his trellis but never could explain how they worked together. Coon, his self-proclaimed disciple, never appreciated the issue. Today nobody calls himself a Polycentrist.

After several attempts at reconciliation,[9] including his own, the modern synthesis was evolving even as Weidenreich was writing his last papers, and in the synthesis was the answer to the contradictions in his work. Weidenreich is said to have approached Ernst Mayr, who was deeply involved in developing the synthesis, saying, "I find this work you young fellows are doing [on the new synthesis] to be interesting, but I am too old to change."[10] All the pity, because the modern synthesis came to address his publications and provided answers for the very problems that many felt invalidated them. Just after Weidenreich died, the famous 1950 Cold Spring Harbor Symposium on "the origin and evolution of man" took place. The geneticist Theodosius Dobzhansky, who had become increasingly sympathetic to Weidenreich's Polycentrism, wrote:

> In general the old anthropological alternative of monogenic versus polygenic descent of man ceased to exist when considered from the vantage point of the present evolution theory. Different populations (races) of a polytypic species may be descended largely from different races of the ancestral species and may differ in some genes in which these ancestral races differed. And yet, a polytypic species may still evolve as a single genetic system. Favorable mutants or gene combinations arrived at in one part (race) of such a species may, under the influence of natural selection, eventually spread to all other parts and thus eventually become a common property of the entire species. Thus, local autonomy of gene pools of racial populations does not preclude retention of the basic unity of the species as a whole. I would like to point out that this view agrees quite well with the conclusions reached by the late Weidenreich on the basis of purely morphological analysis of pre-human populations. This is worthwhile stressing because Dr. Weidenreich has sometimes used expressions that seemed to put him close to the old-fashioned polygenist camp, which he actually rejected absolutely.[11]

But in 1963, commenting on Coon's hypothesis, Dobzhansky's response was quite different, and little wonder, as Coon's hypothesis was quite different. The two points Dobzhansky addressed with the greatest skepticism were Coon's contentions that

1. The races of *Homo erectus* independently crossed a "sapiens threshold" to become the races of *Homo sapiens* independently and at different times
2. The races of *Homo sapiens* are caught, today, at different evolutionary stages, as those that crossed the "sapiens threshold" earliest are the most evolved

The title of Dobzhansky's essay,[12] "The Possibility That *Homo sapiens* Evolved 5 Times Is Vanishingly Small," tells it all.

Coon really believed there was a threshold, a firm, definable border of humanness that the races crossed at different times. Responding to vanishingly small probabilities, his final solution, posthumously published,[13] was to assume that the relevant mutation allowing a population to leap across the threshold because of larger brain size only happened once and spread from population to population because of its advantage. None of these explanations was widely accepted. The threshold was not only one for *Homo sapiens*, but also a threshold for the acceptance of Coon's ideas, and this was a Rubicon that was never crossed. But what if there was no such border? No threshold between the races of *Homo erectus* and a new species. What if *Homo erectus* did not exist?

WHAT IS HOMO ERECTUS? WHY IS HOMO ERECTUS?

Coon could have avoided at least *this* problem if he had accepted one other idea from Weidenreich: his contention that *Homo erectus* is not a valid species. Remember that he did not regard the australopithecines as hominids; Weidenreich wrote: "I believe that all primate forms recognized as hominids . . . represent morphologically a unity when compared with other primate forms, and that they can be regarded as one species."[14]

He regarded humanity, past and present, as a single species from the all-important genetical perspective; he thought they could potentially interbreed and have fertile offspring. But this was not widely accepted, possibly because Weidenreich himself continued to use the old terminology, despite being explicit about the fact that the terms had no taxonomic meaning. He reasoned that it was impossible to go back and change what was already published, and a shift in terminology would just create confusion. The problem is that continuing the old terminology created even *more* confusion. Since then others finally came to agree with Weidenreich's idea. Several paleoanthropologists have made the same proposal, that *Homo erectus* and *Homo sapiens* are not different species. Let's look at the reasons why.

Helmut Hemmer, a postwar German evolutionist, reached a similar conclusion some 20 years later, for similar reasons. He studied the allometry of hominids, the proportional changes that came as the results of size differences. For instance, if one person is twice as tall as another, the former's legs must be relatively thicker. This is because when height doubles, the weight difference will be more than double. Hemmer[15] concluded that the fundamental design (*Bauplan*) of the hominid cranium

> is the same from the earliest definite members of the *erectus* group (e.g., "Pithecanthropus IV" from Sangiran, Java) to the modern races of *H. sapiens*. . . . Since these features vary among the recent races no less significantly than between different fossil groups, or between fossil and recent populations, it is impossible to draw a line anywhere for species delimitation unless one intends also to split up recent man into several species. Therefore it seems necessary to include all of these fossil and recent groups in the single species *H. sapiens*.[16]

Like Weidenreich, he found no fundamental structural differences in the human fossil record, nor variation that was greater than the variation between living races. Since he believed the idea of separate species for living populations was unacceptable and clearly incorrect, he used the present to interpret the past. Similar reasoning was advanced by John Robinson, the South African australopithecine expert. He addressed the *Homo erectus* question as part of his general revision of hominid taxonomy.[17] Robinson argued that in the broad view of human evolution, "most of the obvious physical change had already occurred" at the time of the appearance of *Homo erectus*. All subsequent human populations were mainly characterized by a single evolutionary trend, in his view, "the realization of the cultural potential."[18] He believed *Homo erectus* and *Homo sapiens* should therefore be subsumed in the single species *Homo sapiens*, because it has priority.

A different aspect of Weidenreich's reasoning was emphasized by the Kenyan paleoanthropologist Richard Leakey. He focused on the multiple origins of recent and modern populations:

> I do not favour the idea that the modern form of our species had a single geographical origin. The fossil evidence from widely separated parts of the world indicates to me that *Homo sapiens* in the modern form arose from populations of the more archaic form wherever it was established;

and that similarly, these archaic forms arose from established populations of so-called *Homo erectus.* There are specific examples that cannot be brushed aside.[19]

Leakey recognized the fundamental similarities of *Homo erectus* and *Homo sapiens* crania, attributing them to the influence of expanded brain size, and added:

> I am increasingly of the view that all of the material currently referred to as *Homo erectus* should in fact be placed within the species *sapiens* [which would] project *Homo sapiens* as a species that can be traced from the present, back to a little over two million years.[20]

Emiliano Aguirre, a prominent Spanish paleontologist, wrote much the same.[21]

Jan Jelínek, a lifelong proponent of regional continuity,[22] focused on the lack of a clear boundary between *Homo erectus* and *Homo sapiens.*[23] He examined the transition in African, European, north Asian, and Indonesian samples and concluded the changes were not all that great. Moreover, he raised the issue of establishing criteria for species definitions, questioning whether global morphology, regional morphology, chronology, or cultural traditions provide the more valid means of separating the hominid species. Taking a global view of the problem, he argued that the anatomical links between Middle and Late Pleistocene populations in each of several regions make it impossible to regard some, such as Ngandong, as *Homo erectus* while other contemporaries are *Homo sapiens.* There was no question of different times for crossing a Rubicon for Jelínek. For him the differences were not great enough to warrant such drama. Jelínek had spent some time in Australia, living with native peoples, and for him it was culture, and not any particular anatomy, that made people human. He wrote:

> Have we any solid scientific grounds on which to consider Middle Pleistocene European finds, with earlier morphological cranial changes, as *Homo sapiens* and the extra-European finds evolving in the same direction but in somewhat different degree and time sequence of adaptation into different conditions as *Homo erectus?* The whole mode and the process of the hominid evolutionary process shows that there are not, and that in the past [there] could not have been differences at the species level, but only at the subspecies level, whether the cerebralisation process—as only one part of the mosaic of evolutionary changes—started earlier or later. The logical consequences of such a situation is to lead us to consider the different African, European, and Asian finds of *H. erectus* type as *Homo sapiens erectus.*[24]

His global perspective revealed another reason anatomical links between Middle and Late Pleistocene populations are important—these links are strongly regional. He knew early on what some of our paleoanthropologist colleagues were only to learn the hard way[25]—modern humans are difficult, perhaps impossible, to define.

> If the differential diagnosis between *Homo erectus* and *Homo sapiens* cannot be other than by convention, and . . . this convention must be different for different geographical regions, then the value of such a difference should be critically considered. . . . It is time to replace the paleontological species with a biological one. . . . Paleontological taxonomy cannot be in contradiction with . . . biological facts.[26]

Milford addressed the *Homo erectus* issue as well. Working with Alan Thorne, as well as friends and colleagues including Jan Jelínek and Zhang Yinyun, he proposed that *Homo erectus* should be "sunk," submerged within *Homo sapiens.*[27] Much of the reasoning for this sinking continues to emerge from the species concept and what it means for paleoanthropologists, and paleontologists in general. The modern synthesis gave biologists a species concept that tied right into evolutionary theory. Designed to apply to living species, a biological species is *a group of populations that can actually or potentially interbreed and produce fertile offspring, and which are reproductively isolated from populations in other species.* This is the "genetic species" that played an important role in Weidenreich's thinking, a species that is the main unit of evolution because genetic change can spread anywhere within a species, but cannot spread from one species to another. Further, many scientists recognize biological species as the only natural biological unit because its definition is based on biological criteria, and its limits define the limits to microevolutionary change. Taxonomic units below the species level (subspecies, races) are much more arbitrary in their delineation, while the higher taxa above the species level (genera, families, orders) are constructs based on how species are related.

The biological species is a protected gene pool. Within a species the morphological integrity of the members—the fact they can usually be distinguished from members of other species—is a *consequence* of the breeding barriers defining it. Species are discrete entities because of these boundaries; at the same time the amount of difference within a species is limited. Sooner or later differences between species must increase because all they can do is diverge from each other.

But because the foundation of the species is in its breeding behavior, there are several problems when species are considered in their true multidimensional character—that is, over time. Many past species can be related to present ones through ancestry. When this seems possible, there is a lineage—*a group of ancestral-descendant species that is reproductively isolated from other lineages.* Looking at these definitions another way, a species is a lineage at a particular point in time. Recognizing the preeminence of the lineage, however, creates the problem of diagnosing its existence. One issue is that in spite of the thinking of paleoanthropologists such as Weidenreich, an interbreeding diagnosis cannot really be applied to populations far removed in time, especially the geologic time we might encounter if we were interested in whether 40,000-year-old Neandertals and modern humans are in the same species. A second difficulty is that fossil remains usually provide no direct information about interbreeding between contemporaries. Although morphological differences between species are a result of gene exchange barriers, marked differences can also occur within species as well—for instance, polytypic species such as ours with a wide range and concomitant geographic variation.

Paleoanthropologists have been concerned with how to recognize past species, and different scientists have each accepted one of several solutions. These solutions come from a more genealogically oriented concept of species, one that considers species in their temporal dimension. The genealogical idea of species is that they are defined by descent; each is a distinct individual with a definite beginning and end (birth and death) and each with a unique ancestry. The importance of genealogy in this species concept is the insistence that all members of the species share common descent. Another way to put this is that genealogical species are monophyletic: a monophyletic group consists of an ancestor and all its descendants.

One reason that genealogical species definitions became popular is that earlier this century some paleontologists objected that biological species criteria cannot be directly applied to the fossil record, and therefore developed the idea of a species definition based on morphological difference: the morphospecies. But morphospecies had no evolutionary standing (their definition was based on anatomical criteria alone), and when genealogical species were clearly defined from evolutionary considerations, they were seen to provide the perfect solution for the paleontologists.

Evolutionary species have proved to be the most useful genealogical approach to the species problem. An evolutionary species is a distinct genealogical group, defined as *a single monophyletic lineage of ancestral-descendant populations which maintains its identity separate from other such*

SPECIES DEFINITIONS

Basis	These two species have a genetic basis	
		This species has a genealogical basis
Species	**Biological Species**	**Evolutionary species**
Definition	A group of populations able to interbreed and have fertile offspring that are reproductively isolated from all other such groups	A single monophyletic lineage of ancestral-descendant population evolving separately and maintaining its identity separate from other such lineages, with its own evolutionary tendencies and historical fate

lineages and which has its own evolutionary tendencies and historical fate. The evolutionary species thus retains the essence of the biological species—reproductive isolation (the reason it can maintain its identity and has unique evolutionary tendencies)—while avoiding many of its deficiencies: lack of time-depth, absence of morphological criteria for diagnosis, and, perhaps most important, emphasis on reproductive ties alone as a major cohesive force. An evolutionary species can be identified morphologically, as it must to be useful for paleontologists.

For evolutionary species the issue is not over the *amount* of morphological variation but its *pattern*—how it is distributed over space and how it changes over time. Species are created and ended by lineage splits or extinctions and are not arbitrarily defined by comparing the variation from their beginning to end with the variation across the range of an extant species. The variation over the *life* of a species has a temporal dimension and beside the factors that control species variation, it is also controlled by the accidents of births (species splits) and deaths (more species splits or extinctions) that may be unrelated to any attributes of the species itself. Variation over the *range* of species and over its life is not comparable. This is why diagnosis of an evolutionary species is independent of the amount of variation it encompasses.

Reproductive boundaries are the underlying basis for the main characteristic shared by the populations of an evolutionary species—its unique evolutionary pattern. Within these reproductive boundaries there is an internal cohesiveness that helps maintain this pattern. The cohesiveness results from three factors:

1. The reproductive boundaries that limit genic exchanges
2. Natural selection in the specific habitats where populations live—species can adapt to only a restricted number of environments
3. Ecological, developmental, and historical constraints that come from past evolutionary history

Evolutionary species can be seen as a biological species extended over time, but they are more than that. This broader concept of an evolutionary species addresses the weakness of the biological species concept that is of greatest concern to paleontologists—the difficulty of translating the main criterion of actual or potential interbreeding into useful models of morphological variation. The evolutionary species has its unique evolutionary tendencies that can be readily interpreted as morphological criteria, and it is this aspect that allows them to be identified.

Milford and colleagues applied the evolutionary species idea to the *Homo erectus–Homo sapiens* problem. Some important points follow from the fact that they are ancestral and descendant species on a single lineage:

1. No species splits occurred when *H. sapiens* is said to originate from *H. erectus;* there was no division of one species into two, and therefore no species birthing
2. No distinct anatomical boundary separates the ancestor *H. erectus* from the descendant *H. sapiens*
3. No single worldwide set of criteria validly distinguishes so-called late *H. erectus* from subsequent samples of early *H. sapiens*
4. Just about every way *H. erectus* differs from its australopithecine ancestors also characterizes *H. sapiens:* virtually no features are unique to *H. erectus*

For these reasons they concluded there was no basis for distinguishing a species called *H. sapiens* from a species called *H. erectus.* Coon's observations of regional distinctions crossing species boundaries do not imply that the distinctions are an illusion, but rather that the species boundaries are. *H. erectus* is but an early version of *H. sapiens.*

CENTER AND EDGE:
HOW REGIONAL VARIATION GETS ESTABLISHED

One of the things Weidenreich did not remark on is that the samples of ancient humans he knew so well, especially those from Java and China, were relatively homogeneous. Not identical by any means, but in each region an anatomical pattern was distinct and recognizable, especially when sex differences were taken into account, and the pattern showed relatively limited variation. It was Alan Thorne who first noted this pattern in fossil humans decades later and explained it with his "center and edge" hypothesis.[28]

Alan was influenced by the thinking of Ernst Mayr, who wrote extensively on the question of why populations at the periphery or margin of a species range differed from the more centrally situated ones. Mayr[29] was interested in fruit fly studies, species that we call "colonizing"[30] because their reproductive pattern lets them rapidly expand their populations and move into new areas. Reviewing many studies of these flies, he concluded:

> The genetic differences between central and peripheral populations can be described as follows. The total amount of gene flow is reduced in peripheral populations, and the gene flow becomes increasingly a one-way inflow of genes near the periphery. Many of the peripheral populations . . . are established by . . . a small group of founders which carry only a fraction of the total genetic variability of the species. . . . Environmental conditions are marginal near the species border, selection is severe, and only a limited number of genotypes are able to survive these drastic conditions. Reduction of gene flow and increased selection pressure combined deplete the genetic variability of the peripheral populations. . . . Contiguous central populations, on the other hand, are in the midst of a stream of multidirectional gene flow and harbor at all times a large store of freshly added immigrant genes.[31]

But what did this mean for the morphology? Polymorphic traits, features that can have many different forms in a population, like eye color or the size of the nose, are affected. In the peripheral populations, Mayr wrote, polymorphism could be reduced, and features could even become singular in form.

Alan realized this very description fit the human fossil record he knew best, the Javan and East Asian one. In the paper he presented at the 1977 Pan African Congress in Nairobi, the one that caught Milford's attention and spawned their relationship, Thorne suggested his center and edge hypothesis to explain certain aspects of human evolution.[32] His proposal was about the pattern of the differences that developed in the earliest human inhabitants out of Africa. He noted that when samples of the earliest inhabitants out of Africa were examined, such as the several crania from Sangiran, Indonesia, or Zhoukoudian, China, the peripheral samples were much more homogeneous than the centrally located African ones. But these two large peripheral samples from Indonesia and China had *different* combinations of homogeneous features, often different character states of the same anatomy.

There was probably no better place than the National Museums of Kenya for Milford and Alan to discuss these comparisons and contrasts, because the African fossils of early *Homo sapiens* there, specimens they looked at every day, were dramatically more variable than the Indonesian or Chinese ones. Africa appeared to conform to the pattern of variation Alan predicted for the center. In Nairobi they began to discuss why.

They realized there were several reasons why more variability could be found in Africa. Perhaps the original center was East Africa, perhaps somewhere nearby, but for long after *Homo sapiens* first appeared in East Africa, the region continued to be an evolutionary center. It was here that *Homo sapiens* first evolved and was initially successful. It is quite likely that this center remained the region of optimal adaptation for a long time, and both genetic and archaeological evidence confirms that for virtually all of human evolution most humans have lived in Africa. Populations there were the first to begin to make patterned stone tools like hand axes in a stone tool industry called the Acheulean.[33] Acheulean tools are the first that appear to have been preconceived by their makers, and the concept of "tool types" can be validly applied to some of the artifacts. In fact, one tool type, the hand axe, is the defining characteristic of the Acheulean.

At the center, where it appears developments like these were taking place, a greater range of anatomical variation will be tolerated within groups because humans were best adapted there and selection played a smaller pruning role. Additional differences between populations might be accentuated by competition between the more densely packed central ones. Furthermore, central populations have relatively higher population numbers and therefore often live closer together and are more often in contact where they can exchange genes. Increased gene flow, as Mayr pointed out, elevates the amount of anatomical variation because with more genes in the mix, more polymorphisms are found. Other factors also accentuate variation. Gene flow from other regions will be multidirectional at a species' geographic center, while at the peripheries the direction of genic exchanges is more limited, often directed mainly outward. Combined, these factors would increase the heterogeneity of the central region as a whole, and there would be a larger number of different genetic combinations.

Yet, human populations eventually left the center because one of the most important consequences of the adaptive shift involved in *Homo sapiens* origins was its change to a colonizing species, a species undergoing habitat expansion.[34] Humans expanded their ecological range into arid and highland-to-mountainous habitats and eventually moved out of Africa to spread across the tropical and subtropical regions of the Old World. During this time many regional differences appeared, and populations at the edges of the human range developed distinctions that in some cases persisted into the Late Pleistocene and even recent and modern times.

Why did people leave Africa? Quite possibly it was because, for the first time, they could. An important prerequisite for colonizations was clearly the new *Homo sapiens* adaptive complex. There were important developments in mobility—large body size, long legs, and physiological adaptations reflecting improved water retention and effective sweating.[35] Paleoanthropologists have suggested several other reasons why colonizations began at this time. There are the consequences of the newly developed adaptation to food acquisition by hunting, as well as scavenging and gathering, a change seemingly linked to *Homo sapiens*' first appearance. Climate changes also may have played a role. It has been suggested that the Sahara desert might have acted like a pump, drawing in populations when it was wet and fertile and spitting them out (presumably right out of Africa) when the climate returned to the aridity of today. In any event, the adaptive changes reflected in the large bodies and long legs of early *H. sapiens* may have been the initial impetus.

We do not believe these physiological and ecological changes alone permitted the successful colonization of the Old World. Certain changes in human social behavior were almost certainly necessary as well. By some half million years after the human species first appeared, some significant events were associated with the spread of the Acheulean industry. There was more effective use made of a wider range of habitats, and additional food resources were made available as humans developed strategies of organized hunting and confrontational scavenging and improved the technology they used for collecting and preparing gathered foodstuffs. But we think that the most important innovations were in the divisions of labor by sex and age, allowing groups to use different resources simultaneously. This increased the carrying

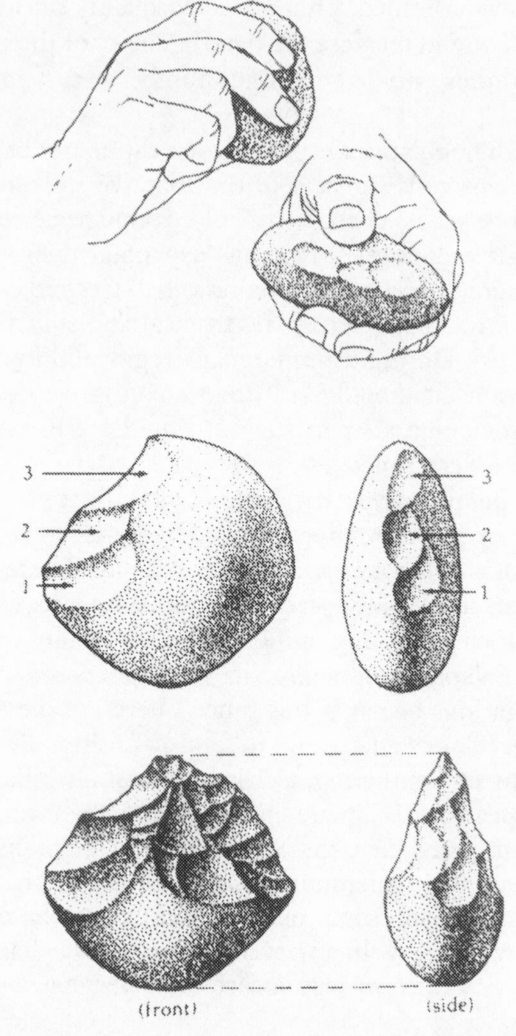

Hand axes are bifacially flaked (flakes are taken off both sides of the stone as the tool is being made), pear-shaped, pointed general-purpose tools of vastly varying sizes. The most common tools of the Acheulean, however, remain the amorphous and ubiquitously utilized flakes like those being struck as hand axes were made. Many Acheulean tools were made from large flakes that, themselves, were first struck from even larger cores. It is quite possible that it was this innovation of producing large *flakes to be worked into tools that underlay or stimulated the technological developments of the Acheulean. The importance of these is perhaps best emphasized by one of the new features found at the earliest Acheulean site. The bones of large animals are covered with numerous cutmarks that suggest they were hunted and certainly show they were systematically butchered.*

capacity of the regions they inhabited and allowed the expanding populations to colonize new and more difficult habitats. If men and women had different foraging strategies and procured different food resources that were later shared, more people could be supported in the same area. This is quite a change from feeding in place, wherever the food sources are found, that seems to have been common in the earlier australopithecines. But early *Homo sapiens* groups, adapting by hunting, scavenging, and gathering, would require a wide yearly range to take best advantage of seasonal resources, because this means moving from place to place. Their increased ranges would have brought groups into contact in regular, predictable ways. From these contacts, and the mate exchanges that must surely have followed, broad social networks developed.

Clive Gamble believes that it was these social networks, as much as the rest of the anatomical and behavioral package we have been describing, that made humans into colonizers.[36] Social networks improved the base of knowledge necessary for populations to move into new habitats and made them less dependent on the distribution of particular resources. As Gamble argues, advantage could be taken of opportunities with fewer risks, and the spread of human populations out of Africa and the successful colonization of other regions was the result.

By a million years ago human populations had begun to colonize the world outside of Africa, inhabiting other regions successfully in significant enough numbers to leave archaeological traces and a fossil record. Populations inhabited the tropics of South and Southeast Asia (including Indonesia, connected to the continent during periods of low sea level, where the earliest Asian remains were discovered), and ranged eastward into central China (finds at Gongwangling and Yuanmou) and as far west as the western edge of Asia where, in the Caucasus, an ancient mandible was discovered at Dmanisi, Georgia.[37]

Colonizing populations were probably small, according to workers such as Joseph Birdsell,[38] who attempted to model Pleistocene population movements from his very extensive knowledge of Aboriginal Indigenous Australians. Small population size usually leads to some genetic variations becoming lost, because when there are only a few parents, it is possible that some genes won't get passed on just by chance. When the population size is small enough, this drift process is called a bottleneck, a graphic term referring to cases when the population is, at the extreme, as small as a single mating pair. These demographic factors can create homogeneity in populations, and no doubt created the homogeneity we can see in the fossil record. The key is that different anatomies become common in different peripheral regions, a reflection of varying population histories.

The differences of various regions may reflect rapid adaptive change which occurred when new habitats were encountered. One effect of rapid adaptive change was the same as small population size. If the requirements of a new habitat created strong selection, bottlenecks could occur this way as well and additional genetic variation would be reduced or lost. These natural consequences of colonizations combined to create relative homogeneity in the more peripheral colonizing populations farthest from the center. The homogeneity was expressed as the loss of some variations and reduced frequency of some polymorphisms in others. This is the center and edge effect, and there were some long-term consequences of the reduced variation in the peripheral populations. Some of the less variant features that were established persisted in those peripheral regions for long periods of time, in some cases until the present. In peripheral regions some of the features that reflect modern geographic variation have been found to have appeared in the initial immigrants.

Therefore the anatomical homogeneity that has actually been observed to characterize fossil human crania from peripheral sites with large sample sizes such as Sangiran in Indonesia (and later, Zhoukoudian in China and Ngandong in Indonesia) reflects a reduced number of genetic polymorphisms in nuclear (and mitochondrial) DNA. More genetic variation has been lost because the characteristics of these gene pools were initially established by the partial isolation of numerous small populations with histories of drift and bottlenecking.

An example of this homogeneity is found in the forehead shape of the early Indonesian sample. Virtually all of the earliest Indonesian skulls have low, flat foreheads with a thickening of the cranial bone called the sagittal keel running right along the middle of the skull and extending onto the forehead. There is only a shallow hollowing in the side-to-side direction, between the front surface of the forehead and the top of the browridges. These bony ridges are continuous, from one side of the face, above the eyes, to the other. They pass across the face in a fairly straight line and are usually well developed at their center, over the nose. Not every skull is identical and each of these features varies in the sample, but *most* of the *combination* of features characterizes all of the early Indonesian specimens. Most of these features appear in the African early *Homo sapiens* remains as well, but the African sample is much more variable in each of them, and they are never found all together.

The second point of the center and edge hypothesis, the different histories of random loss and intense selection, is also evident in comparisons of the fossil remains. Because we would expect different losses and other genetic changes in each homogeneous peripheral group, they would come to differ in adaptive potential. Alan Templeton argues these conditions are optimal for the rapid appearance of adaptive divergences.[39] When populations are small, even if selection is intense, additional differentiation of relatively neutral variations can be expected as well. This is because the advantageous features may be linked to other characteristics, whose genes may be nearby on the same chromosome. Individuals who proliferate the advantageous genes can also spread linked ones, as long as these hitchhikers are not disadvantageous and do not offset the advantage of the favored genes. In all, this means many features would be expected to vary between populations; of course, some of the variations would reflect differing adaptations, but others may be more neutral and differ because of the different colonization histories. An example of this heterogeneity found *between* peripheral samples can be found in the face. The Sangiran faces from Java, best known from the most complete specimen Sangiran 17, have cheeks beginning well to the rear of the front teeth, with their base over the anterior molars, and their sides angling backward so they face to the sides, not anteriorly. The Chinese

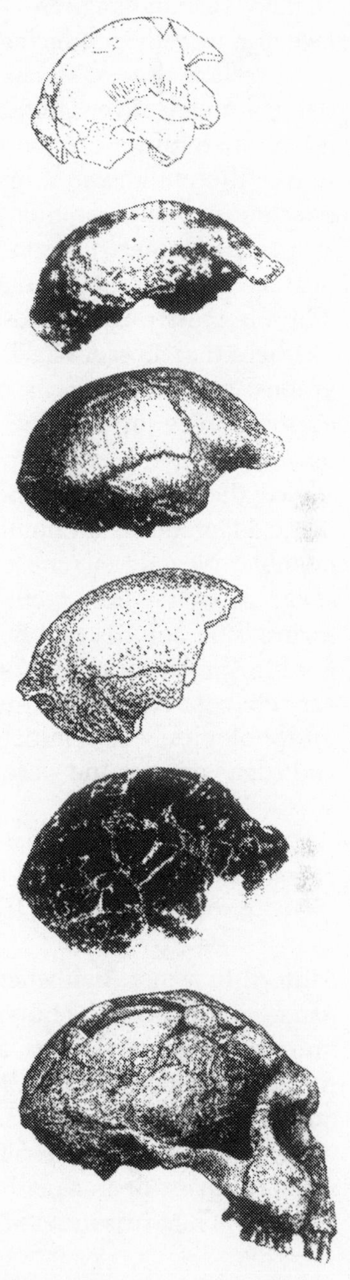

Indonesian crania from the Kabuh formation, found near the village of Sangiran and from Trinil.[40] From top to bottom these are the females Sangiran 12, Trinil 2, and Sangiran 2, and the males Sangiran 10, Sangiran IX, and the most complete, Sangiran 17.

Gongwangling face, of similar age, is quite different, as are the later Chinese faces. The base of the Gongwangling cheeks is much more anterior than in Sangiran 17, positioned just behind the nose, and they face horizontally, producing extreme facial flatness.

These are alternative character states for facial features that probably had no adaptive significance at the onset. Later, though, their different anatomies were the basis of local adaptations that affected the face. Different ancestral morphologies may influence the ways adaptive requirements are met; for instance, skulls and faces of different shapes may require different forms of buttressing in dealing with masticatory stress. History may dictate alternative adaptive responses. Thus, because they are most strongly expressed in living Eskimos, the flat faces that most Asians have today could be explained as cold adaptations, or as adaptations to maximize the leverage of vertical bite-muscle forces through the incisors. But both of these requirements have been met in different ways in non-Asian populations. For instance, the European Neandertal faces have also been explained as climatic adaptations, although they are about as structurally different as possible. Neandertals must have had strong forces through their incisors as well—these teeth are as big as incisors get in any human forms. The point is that both Eskimo and Neandertal anatomies are used in similar climatic adaptations, to deal with the consequences of very cold conditions, but they evolved from quite different preexisting morphologies, whose initial variations were established by the "center and edge" process and were not adaptive at all.[41]

IMPORTANCE OF THE CENTER

Milford first met Alan when he presented his first "center and edge" paper in Nairobi. The human fossil sample they studied together had no peripheral populations, and their discussions came to focus on the role of the center, first in the pattern of origins but later in the subsequent evolutionary process. What was interesting about the center, meaning Africa below the Sahara, is that since it is the area of the world where humans first arose, it is therefore the area where they are best adapted. They reasoned this should have a number of consequences:

1. Populations in the center should be heterogeneous because selection weeding out variations should be less intense where humans are best adapted. Furthermore, with denser habitation

more populations were in contact, leading to more genetic exchanges and variation.

2. Distinct features relating ancient populations to modern ones should be most difficult to find because there was no history of bottlenecking or especially small population size.
3. Local population sizes and the size of the total population should be much greater at the center.

The first two consequences were reflected in the African fossils they studied, but the third was not as evident at the time. It became important later, when geneticists began to interpret their evidence of greater variation in Africa to mean greater time depth and an indication that a single origin of *modern* humans was there. Genetic variation seemed greater in Africa than in other places; and although as Templeton reminded his colleagues,[42] the differences in variation between African and other populations was not great enough to be statistically convincing, the greater variation was assumed anyway and argued to mean that modern humans lived there longer. The implication taken was that modern humans had a single origin in an ancient African population.

But center and edge *predicted* greater variability in Africa for quite different reasons, because *humans* (not just modern ones) originated there. The predictions of much larger populations in Africa have an impact on this problem because large population size is an additional explanation for greater genetic variation. In addition, when population size is expanding, the magnitude of genic exchanges between African populations and the genic exchanges between Africa and other regions can each impact the genetic variability shown by African populations.[43] Turning to the cytoplasmic genes, "The greater mtDNA diversity in African populations may reflect greater effective population size in Africa during the past."[44] So *either* deeper genetic roots *or* larger population size can have the same effect on genetic variation.

At the very beginning in Nairobi when Alan and Milford were hashing out the details of center and edge and how it applied in human evolution, based on their differing experiences with ancient human remains, they were struck by how well it fit the always-scanty fossil record. And it seemed as though the basic pattern worked for some time. Distinct regional skeletal features in African populations are not obvious, and have little time depth. Coon's comments in *The Origin of Races* reflected this. In answer to being asked by a black neighbor, "Who were my ancestors?" he replied, "I [do] not know. . . . The origin of the African Negroes and of the Pygmies is the greatest unsolved mystery in the field of racial study."[45]

Of course, what he *meant* was that the *characteristics* of modern Negroes and Pygmies could not be found because they did not extend far into the past; their *ancestors*, of course, were there all along. Coon seemed to stumble on his tongue continually over this issue. The absence of modern African features early on that Coon observed was not because the Africans had crossed some kind of a threshold last, it was because African populations did not have the history of drift and bottlenecking and adaptations to extreme climatic and habitat conditions that many of the colonizers suffered. There was less heterogeneity at the peripheries, more at the center.

This is one idea that did not die with Coon nor disappear with the general disregard of his work. G. P. Rightmire, a student of Howells', has developed an expertise with the African materials over the past decades. He saw considerable evidence of evolutionary continuity in Africa, and was among the first to champion the idea that modern humans could be found there earliest.[46] Yet, he had continued difficulty in finding unique links between the Middle and Late Pleistocene populations. He speculated several times that the Late Pleistocene, early modern human groups could well have evolved outside of the continent and later migrated there. While evolutionary continuity between Middle and Late Pleistocene populations was obvious and well established in eastern Asia, China and Australasia, in contrast it was not evident at the center, according to Rightmire. The early moderns may well have evolved there—no evidence shows this to be incorrect, but there were no unique links suggesting its correctness either. Unlike the peripheries, regional features were not there at the center.

Alan and Milford clearly saw this pattern of central and peripheral contrasts in variation within the fossil record. What they did not know then is that the pattern would be found to hold right up to modern times. In spite of the population movements of the Holocene, in spite of the dramatic population expansions of the last 50,000–60,000 years, it appears that there generally remains more genetic variation in African populations than elsewhere. And why not?[47]

- People have lived there the longest.
- For most of prehistory Africa had by far the largest population.
- With the highest population densities there was less drift and bottlenecking.
- There were fewer population fluctuations created by the shifting environments of the Pleistocene.
- It is a very large continent, of some 11.7 million square miles.

- Africa spans the greatest range of climatic variety because it extends poleward for almost 40° on both sides of the equator and therefore has the most adaptive variation.

Moreover, in discussing the variation in African populations, there has been a tacit assumption that while Africa has been the source of genes of all kinds for other areas, it has somehow remained genetically pristine and isolated, and that today's comparisons sample ancient patterns. But certainly in Holocene times, genic exchanges involving northern Africa, both across the Straits of Gibraltar and from the Levant, have modified gene pools across much of northern and eastern Africa. To the southeast, the island of Madagascar was settled by sea from Indonesia some 1,000 or more years ago, and there is evidence that this migratory process also resulted in landings on the east African coast, yet another influence on African genomes. Arab traders have long traveled up and down the east coast, leaving genes as well. These are just a few examples from historic times, and it is overwhelmingly likely there were other contacts in the more distant past.

The earliest good evidence for what might be a distinct pattern of regional evolution in Africa is an example of center and edge in a more limited application. This evidence may well be at an *African* periphery—in this case, along the southern Cape. When the Dutch settled the southern Cape, they identified two native groups: San: "Bushmen" hunter/gatherers and "Strandlopers" who gathered marine resources; and Khoikhoi: "Hottentot" pastoralists who herded sheep and cattle. The distinctions were more economic than biological, and together they are referred to as Khoisan. These were among the smallest African peoples, second only to Pygmies, and the understanding of Khoisan origins is tied to theories about all sub-Saharan populational origins, and also to racial politics in formerly apartheid South Africa, over the issue of exactly who the native South Africans were. One major dispute is whether these small peoples evolved locally or are the sole remnants of a once Africa-wide hunting/gathering population. This is the type of question that a good fossil record can help answer.

The best dated sample of fossil human remains that may be a direct ancestor of a modern African population is from the early Upper Pleistocene of the southern Cape, at the Klasies River Mouth Cave.[48] There are a number of fragmentary remains from this cave, cranial bits, frontal and zygomatic pieces, maxillary and mandibular remains, and postcranial bones (a vertebra, parts of an ulna and clavicle, and a foot bone) spread through several layers and associated with a Middle Stone Age industry (a Paleolithic industry manufactured using the same prepared core techniques as industries associated with European Neandertals). Most of the specimens come from a level in the cave dated to 90,000 years ago, but two maxillary fragments are older, by perhaps as much as 30,000 years. Fragmentation of bones happens without human activities, but Tim White identified cut marks on the frontal bone, right across the forehead, and suggested cannibalism may have contributed to the broken-up condition of the specimens.[49]

As we discuss in Chapter 11, this ancient sample is often considered modern because many of its constituents are very small. Small body size is also of interest because of the recent inhabitants of the Cape[50] and the question of their ancestry. One issue raised by Klasies is about the evidence for local antiquity. According to Rightmire and Deacon,[51] the Klasies postcranials are no larger than recent San. The proximal ulna is small and slender, with relatively large joints (the postcranial specialist Steven Churchill, writing with colleagues,[52] regards the anatomy of the elbow joint as archaic, and his metric analysis associates it with archaic and not modern samples). The ulna is about the average length for a San male, and a foot bone, the first metatarsal, is from an even smaller individual. In an evolutionary context, Klasies is the earliest African population that was really small. If this goes a long way toward explaining its gracility (small usually means lightly built), it also provides an anatomical link with the modern San, however tenuous. It is true, as Alan Morris puts it, not actually meaning to pun, "The actual physical evidence for the linking of the Klasies specimens with any extant regional morphology is actually extremely thin."[53] But small body size may be enough, and Deacon adds the argument that the archaeology suggests there was isolation for these populations from the southern Cape, possibly during the time the San evolved their characteristic features (there are no human skeletal remains for the southern Cape between the time of Klasies and the end of the Late Pleistocene). If it was true that these earliest Late Pleistocene folk made a significant contribution to modern San it would explain why Linda Vigilant and colleagues have observed that their mitochondrial variations seem so distinct from other African populations.[54]

AT ANOTHER PERIPHERY:
THE NEANDERTAL QUESTIONS

At first Milford found regional continuity hard to swallow because he had learned that racial features are much more ephemeral than the characteristics of the main evolutionary stages. It is possible that a flat face or a large nose persisted for some time in one area or another, but how could this hold a candle to the dramatic Pleistocene expansions in brain size? In Milford's early paper on Neandertals, written with his Case Western Reserve University colleague, David Brose, his focus on stages rather than regions is apparent.[55] In it Milford had used G.H.R. von Koenigswald's phrase "tropical Neandertals"[56] to refer to the Javanese specimens from Ngandong and was almost immediately criticized by David Pilbeam of Yale University, who likened it to calling today's Javanese "tropical Europeans."[57] Pilbeam was quite right, but it took Milford a long time to realize why and shed his ideas of a "Neandertal phase."

The Neandertal issue was central to much of Milford's thinking, and in the exchanges of ideas with Alan, he used it as a pry-bar to open up his own reasoning and examine its deepest assumptions. He began to pay more attention to the homogeneity of Neandertals; in spite of the temporal changes he focused on in the Vindija paper, and considerable individual variation they exhibit, Neandertals on the whole are very much alike in several characteristic features. Milford began to see Europe as a region, not a manifestation of a worldwide stage. Europe was an edge, at the very periphery of the human range, and Milford, as we discuss in the next chapter, now understood the evolutionary continuity in Europe as a consequence of genic exchanges balanced with selection and drift, not a consequence of isolation as he had thought. And once he abandoned the idea of stages, he accepted that gene flow was involved in *in situ* change. Multiregional evolution meant ubiquitous gene flow didn't challenge his views on the evolutionary pattern of later humans. Rather, it supported them.

Clearly, with this shifting evolutionary paradigm, the question of the fate of Neandertals was also becoming much more complex. When the issue was framed as replacement versus in situ evolution, its solution seemed simple, at least in principle, because it was easy to disprove the extreme position of total replacement. But the precept that genic exchanges, local adaptations, and an evolving cultural system all played roles in Late Pleistocene human evolution makes it more difficult to isolate an extreme position that could be easily dismissed (only a few paleoanthropologists, such as Chris Stringer,[58] support the notion that Neandertals are a different species).[59] As the questions turn to the specific details of what happened, a much wider range of data becomes relevant, and we have come to believe that it is finally time to wonder whether even any of the old *questions* are appropriate any more, let alone the answers that have been provided for them.

10

MULTIREGIONAL EVOLUTION

IF AN EVOLUTIONARY BIOLOGIST from some remote galaxy (or any other place that rendered her ignorant of the peculiar ways of men and nations) arrived on earth to study the origins of that strange and dangerous species *Homo sapiens*, she would be very surprised to learn, after perusing the literature on human evolution[1] and talking to the earthling prehistorians, of the central role in the human origins issue given to Neandertals by the local informants. It is indeed very peculiar that Neandertals are considered so important in understanding modern human origins, because they have very little to do with it—in a biological sense, that is. In fact their only relevance is to a very small section of the world at the west end of the Eurasian continent, where they inhabited the hilly and mountainous regions from Uzbekistan to Wales in very small numbers. However, historically, politically, and psychologically they have everything to do with the human origins issue. They are at the root of how Western science defines humanity; they are the "other" to which humanity is compared: they are what a "modern human" is not. Replacement theories explaining modern human origins often come from ideas about the fate of Neandertals, Eurocentrically applied to interpretations across all of human evolution. How did this small segment of archaic humanity achieve such importance?

NEANDERTAL INTERPRETATIONS

For most of the last quarter million years, Europe and a good part of western Asia have been home to this enigmatic and troublesome race. Neandertals were first discovered, or in some cases recognized in museum collections, in the middle of the past century: the Engis child from Belgium (1829), Forbes Quarry from Gibraltar (1848, but not recognized as particularly special until 1865), and of course the namesake, the Neandertal skeleton from the Feldhofer Cave on the Neander River in Germany (1856). The "Neandertal question"—the place of the Neandertals in human evolution—is paleoanthropology's oldest problem, since the first archaic human fossils to be discovered were Neandertals. The unusual anatomy of these folk and, soon it was also discovered, their stone-age culture rubbed the question of a savage and primitive European ancestry right in the face of the European scientific world. And as luck had it, the Neandertal practice of living in caves and burying their dead gave the scientists a lot to talk about, once paleoanthropologists realized where and how to look.

As these fossils were discovered, they were placed into schemes of human evolution developed to straighten out the relationships of the human races. The first phylogenetic tree specifically addressing humans was constructed by Ernst Haeckel,[2] who innovated many of the concepts of phylogeny, including the word itself. The human phylogeny he created postulated hypothetical ancestors between the ape and human species—*Pithecanthropus alalus,* the speechless ape-man. This was the "missing link" his disciple Eugène Dubois sought to find in Borneo and Java. The position for Neandertals on this tree was taken up by another species Haeckel called *"Homo stupidus."*

Neandertals were quickly attributed to a "barbarian race of European natives," but the most popular prevailing view soon came to be that the apelike features showed that "Neandertal Man" could not be a human ancestor. This was formalized by the Edinburgh anatomist William King (1809–1886), who in 1864 was the first to apply a formal terminology to express his contention that the Neandertal remains should be classified as a distinct species—*Homo neanderthalensis.*[3] King's earlier use of the term explains why the species is written in formal taxonomies as *Homo neanderthalensis* King, and not *Homo stupidus* Haeckel. However, Thomas Huxley emphasized its humanity and attributed the Neandertal findings to an extreme variety of modern human, denying that it could be considered a link between apes and humans (an unusual position, given how many evolutionists of the time considered some of the human races as links between apes and humans). Thus, at the very time of its first discovery, the question of whether Neandertals were race or species was there.

With the subsequent discoveries of several complete Neandertal burials—two from Spy (discovered in 1887), La Ferrassie with two adults and children (1909–1921, and 1973), and La Chapelle (1908), the French paleontologist Marcellin Boule epitomized the most widely accepted ideas about them in his contention that the Neandertals were "a homogeneous type, which differ greatly from all living types."[4] He accepted them as a distinct species, lying somewhere between chimpanzees and Europeans, but as a separate side branch. His detailed description of the "old man's" burial at La Chapelle cemented this view in the minds of most scholars. Yet, Boule's view was more complex than is generally appreciated. He wrote:

> Whereas the Neandertal man of Europe occupies the position of a type apart, . . . which seems according to all the evidence to have vanished without issue, the little we know about this type in Asia shows it as included within a regular evolutionary sequence.[5]

European Neandertals were different from the other fossil humans—perhaps the quintessential peripheral population. They existed in Europe for about 200,000 years, quite a long time. That they are the best known of all Ice Age human fossil remains is ironic because Neandertals are very disproportionately overrepresented in the fossil record. Europe is densely populated today and is a region with intense use of the landscape, with construction, farming, and road building. Much of the land has been worked and many fossils discovered in the process. The European fossil record was further exposed with improved knowledge of where to find human remains. Neandertals lived in south-facing caves and often buried their dead there. But finding a large number of Neandertals does not necessarily mean that a large number of Neandertals lived in Europe. In fact, at the extreme western end of their range the Neandertals occupied a geographic and ecological margin. For most of their existence, they probably lived in small, isolated populations surviving in temperate to arctic conditions. The total population of glaciated Europe may well have only been in the tens of thousands.

These peripheral folk remained an extinct, primitive race in the hearts and minds of most paleoanthropologists. Perhaps a real key to understanding the popularity that Neandertal extinction sustained in western Europe is the way in which the Neandertal extinction became a metaphor for all colonial invasions.[6] For instance, in *Everyday Life in the Old Stone Age*, a popular book from the middle of this century the Neandertal is described:

> His large head, with the thick frontal bones, must have been very good for butting a brother Neanderthal, but it was no use against the stone wall of advancing civilization, and like the Tasmanian and Bushman, the Red Indian and Australian of nowadays, he fades out of the picture and his place is taken by a cleverer people.[7]

Looking back, we can see that knowledge was not growing, even though the number of Neandertals increased. Moreover, as Trinkaus and Shipman put it:

> Despite a major change from the 19th century emphasis on progress and betterment to the 20th century emphasis on isolating ourselves from all but our most remote near-human ancestors, the themes in Neandertal studies have remained relatively constant.[8]

Weidenreich, as we noted, continued to promote the idea of some form of Neandertal ancestry for Europeans in his debates with Howells. The position he finally arrived at was that the European populations were isolated and cold-adapted,[9] and that the thread of ancestry from Neandertals to modern Europeans mainly passed through the Levantine Neandertal populations, as represented at the Mount Carmel cave of Tabun from Israel. For him it didn't matter *which* Neandertal might have been ancestral because of the complex pattern of ancestry for all populations depicted on his trellis. But this was, by far, the minority view in the paleoanthropological profession.

The Mount Carmel remains, first published in detail in 1939,[10] came to play an important role not only for Weidenreich but also in how the modern synthesis of the 1940s changed the rules of the game and the details of the Neandertal debate. These caves, Tabun and Skhul, are nearby each other, in a small wadi (gully) near the Israeli coast, the Wadi-el-Mughara of Mount Carmel. Discoveries between 1929 and 1934 of human burials in the caves provided the first large sample of fairly complete Paleolithic humans. When they were first described, the astounding realization was how variable the individuals were. At the time they were thought to all date to the same age, about 100,000 years ago,[11] and for years paleoanthropologists grappled with how to explain what appeared to be a mixture of Neandertal and more modern features:

- Were they a hybrid population of the two? (But that would seem to imply once-pure races.)
- Were they in the throes of evolutionary change? (But when there is selection the populations get less variable, not more variable.)
- And if they were caught in the act of evolving, was it Neandertals into moderns, or moderns into Neandertals as McCown and Keith thought?

With our 20-20 hindsight, it now seems clear that even though the sample of fossil human remains increased, no reasonable understanding of Neandertals could ever be achieved without applying evolution

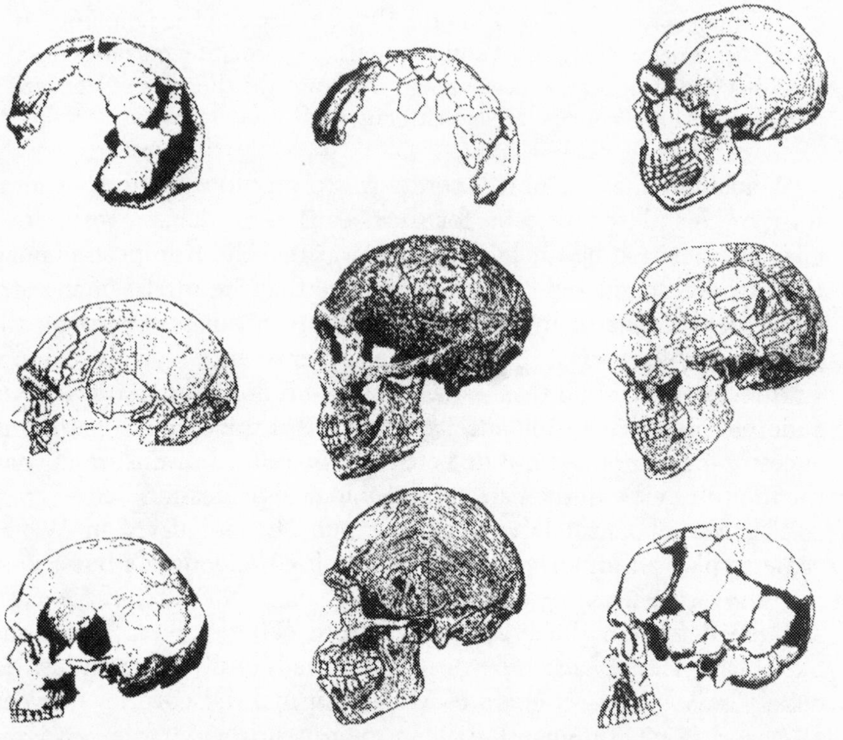

Lateral views of Levant crania.[12] *The specimens* (top row, from left to right) *are Qafzeh 3 and 5, and Tabun;* (second row) *Skhul 9, Amud, Skhul 4;* (third row) *Qafzeh 6, Skhul 5, Qafzeh 9. Of these crania only Tabun and Amud are classically considered Neandertals, but as can easily be seen, the differences in anatomy within the sample are gradual, and not large. This has always been the problem with their interpretation: where do Neandertals end and other races begin? The Mount Carmel sites are Skhul and Tabun. Their variation is an accurate reflection of the variation in the larger sample now known from the region.*

to their interpretation, and it took the modern synthesis to do so. The Russian geneticist Dobzhansky first did this, taking an evolutionary approach to the question in 1944 when he reanalyzed the Mount Carmel remains. The resemblance of the human remains in this collection to both Neandertals and moderns, to differing degrees, had been interpreted by the scientists who described it (Theodore McCown and Arthur Keith) as the result of the population being "in the throes of evolutionary change." Dobzhansky took issue with this interpretation, arguing it is much more probable that the variation was the result of intermixture between Neandertals and less specialized populations. For Dobzhansky and many later workers (for instance, Mayr, in his famous 1950 reevaluation of human taxonomy), this proof of mixing showed there was a single species at Mount Carmel, and Neandertals became a *Homo sapiens* subspecies (*Homo sapiens neanderthalensis*). But there was still confusion about what that meant, in particular the continued confusion of temporal subspecies with geographic subspecies (was there a Neandertal *stage* or a European Neandertal *race*?). While Neandertals became, for many, the European version of archaic *Homo sapiens*, it was never clear exactly what that meant.

F. Clark Howell was the first paleoanthropologist to put Neandertals in the evolutionary context of the modern synthesis. His master's thesis at the University of Chicago, submitted in 1951, was on the Ngandong hominids and he earned his Ph.D. with Washburn in 1953. From his Ngandong M.A. dissertation and his interest in Neandertals, Howell was familiar with Weidenreich's work, and proceeded from his assumption that features of the western European Neandertals were the result of cold adaptation in an isolated population. But Weidenreich's observations and speculations about Neandertals needed to be placed within the frame provided by the modern synthesis, and Howell did this.

The influence of Weidenreich and Washburn combined in his earlier work, in a seminal paper on the evolution of Neandertals in Europe and western Asia, that Howell published in 1951. In it, he examined the evolutionary trends revealed by comparing earlier and later parts of the Neandertal sample. Because the direction of change in the western population seemed to be away from modern humans, especially when comparisons were made with the Skhul hominids that were regarded as being earlier than most western European Neandertals, he concluded the "Classic Neandertals" of western Europe were peripheral to the mainstream of human evolution. As the most extreme of the Neandertal groups, their unique features precluded the possibility that they were ancestral to later Europeans. Howell argued that while the Classic Neandertals evolved in western Europe, "at the same time modern man was developing further to the east."[13] Howell reasoned the evolutionary direction leading to so unique a set of features could only develop with genetic isolation, and he sought to show that the Neandertal populations of western Europe were indeed isolated, by the pattern of glacial advances during the earlier Würm glaciation.[14] In a more general restatement of his arguments Howell examined a larger Neandertal sample and buttressed his conclusion that modern populations evolved only from earlier Neandertals and not from the late Classic Neandertals of western Europe:

> East-Central and Eastern Europe were not particularly isolated during the Early Last Glacial . . . there was broad racial continuity at least as far east as the Crimea and southward into the Levant.[15] . . . Important structural changes were taking place in the human populations of the Levant during the Early Last Pluvial . . . these were of primary significance for the evolutionary origin of subsequent anatomically modern peoples of the European Upper Paleolithic.[16]

Howell developed what has become the most commonly accepted precept of the place of the Neandertals in human evolution, advocating a general Neandertal ancestry for modern Europeans that excluded the Classic western populations. He thought that Neandertal specimens such as Ehringsdorf could be ancestral to both the extinct western Neandertals and the Mount Carmel population such as found at Skhul, who in turn gave rise to the modern Europeans. The current understanding of dates for these sites still allows for this possibility, and it would be fair to say that many paleoanthropologists continue to believe some of the Neandertals contributed to the modern European gene pool.

But others do not, and if anything, the Neandertal controversy is more heated now than ever. Why is there so much focus on the Neandertal problem once again today? Three reasons, really. The first is that there are so many of them. As Chris Stringer put it:

> If we cannot resolve the Neanderthal problem and thereby arrive at an understanding of the relationship of Neanderthals to "modern" *Homo sapiens*, there would seem little hope of resolving any of the more complex issues concerning human evolution.[17]

The second comes from how Neandertals have attained a critical role in defining modernity. If we cannot be sure what modern humanity *is*, it is increasingly evident that we can agree on what it is *not*—Neandertal! With native peoples no longer available for the role, it is the Neandertals who have become "other." How better to define ourselves than in comparison with these folk? Look at their use in fiction:

- "with a big face like a mask, great brow ridges, and no forehead"—Wells's *The Grisly Folk*[18]
- whose "mouth was wide and soft and above the curls of the upper lip the great nostrils flared like wings, there was no bridge to the nose and the moon-shadow of the jutting brow lay just above the tip"—Golding's *The Inheritors*[19]
- with a child who was "the ugliest little boy she had ever seen . . . from misshapen head to bandy legs"—Asimov's *The Ugly Little Boy*[20]
- or with a face featuring "a large beaky nose, a prognathous jaw jutting out like a muzzle, and no chin. Her low forehead sloped back into a long, large head"—Auel's *The Clan of the Cave Bear*[21]

The only thing that has changed over the years is the assessment of how these misshapen, ugly folk behaved.[22] The earlier versions were people you would not want to meet in a dark alley, but the later ones? Asimov's ugly little boy learned to speak perfect English, and won over the heart of his nanny by saying to her, "I know your name is Miss Fellowes, but—sometimes I call you 'Mother' inside."[23] Golding's Neandertals were telepathic, and Auel's had "big, round, intelligent, dark brown eyes.[24] They became more like the natives whose role they inherited by becoming the "other." Neandertals provided the clearest picture of what we were *not*, and perhaps some insight into how we came to be successful at their expense.

The third reason for the renewed focus on the Neandertal problem is the importance Neandertals have attained in the conflicts over Multiregional evolution—their interpretation is perceived to be a weak point in the Multiregional evolution model. The Eve theory brought a different perspective to the understanding of Neandertal, once again interpreting these folk as a separate, unique species. This is because if Eve is valid, Neandertals must have been replaced by her descendants. In fact, some scholars regard Neandertal extinction as a refutation of Multiregional evolution, because it would mean there was no evolutionary continuity in Late Pleistocene Europe. Eve has made Neandertals critical to modern human origins theories by denying them any role in it. This importance is asymmetrical, though. If Neandertals could be proved extinct in Europe, without any mixing or contribution to later Europeans, it would not prove Multiregional evolution wrong, but only that replacement was the mode of Multiregional evolution in Europe. It would take a demonstration of replacement everywhere else, except the place of origin, to invalidate the model. *But*, if it could be shown there was any continuity between Neandertals and later human populations in Europe, that would be more than sufficient to disprove the Eve theory.[25]

The link between Eve and the Neandertal issue is nowhere more clearly reflected than in the popular press, through the way these prehistoric Europeans are considered by science writers who are certain the Eve theory *must* be true. Look at two almost consecutive issues of *Discover* recently published. In March of 1995 there was an article on cystic fibrosis,[26] a debilitating, deadly disease that, untreated, leads to death before reproductive age is attained. The disease is caused by a gene, discovered in 1989, that must appear in its homozygotic form with the same allele from each parent. For such a deadly allele to be retained there must be an advantage for the gene in its heterozygotic form (the deadly allele from only one parent, the normal allele from the other). Otherwise it would disappear. Research by Xavier Estivill and colleagues was cited that suggested this heterozygotic advantage may be in the reduction of diarrhea, an effective adaptation to cholera.[27] The gene is by far most common in Europeans, where cystic fibrosis is the most prevalent fatal genetic disorder, but uncommon in Africans and virtually unknown in East Asians. The Estivill study estimated a date of *at least* 52,000 years ago for the time of origin of this gene in Europe.[28] Even the most recent date possible, according to this study, would seem to make the gene originate in European Neandertals, and show them to be ancestors of modern Europeans. This is so unacceptable, the Eve theory is so uncritically assumed, that the alternative explanation provided in *Discover* is that the mutation must have arisen in the Levant, in Eve's descendants, but "before the settlement of Europe by modern humans [although] it must have originated after humans spread out of the Near East in other directions, sometime after 100,000 years ago; otherwise cystic fibrosis would be as prevalent elsewhere as it is among Europeans."[29] In Africa, a gene with a distribution and early date of origin like this would be taken as proof of the Eve theory. But for Europeans, it is more credible to posit Palestinian pedestrian police, keeping those modern European ancestors out of Europe long after the ancestors of the other modern populations were allowed to leave.

In a *Discover* soon following, another science reporter also grappled with Eve and the Neandertals, and maintained Eve's descendants replaced those ancient Europeans without any mixture.[30] Here, a different difficulty with the Eve theory is resolved. The notion that Neandertals and moderns lived side by side in the Levant is accepted,[31] where, the author admits, they made the same tools in the same ways, hunted the same game, butchered it the same way, and even practiced the same ceremonies, at least when it came to burying their dead. The problem for the Eve theory is how this separation could be maintained in the face of all these behavioral (dare we say cultural?) similarities?

The article cited the paleoanthropologist Clive Gamble as saying:

> Cook's men would come to some distant land, and lining the shore would be all these very bizarre-looking human beings with spears, long jaws, brow ridges . . . God, how odd it must have seemed to them. But that didn't stop the Cook crew from making a lot of little Cooklets.[32]

But nothing was learned from that quote. If moderns and Neandertals were making little Neandertalets for 50,000 years, the Eve theory would be wrong. But that can't be, so instead it must be that they didn't have the urge to merge. This interpretation makes Neandertals a different species that didn't mate with the moderns because they found them unattractive; the Neandertals looked for different facial features to identify their mates, as presumably did the moderns. According to this reporter, this allowed the two to "coexist through the long millennia doing the same human-like things but without interbreeding, simply because the issue never came up."[33] These moderns must have been different humans than the ones Gamble was talking about, certainly quite different from ourselves, who have the need to put words like "sodomy" in our dictionaries, and in our holy books.

On our more pessimistic days we believe the Neandertal problem will never be considered solved. Our dear friend and colleague David Frayer often seems much more right than wrong when he writes:

> Recent publications have produced a set of disjointed interpretations about what European Neanderthals represent morphologically and what they were capable of behaviorally. Trying to piece together all that has been recently written . . . results in a picture resembling postmodernist art, where a series of incongruous, completely unrelated images are combined together in the same scene producing a phantasmagoria. For example, while there is still no human fossil evidence which supports the co-existence of Neanderthal and Upper Paleolithic forms in Europe, we now have a series of models and speculations about the details of this co-existence and why one group replaced the other, be it linguistic incompetence, spousal inattention, or inferior hunting practices. . . . There are

> also suggestions that Upper Paleolithic groups are directly derived from African migrants, despite the complete absence of any supportive analysis for the presence of ancient or modern African features in the earliest Upper Paleolithic humans. In short, the study of European Neanderthals has reached a state in paleoanthropology where the [functions of the] fossils themselves [as evidence for human evolution] have been supplanted by speculations about them.[34]

But we believe that even if the Neandertal issue is never resolved to everybody's satisfaction, Multiregional evolution will be.

THE MULTIREGIONAL MODEL

Multiregional evolution is a model of population variation and evolutionary change in a widespread, geographically diverse species that is internally subdivided. It is a description of the pattern of human evolution and an explanation of how evolutionary processes created it. Multiregional evolution characterizes humans while they evolved as an interconnected polytypic species from a single origin in Africa some 2 million years ago. In the last chapter we described how small population effects during initial colonizations, as humans subsequently expanded out of Africa, helped establish regional differences which were maintained through isolation-by-distance and adaptive variation. But this was only the first of many "out-of-Africas" and "into-Africas" as well.

The human species has been a dynamic evolving interconnected entity in which advantageous changes could spread through genic exchanges. Technological innovations and new ideas could also spread through the network because of the common background of humanity's evolving cultural system. Most modernizing features arose at different times and places and diffused independently. Evolution was gradual, in the sense of not involving new species, but was neither constant nor at an even rate. Its pace was punctuated by periods of rapid change and other periods of stasis. In the human network of populations interconnected by lines of genic exchanges, and increasingly by cultural communications, the pattern of evolution in one region cannot be understood apart from the others.

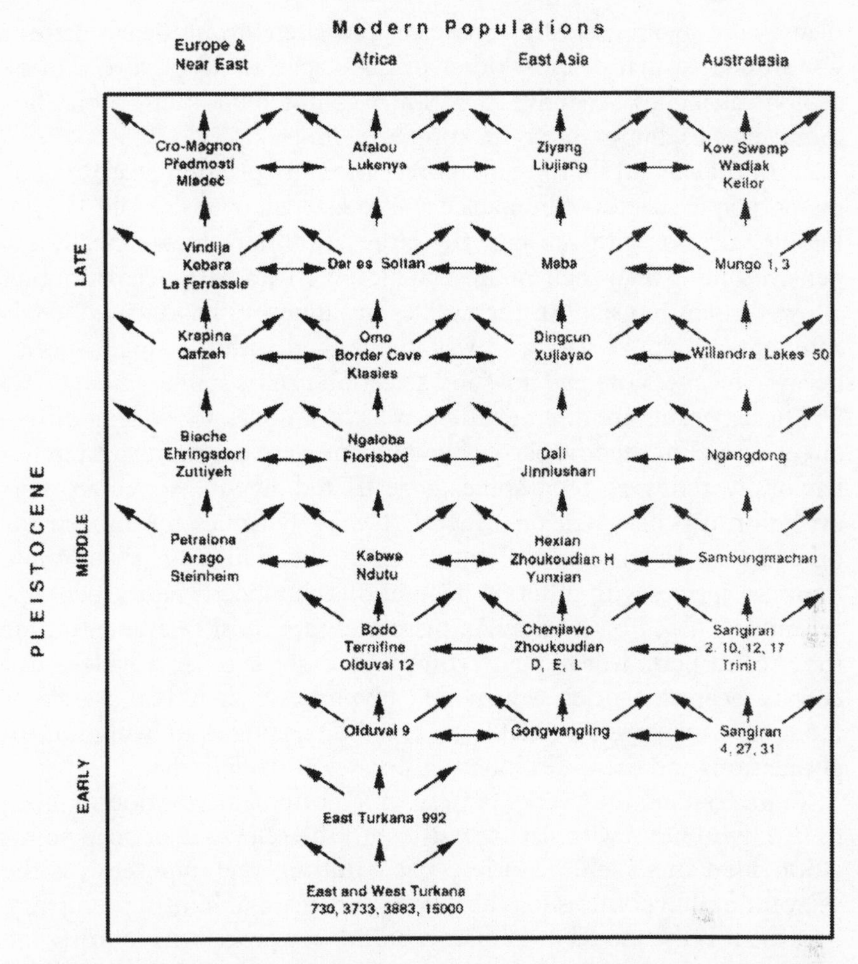

Multiregional evolution,[35] as it might be used to model the relations among human fossil remains. The specimens are all that remain of populations ancestral to or related to each other as shown. A diagrammatic rendering of the model such as this one is useful because it frames the fossils in time and space, but at the same time is necessarily misleading for two reasons. First, it shows all of the arrows of genic exchanges with equal magnitude; this surely is not correct, but the fact is that we have no real information about the actual magnitudes, except that they were all different and varied over time. Second, it shows no population extinctions, although there undoubtedly were many.

TWO PATTERNS

There are two very distinct periods of human evolutionary history, and the general patterns are quite different for each.

1. In the longer part, for the first two-thirds of their evolution hu-

mans were geographically restricted and their evolutionary pattern was similar to that of most other primate species, adapted to limited ecological niches. Adaptive variation was not found within one hominid species, but in different, competing ones.

2. In the second shorter and more recent colonization phase[36] a single polytypic species of humans came to exist all over the Old World. People, already with substantially larger and more metabolically expensive brains than their predecessors, left Africa with behaviors that allowed them to establish themselves in numerous habitats and environments. Adaptive variation was maintained within a single species because balances of gene flow and selection can be stable.

These periods are discrete and can be best understood using different criteria. For the first part, when humans were less sophisticated behaviorally and were geographically restricted, hypotheses about their evolution can be based on general species dynamics formulated for other species in similar ecological conditions. This is when multiple hominid species with different adaptations abounded and the ecological models based on nonhuman primates[37] are most relevant. But for the second period of human evolution we need to use a widespread culture-bearing species as a model. Hypotheses such as those about modern human origins need to be based on analogies to living human populations and the evolutionary processes at work in them.

To understand the overall pattern of later human evolution we need to test hypotheses with data that attest to the relevance of both population adaptations and histories. The Multiregional hypothesis is the relevant null hypothesis for the colonizing phase of human prehistory, because it is the model of human evolutionary processes requiring the fewest number of assumptions to account for the paradox of global change and local continuity. In fact, the main assumption it requires is that the evolutionary processes at work in relatively recent modern human populations were in operation in the past. Our species is unusual and difficult to model because it is polytypic, with extremely broad geographic and ecological ranges. Most polytypic animals are so because they occupy an ecological niche that is broadly distributed, but the human pattern is of a widespread single polytypic species with many different ecological niches. It is a very rare one.

The difference is human culture, the unique human capacity to transform experience into symbols for information storage and transmission and complex communications. Culture allows unprecedented behavioral flexibility, providing critical elements of what is necessary for human populations to occupy so many diverse niches. Culture affects evolutionary processes in many other respects; it influences the demography of populations and plays an inestimably important role in shaping the evolutionary histories of human populations. While we do not think that culture appeared suddenly in its present form—its definition and evolution are obviously complex—the behavioral flexibility it allows was clearly significant during the colonizing phase in human evolution. Humans in this later phase should not be expected to follow the species proliferation or extinction patterns of other fauna that lack this degree of flexibility. This is why we think Modern humans themselves, more than any other living species, represent the best model to base the reconstruction of the Multiregional pattern of human evolution.

MULTIPLE DESCENDANTS, MULTIPLE ANCESTORS

Ethnologists such as John Moore argue that populational, linguistic, and cultural relationships between populations today are both multiple and complex. Each population descends from, or is rooted in, "several different antecedent groups."[38] An American, for example, may speak English and understand German, have biological roots in Britain, Holland, and Germany, and grow up in center-city Philadelphia, with all its subcultural modes and mores. But this pattern is not limited to a putative "melting pot" like the United States—it is a worldwide one. Roger Owen studied a series of native tribes across the world, from the American Great Basin to Australia to Tierra del Fuego, and everywhere he looked, there were no ethnically isolated groups. Instead, marriages across what seemed to be the most profound ethnic boundaries were common, and multilingualism prevailed.[39] Susan Sharrock surveyed Native American tribes from the Northern Plains and concluded that "the ethnic unit, the linguistic unit, the territorial coresidential unit, the cultural unit, and the societal unit did not correspond in membership composition."[40] Populations, languages, and cultures continue to change, divide and merge in a matrix of changing patterns in which a population (or language or culture) can have several ancestors and several descendants. Moore likens this to the channels in a river that can separate and recombine numerous times. He calls the process "ethnogenesis" and notes that the continued population contacts and exchanges it implies means there is *and can be* no link between language, culture, and biology.[41] Assuming such a link is part of a different, opposite model of population variation which is based on the idea that variation is created by continued branching and divergences, Moore writes:

> The ethnogenetic perspective argues that humans have evolved as a species and not as unconnected regional populations, so that there always have been admixed populations. . . . The ethnogenetic perspective takes the multiregional position, arguing that human history has always been characterized by interaction across profound ethnic and cultural boundaries, by the amalgamation of linguistic traits, and by the recurrent hybridization of cultures.[42]

Multiregional evolution addresses the same process, with the time depth that must be added to understand how it led to the present distribution of human variation. It attempts to model how evolutionary change continued in the face of geographic dispersion without speciation—the pattern that seems to characterize at least the last million years of human evolution.

The alternative model for human evolution is speciation, new species appearing and replacing previous ones, contemporary species competing and one prevailing. Multiregional evolution suggests a pattern differing from species-level evolution in one very important way: it does not involve branching, or treelike divergences. Weidenreich's network or Moore's river channel model is a vivid, and much more appropriate, analogy. Population geneticists have long known that the genetics of human populations cannot be validly modeled as trees.[43] Weidenreich, who was quite explicit in rejecting trees as models of human evolution, realized there were never purer, less mixed populations. Whether Doberman pinschers or Chinese, Weidenreich knew that today's races and the races of the past had multiple ancestors and could never be considered "pure." Moore realizes much the same thing, and applies it to sociocultural anthropology through his ethnogenetic model: "This distinction between admixed and unadmixed populations is illusory and a product of our ignorance of the ethnohistory of most of the world."[44]

Certainly over the span of the Pleistocene, 1.75 million years ago to the present, the human species has evolved as a whole, as various localized advantageous changes spread widely and persisted because populations mixed, diverged, mixed again, and continued to do so. The two key elements of the model are:

1. The historic and adaptive processes that created and maintained the pattern of variation
2. The dynamics of reproduction, communication, and population movements linking local populations and providing the network

for advantageous changes—whether these are based on new gene combinations or new ideas—to diffuse throughout the species

The Multiregional model accounts for the combination of long-lasting diversity and species-wide evolution by examining the consequences of the species' internal subdivisions—widespread diversification linked by gradations of continuously varying features that reflect gradients in selection, genic exchanges between adjacent populations, or both. These gradations persist when they balance the forces that create differences. For instance, directional genic exchanges often extend from the center to the edge of the species range, and local selection is often most intense in more peripheral populations.

The Multiregional evolution model begins with the obvious—humans are a single widespread polytypic species, with multiple, constantly evolving, interlinked populations whose dynamics can be partially explained by evolutionary processes that pertain to other widespread polytypic species. These processes are framed by two aspects of variation:

1. Clines (continuous gradations) created by the interactions of evolutionary forces and their variations
2. History: the consequences of population placements relative to other populations—the center and edge mechanism

These processes have affected humans since *Homo sapiens* became a widespread species, and therefore they can explain the patterns of morphological variation that we see in prehistory. As early as when humans first successfully colonized regions outside of Africa, evidence of regional continuities in different places became apparent and convincing.

GEOGRAPHIC DIFFERENCES

Humanity is a polytypic species, a species with geographically distinct groups of populations (or races). "Distinct," in this context, does not mean every individual can be distinguished from all others, or that all the members can be uniquely described with one or even many characteristics. We know this isn't true from our experience as forensic anthropologists, trying to identify human remains, and from the determination that only about 10 percent of all genetic variation is found between populations, most variation is within them. Regional distinctions are expressed as average differences, even while specific features are extremely common in nearby populations. It is interesting that most people derive clues about origin from these features. Ideas about "self" and "other" are partially based on physical characteristics in many, if not most societies around the world. Traits that can help distinguish at least some and perhaps many members of one population from another occur in complexes that include features that vary regionally. These may potentially provide information about regional histories. Many of them have a skeletal basis, and some occur in the skeleton alone, with no external expression. This neither means that the features are unique to specific regions, nor that they characterize all individuals from a region. But just like the phenotypic characters that allow someone standing on a busy corner in Philadelphia to make up hypotheses about the ancestry of the people she sees, these too exist in complexes that tend to occur in higher frequency in some regions than in others. But we are not interested in using these characters to place individuals into categories. We are interested in understanding them as one aspect of human variation within and between regions over time.

The two fundamental reasons why geographically disperse human populations differ are adaptation and history. Some of the consequences of history were outlined in considering the effects of the center and edge process during regional colonizations. In particular, at least at the beginning, more peripheral populations were found to be homogeneous for a number of features. Certain of these were adaptive, important elements that helped populations meet the requirements of their environments. Adaptive features appeared in high frequency in response to changes in the conditions that made them helpful, not always at the time of colonization. For instance, short arms and legs are adaptive in cold conditions. Climate change, or population movement into cooler regions, can create selection for reduced limb length if other diverse functions of limb length such as locomotion or throwing are not adversely affected.

Other features, however, were established in high frequency when small population sizes and locally intense selection were parts of the colonization process. These features are commonly occurring variations of anatomy that can be called "equivalent alternatives." As an example of equivalent alternatives, broad noses might have been established as the most common variation of nasal form in one region, narrow noses in another. If there was no selection on nasal breadth—no advantage of one nose form over another—they are equivalent alternatives because both nose forms function as well in respiration and olfaction.

Some of these once-common features disappeared but others persisted. Why? Of course, when a feature became advantageous, selection would help it persist. For instance, broad noses might be important in a particular climate. But other cases of this persistence reflect a kind of evolutionary momentum. Features that do not improve adaptation, nonadaptive features that were initially established at high frequencies, may remain unaltered if no evolutionary forces act to change them. There is no entropy effect on the evolutionary process: things do not disappear if they are not used. Some cases of continuity, then, involve nonadaptive characters that never became maladaptive.

However, there is a second reason for persistence. Features that were established by chance, or for reasons that are no longer relevant, can become incorporated into new adaptations. These are examples of exaptation, adaptive features whose origin was unrelated to their later adaptation.[45] Exaptations are common because the evolutionary process can only work with what is at hand. The keen eye and accuracy of a pitcher did not evolve for pitching, or possibly even for throwing. Hand and eye coordination and accurate spatial location are consequences of a past arboreal life and an even earlier adaptation for grabbing insects out of the air.[46] Because exaptations were not initially established to promote the adaptation that was important later, they may differ substantially between various populations, while meeting the same adaptive requirements.

In other words, when there are exaptations, populations may meet the same adaptive challenge in differing ways because of past population histories. In these cases the different characters promoted in various populations do not reflect different adaptations. Again turning to the nose, there is an example Milford is rather proud of, his first published piece of research. The nose is most efficient in warming and moistening inspired air[47] (or gathering moisture from the expired air[48]) when the nostrils are not circular. This is because circular nostrils have the *least* amount of surface area relative to the volume of air passing over the surface, and this surface is where the action takes place. But how to deviate from nostril circularity? The ancestral condition is important here. Before the adaptation to dry or cold conditions the nostril form was influenced by anterior tooth size: broad front teeth created a broad nose because the canine roots are at the sides of the nose, and small front teeth a narrow one. These differences, unrelated to climate, were different exaptations for populations changing their adaptation. Broad noses were incorporated into an adaptation using broad nostrils, while narrow noses were incorporated into an adaptation using tall nostrils and thereby a more projecting nose.

Features such as these do reflect past "race"-level differences. The races themselves are transitory, but in many cases the features have not been. Whether they were first established by adaptation or by other historic processes such as past bottlenecking or drift, the observation that geographically differing characteristics persisted for long periods of time, in some cases up to and through the Late Pleistocene and some even to recent and living populations, is called regional continuity. The *explanation* of regional continuity in *Homo sapiens* is found in the Multiregional evolution model.

A PARADOX

The contradiction between the isolation of populations that seemed to be required to establish regional distinctions, and the continued hybridization between them and the absence of any "pure races" that his observations revealed, is the second of the two problems in Weidenreich's Polycentric theory that he was unable to resolve. How *did* populations retain geographic distinctions and yet evolve together? This is the apparent paradox that Multiregional evolution addresses. If it is true that isolation is necessary for long-term geographic differentiation, wouldn't that eliminate any common patterns of evolution unless they were purely fortuitous? If genic exchanges are required for common evolutionary directions, wouldn't they eliminate geographic distinctions? How could we possibly expect these two contradictory processes to be of just the right magnitude to balance each other and allow both persisting regional distinctions and common evolutionary changes?

These are false choices. Part of the resolution of the paradox lies in recognizing their incorrect assumptions.

One invalid assumption is that genic exchanges are necessary for there to be common directions to evolutionary changes in different human populations. Exchanges create these commonalties, of course, as advantageous alleles spread widely because of the benefits they confer, and thereby result in the same evolutionary changes in different places. But human populations are also particularly receptive to a second process, as communication systems disseminate the behaviors that make changes advantageous, such as new technologies, improved organizational skills, or important changes in the communication systems themselves. The exchanges of ideas, information, and technology are independent causes of common evolutionary directions. As ideas disseminate widely, they produce the same changes in selection in different regions, and populations with similar gene pools could respond in the same manner. What might appear to result from genes spreading might instead be the result of the forces that cause their frequencies to change. After all, the similarities between human gene pools are more extensive than one might imagine. Most human genetic variation is within populations and not between them, and it is commonly quipped that any two fruit flies have more genetic differences than the most extreme comparison of two people would show. There is ample opportunity for similar selection to cause similar changes.

Actually, of course, spreading advantageous gene combinations *and* spreading advantageous ideas are both important in explaining species-wide changes, and they are not really independent of each other. When ideas and artifacts spread, genes frequently disperse as well. This is part of a more general framework of balances controlling the pattern of variation in subdivided widespread species.

The other invalid assumption made when regionality and common evolutionary changes are considered contradictory involves the role of genetic isolation. Isolation is not essential for maintaining geographic distinctions. Gradually varying distributions of a feature can develop when the source of selection the feature responds to varies gradually. Many characteristics that vary geographically are found across broad gradients, responding to selection that differs over a wide geographic range. For instance, skin color corresponds to the amount of solar radiation skin is exposed to, and it varies from the equatorial regions toward the poles. This pattern may be seen in skeletal features that can reflect climatic adaptation, such as relative limb lengths or nasal form. However, the interpretation of the pattern of variation shown by features such as these can be complex, because many, perhaps most, adaptive characteristics *function in several different adaptive systems.* Therefore the distribution of these features may respond to several different evolutionary pressures. Relative limb length, for instance, is highly correlated with climate, but also is important in different patterns of mobility. Long limbs aid long distance travel, shorter limbs are useful in rapid movements across irregular terrain. Limb length would be a compromise in a highly mobile Arctic population. But behaviors can vary over space as well, and the point is that when the source of selection varies gradually over space, the adaptive response may do so as well and there will be geographic distinctions.

In addition to variation in sources of selection there is another, more complex source of clinal variation, in this case incorporating genic exchanges in the explanation of how regional features can be maintained. Perhaps the most critical insight of Multiregional evolution is that there are gradients formed between contradictory elements—genic exchanges *versus* selection and/or drift. These do not somehow cancel or oppose each other, but interact to create clinal variation. Milford learned this early in his education, although he only realized its importance much later. As a graduate student, he wrote a paper applying the idea of competitive exclusion to human evolution.[49] Competitive exclusion is a way of generalizing the results of competition between species when it is for limiting resources whose abundance affects population size. When this takes place, the competitive exclusion principle predicts that one of the species will move away to avoid the competition, change its adaptation so there is no longer competition, or become extinct as the result of the competition. In other words, the competition creates an unstable relationship between the species.

But every principle seems to have its exceptions, and Milford was brought to realize this when taking his Ph.D. examinations. Charles King, an ecologist on his committee, had read his paper and now asked him, "Under what conditions will competitive exclusion not apply to a competition for limiting resources?" Milford determined the answer was the case in which the competition was at the fringes of one of the species' ranges. That species could be the inferior one when competing, but losses in its numbers could be made up with migrations from the more densely occupied center of the range. A balance of genes lost to genes reintroduced would form, and as a result, the competition could be stable in spite of competitive exclusion. This was an example of the kind of balance of opposing evolutionary forces that would become an important part of the Multiregional model, many years later.

Clines will form when opposing evolutionary forces affect a feature: for instance, genic exchanges and selection. Populations are usually more numerous and dense toward the center of a species range, sparser and less common toward the edge. Population movements are often from dense to more sparsely inhabited regions, and advantageous new genes or gene combinations may first appear in the center where there are bigger populations, then spread outward because of the advantages they confer. Toward the peripheries, however, there may be opposing evolutionary forces. Genic exchange acting on a feature in one direction and selection in the other—for instance, genic exchanges introducing a new gene but local selection acting against it at the periphery—will create a gradient, even if the source of selection is not distributed along a gradient. This is because the intersection of two opposing forces that vary according to their position is always a third force that varies positionally, and morphology reacting to it will vary continuously over space. One of these forces will not overpower the other. The gene being introduced by genic exchanges, for instance, will not achieve high proportions at the periphery, but conversely the selection against it will not make it disappear.

Instead opposing forces of genic exchange and selection will invariably form a balance. No other possibility exists. Genic exchanges and genetic drift can also oppose each other. Particularly when the genic exchanges come from gene flow (individual or population movements), many genes introduced into more peripheral populations by successful immigrants are minimally advantageous or neutral to selection in a peripheral environment. Disadvantageous genes, of course, would disappear. But because peripheral or ecologically marginal populations are more subject to drift, a second possibility exists for balances to form. This is because drift often leads to the loss of rare or infrequent genes, even when they are neutral, and genes introduced by genic exchanges and lost in peripheral populations are distributed along a gradient. Thus there are two sources of clinal variation in features: gradients of selection and gradients created by opposing evolutionary forces. These control just how different interconnected populations may become.[50]

Clinal models both explain how multiregionalism works and address an issue raised in its criticism: how could one expect a fortuitous combination of genic exchanges with just enough isolation for differentiation, just enough gene flow to prevent speciation?[51] But clines formed by a balance of opposing forces are in a stable equilibrium, not an unstable one. Fortuitous combinations are not required because *any* combination will form a gradient. The most important characteristic of these clines is in how they relate the conflicting causes of variation. The *balance of forces* defines the *steepness of the gradient*. Just as a seesaw can balance children or cheeseburgers, clinal balances are independent of the absolute magnitude of the forces and depend only on their relative sizes. The relative magnitudes of balancing forces are important because they create the slope of the gradient—whether a feature varies a lot or a little from one place to another. Once conditions for a gradient are met, it means one of the forces cannot overwhelm or swamp out the other—they will always form a gradient.

Long-lasting genetic differentiation created by clinal balances is commonly thought to be the main cause of human racial variation.[52] Multiregional evolution's central contention is that clinal balances always characterized humanity for the entire history of *Homo sapiens*. While the populations changed and the details of the balances varied with the ongoing process of evolution, local continuities for certain features lasted for long periods of time. *These account for the observations of regional continuity, and at the same time illustrate a historical as well as a clinal dimension to modern population variation.* But why should some clinal balances be long-lasting? Why is there long-term continuity in some features, especially at the peripheries? Three factors come into play.

The first of these is the homogeneity that is

- more common in subdivided species
- more prevalent in small populations because of genetic drift
- more often characteristic of marginal populations, which are smaller than the more central ones
- more often found in the colonizing populations and their descendants

Homogeneous features create continuity because, once they are established, they will not change unless they become disadvantageous.

The other two factors creating regional continuity are long-lasting adaptations and exaptations. They differ mainly over the role history plays. Exaptations rely on the anatomy already present and can only be readily identified when there is the potential for equivalent adaptations. For example, Tracey Crummett has shown there are three factors that contribute to the coal-shovel-like form of shovel-shaped upper incisors: the size of their marginal ridges, the extent to which the flat surface of the crown curves side-to-side, and the size of the tubercle, a lump on the base of the inner crown surface.[53] Populations of western Europe and eastern Asia probably came to differ in these elements contributing to maxillary incisor shoveling quite by accident; that is, as a consequence of the colonization process or of small peripheral population effects. According to Crummett, in Europe crown curvature became a more important element than marginal ridges, while in Asia the marginal ridges were more prominent and the crowns straighter. When each came under selection to increase incisor strength in a space limited because of decreasing jaw size, European incisor crowns became

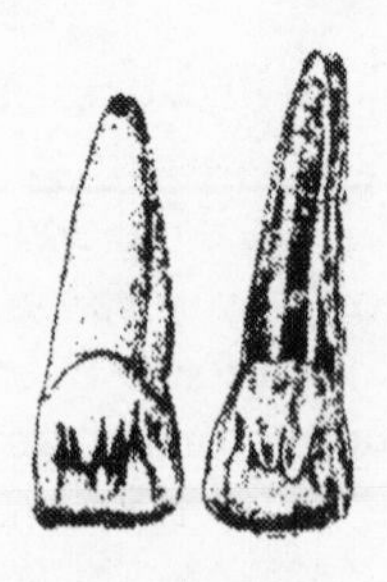

Upper central incisors from Zhoukoudian[54] (left) *and Vindija Cave[55] (Croatia), accomplishing shoveling in different ways. The Asian version, to the left, has a straight crown, the Aurignacian-associated European has a Neandertal-like curved crown.*

more curved and Asian crowns more heavily ridged. The contrasting exaptations are important markers of regional continuity in the two areas.

HOW DO YOU IDENTIFY REGIONAL CONTINUITY?

Regional continuity may be expressed in both adaptive features and nonadaptive ones, but the evidence used to establish the existence of continuity must be treated quite differently in these cases. In developing the Multiregional model, Milford, Alan, and Wu realized that long-lasting adaptations may persist because the need for them never changes. For instance, the Bergman/Allen rule predicts that cold-adapted humans need relatively short limbs to retain body heat. A succession of unrelated populations over a long period of time may retain relatively short limbs in a cold climate because of this adaptation alone. A succession of related populations may *also* retain short relative limb length both because of the requirements of adaptation and the influences of history. The problem is that one cannot be distinguished from the other, and therefore some clearly adaptive features may *reflect* an evolutionary continuity *but they cannot be used to prove it.*[56]

Nonetheless, continuities in adaptive characteristics are an expected product of the evolutionary process and may well be the most common form of regional continuity. They are often expressed as exaptations, which by their very nature depend on history as well as the adaptive process. The adaptations influenced by already existing morphology have the potential for equivalence—the same requirements may be met in different ways. For instance, in the examples already noted different nasal shapes may all function in adaptation to cold or dry conditions, and the various manifestations of upper incisor shovel shaping are all equivalent means of expanding incisor size when there is only limited space in the mouth.

When populations adapt differently they can become separate species, and these species can preserve different adaptations because they do not interbreed. Our problem is to understand how population differences can occur and persist within the same species. Alan Rogers describes the following model. Suppose, he suggests, we envision an adaptive landscape in the Pleistocene with several different adaptive valleys representing different stable adaptations. The idea of an adaptive landscape comes from the work of Sewall Wright,[57] who examined the consequences of the fact that it is individuals and not traits that survive to have greater or fewer numbers of surviving offspring. The genomes that evolve are therefore compromises between many pulls and pushes by the forces of evolution acting on different features. Furthermore, each of these features may itself be a compromise, as the gene pool of a population is affected by adaptations and adaptive strategies throughout each individual's entire life history, and what is advantageous at one age may not be at another. As these compromises change, the requirements of one adaptation can negatively influence the ability to perform another, and different individuals may have different strategies. Adaptations are not "perfect" ways of dealing with a given selective agent and many represent "trade-offs" to allow for other adaptations and the constraints of structural and genetic history.

Wright developed the concept of a genome that incorporated these compromises, expressed as sets of what might be thought of as co-adapted genes. He suggested the best of these compromises can be likened to a series of dips across a flat landscape that might be thought of as an adaptive plateau. (Actually Wright wrote of them as peaks, not

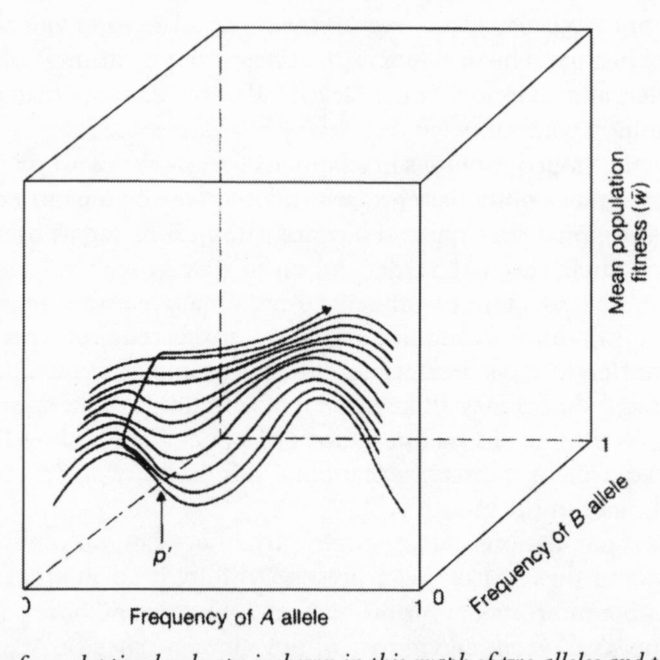

The idea of an adaptive landscape is shown in this graph of two alleles and their combined fitness (vertical axis).[58] *A local fitness peak, where a single allele such as A shown here seems most advantageous (for instance, at the high point of the closest curve, P^1, which is a graph of the fitness of allele A calculated without taking B into account), may be seen quite differently when the effects of advantages of another allele are taken into account. The peak value for the A allele is actually a local minimum value for it when both alleles are considered.*

dips, but the idea of stable configurations is better expressed by dips.) Species, or geographic races within them, have gene pools with frequencies that gravitate toward these stable adaptive valleys. The frequencies reflect effective genetic compromises for specific adaptive niches. Gene pools that are between these stable dips will quickly fall into one or another. He described the evolution of these gene pools in his shifting balance theory.

Each dip in the adaptive landscape represents a distinct adaptive configuration for the entire gene pool, and for the best adaptive compromise to change, to get from one dip to another, gradual change of individual features may not be enough. What is required is a more drastic genetic reorganization—what Ernst Mayr has called a genetic revolution. The process of speciation may provide the opportunity for such reorganizations, especially when it involves the isolation of small populations at the species' edge, where selection is strong because the habitat is least desirable, and where the potential for drift is greatest. Drift can play a particularly important role in this process because of its potential to unlink stable compromises in small populations, a prerequisite for new adaptive configurations to appear. As Wright envisaged the process, adaptive evolution without speciation could also readily respond to ecological changes in species with a subdivided internal architecture. Geographic divisions within a species can be stable as well and also thought of as inhabiting adaptive valleys, in a series of dips that can be spanned only with difficulty. These dips, though, are shallow compared with the adaptive valleys of species.

Rogers suggests how the same adaptive landscape model might help us envision evolution within our subdivided species. Within a species the adaptive topology has shallower dips than the adaptive valley occupied by the species as a whole. Within a species the valleys often have multiple low points, and the different dips are closer together. It is much easier to move between these different adaptive configurations, and they are neither especially stable nor necessarily long-lasting relative to the life of the species.

For example, Rogers suggests that perhaps 0.75 million years ago an Asian population became stuck in one valley, with upper incisors that have straight crowns and large marginal ridges, while a European population was stuck in another, with incisors with curved crowns and only moderate marginal ridges. Both crowns accomplish the same purpose and they can stay this way for a long time, because selection keeps pushing them down in, even in the face of genic exchanges, assuming it is important in both populations to have large incisors in small spaces. In Asia the anatomy persisted until today, but in Europe changing selection on the anterior teeth and the influence of populations entering the continent between glacial advances moved the incisor anatomy to another nearby dip, and by 20,000 years ago teeth with marginal ridges and curved crowns were virtually nonexistent.

Long-term equivalent exaptations like these incisors are sure markers of regional continuity and provide the best evidence to establish its presence because they can readily be maintained by a genic exchange-selection balance.

There is also regional continuity in nonadaptive features. Nonadaptive features can persist after they are established at high frequency, when no evolutionary forces act to change them significantly. Because of their nature, they are very unlikely to persist if the history of a region is marked by population replacements. One of the best arguments for a significant European Neandertal input into the gene pools of later Europeans is the persistence of nonadaptive traits.[59] Features such as the shape of the hole for the mandibular nerve to enter the mandible, or the presence of a small bump on the outer edge of the femur shaft, near its top, are difficult to explain any other way.

The mandibular foramen, for example, is an opening on the inside of the vertical part of the mandible for the branch of the mandibular nerve that reaches the teeth. This is the uncomfortable spot a dentist tries to reach with a nerve block for the mandibular teeth. In the H-O form the rim of the opening has an oval shape with the long axis of the oval oriented horizontally. Alternatively, in the normal form the rim

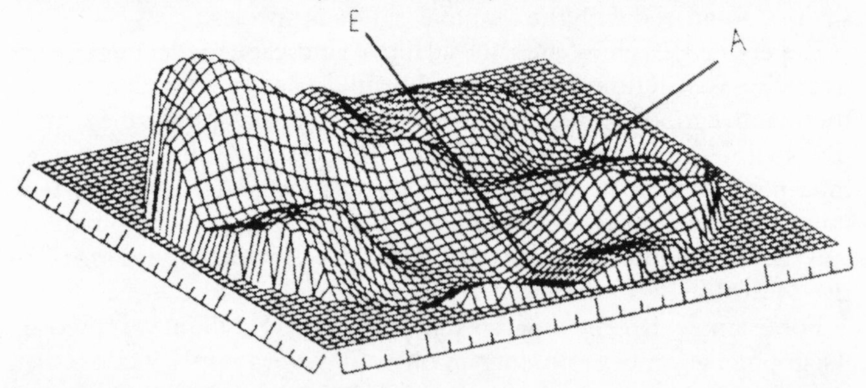

An adaptive landscape within a species has multiple shallow valleys of optimum adaptive compromises. This figure shows how two different I^1 anatomies, an Asian (A) and European (E) one, can be in different local adaptive optima, low spots on the adaptive landscape. Each represents a range of possible anatomical combinations and not a single "best" adaptation (the valley bottoms are broad). The optima are separated by less desirable adaptations that are not strongly disadvantageous (the ridges and peaks between them are not tall).

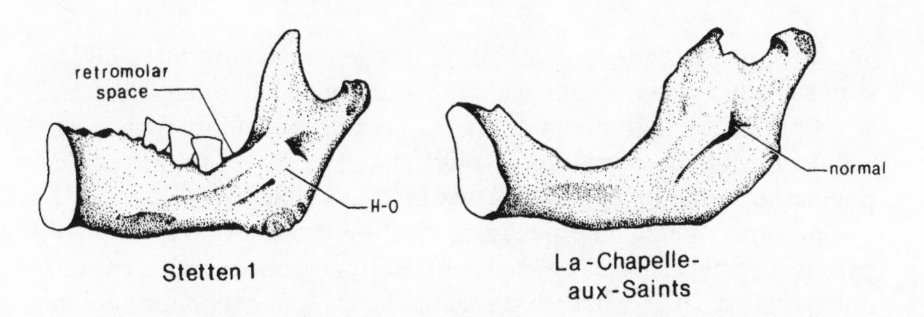

Normal mandibular foramen form in a Neandertal, La Chapelle, compared with the horizontal-oval form in a post-Neandertal European from the German site of Stetten. These specimens are used to emphasize the fact that H-O foramen forms, while common, are not the only form found in Neandertals. The anatomy persists in many of their successors. The retromolar space, a gap between the back of the third molar and the front of the ramus, has a similar distribution (it is in both samples), although it is by far the most common Neandertal anatomical variant.

MANDIBULAR FORAMEN FORM IN LATE PLEISTOCENE AND RECENT EUROPEANS[60]

European Sample	Horizontal-Oval Foramen Frequency (%)	Normal Foramen Frequency (%)
Neandertals	53	47
Early Upper Paleolithic	18	82
Late Upper Paleolithic	7	93
Mesolithic	2	98
Medieval	1	99

may be broken, along its lower border, by an unbridged vertical groove. The broken rim is the usual form in living populations.

The horizontal-oval mandibular foramen is virtually unique to European fossils. It is found in almost no other remains, including Late Pleistocene Africans and the Skhul/Qafzeh sample, the putative alternate ancestors of the post-Neandertal Europeans. But the horizontal-oval foramen has a significant frequency in the subsequent post-Neandertal populations of Europe and only decreases to rarity in recent Europeans. The exact form of the foramen opening is an example of non-adaptive equivalents. It is important that the foramen be there (the nerve must enter the mandibular body), but it makes absolutely no difference which form its rim has.

It is unlikely that any explanation but genetic continuity could account for this pattern in the mandibular foramen and other features. It strains the imagination and defies the laws of probability that parallel acquisition of the same nonadaptive structure by an unrelated later population could be a valid explanation (but see Stringer[61]).

One other point. The persistence of individual features over long periods of time reflects regional continuity, and observations of continuity are based on such features. However, evolutionary continuity can probably never be convincingly *demonstrated* that way, e.g., by examining single traits or features one at a time.[62] There is simply too much normal variation for such an approach to be valid, just as a forensic anthropologist needs more than a single feature to establish "racial" identity in a court of law, even when the identification is as broad as European, Asian, or African. It's that partitioning of genetic variation again: there is much more variation within human populations than between them. In fact, if single traits clearly reflected the evolutionary sequence in different regions, it would be a *disproof* of Multiregional evolution because it would suggest there was no network of genic exchanges linking populations. To the contrary, as far as the *demonstration*

Regional features are found in at least some fossil skulls. Here a Chinese specimen just over 100,000 years old (Maba, middle*) is compared with a just slightly younger Neandertal from Europe (La Chapelle,* above*) and an earlier Chinese skull from the 400,000-year-old Zhoukoudian site (skull 12,* below*). Maba is sometimes called a "Chinese Neandertal," but this comparison suggests that such an assessment is unreasonable. Maba's facial morphology is different from the European's; this can be seen in features such as the nose, which is smaller and less projecting, and the cheek, which faces forward rather than to the side. The significant resemblances are with the earlier ZKD specimen.*[63] *These Asians share a* combination *of features including similar rounded foreheads set behind projecting supraorbitals that curve over each orbit, a flat mid-face, and a nose with a low nasal angle.*

of regional continuity is concerned, combinations of several features must be examined to focus through the blurred picture that normal variation creates. This is a natural consequence of continued interconnections between the populations.

SOCIAL FACTORS

One explanation for long-lasting combinations of features could be long-lasting selection. But what could promote such long-term stability for selection? Certainly not climate, as in all but a few areas the Pleistocene is a period of dramatic climate changes. Specific culturally oriented adaptations are unlikely for long periods as well. We believe social factors, such as kin identification and mate choice, may provide local sources of selection for unique features that have the potential to produce long-term continuity.

Mate choice is tied to kin recognition in a complex way. *Within* our social species criteria for mate selection are defined by complex social issues, including relatedness.[64] Humans have elaborate mating "rules" that are largely based on interpopulational alliances and their underlying social basis in kinship. Physical characteristics, particularly facial features, can reinforce and underscore social elements of interpopulational relations. They may help make a socially acceptable mate "handsome" or "sexy" or otherwise more attractive. Of course, individuals persistently break marriage "rules" and frequently find mates that are socially prohibited. But facial features can underscore linguistic and other cultural factors that influence mate choice, and also may be indicators of social identity and important factors in kin recognition.

The role of kin recognition in internally subdivided species has never been carefully or systematically examined. Certain mechanisms of potential mate identification, balancing learned behavior and genetic predisposition to recognize key resemblances, have evolved to meet this problem in polytypic species where migrations and mate exchanges provide special opportunities for interactions with unrelated individuals. The complex nature of mate choice derives from the fact that picking mates based on simple similarity to one's self would maximize matings with sibs or other very close kin, as these are most likely to be similar. Yet ignoring any similarities could result in mate choices outside important related alliance groups. How a balance between these extremes is reached is suggested by P. Bateson's study of a simpler problem in Japanese quail.[65] These birds show a clear preference for first cousins, remaining in their proximity significantly more often than in proximity to birds with other degrees of relationship and to unrelated birds. In fact, the quails spent the *least* time in the proximity of their sibs. Time spent in proximity is directly related to mate choice in this species. Bateson posits that they are poised to prefer mates who are slightly different from the individuals they are familiar with in early life.

In humans there are more complex problems of interactions between related or potentially related individuals. Recognition of kin comes to play a double function in humans, where social systems of kinship and alliance are of significant importance, not only in mate choice but also in other interactions where shared genes could play a role. Behaviors that provide aid to related individuals can help one promote one's own genes, because relatives are more likely to share some of the same genes than unrelated individuals are. Mark Flynn suggests recognizing related individuals might be of particular consequence in peripheral populations, during periods when there were influxes of new people. Many scientists have proposed the idea that human social evolution is strongly influenced by the special behaviors of individuals toward their kin. Human society helps related individuals help each other.

Kin recognition in the quails is based on their plumage. In humans visual identification of individuals related by systems of kinship or alliance is largely based on cultural clues. But for reasons of technology, organization, and evolving complexity, cultural signals can be ambiguous. For much of human evolution, and to some extent even today, features of the face have helped provide this information as well. It is almost certainly not a coincidence that the face is where many of the externally visible features showing regional continuity are found. The role played by facial features that promote recognition of related individuals is therefore potentially important. Humans have neuroanatomical adaptations to help attain the maximum amount of information for such identifications. At the structural level of brain integration, the specific way in which the neurons are wired together, there are mechanisms that accomplish facial recognition that are quite complex and distinct. Information is extracted from views of faces by specific parts of the brain, mostly in the right hemisphere (the nondominant hemisphere for right-handed people). There are three important regions of the brain that are involved in the process of facial recognition. At the base of the visual cortex at the back of the head, one area processes the visual input and keys onto the unique features of the face. The second region, at the tip of the temporal lobe, stores information about individuals. Names are stored separately from other biographical information, which is why it is common to recognize individuals and associate them with particular times or places but not remember their names (the scourge of college professors!). A third area, positioned between these, is an association area that links the recognition and information storage regions. It is here that the question of "familiar or not?" is settled.

The neuroanatomy of this facial recognition system is unique to humans and seems to be the consequence of the kind of neural reorganization[66] that distinguished hominid brain structure from other primates, beginning early on in human evolution, well before *Homo sapiens*.[67] This is a neural processing system with unique importance in human evolution because its early appearance was one of the important adaptations that allowed successful colonization of regions outside of Africa[68] and the subsequent Multiregional pattern of human evolution.

GENETIC BASIS OF MULTIREGIONAL EVOLUTION

Humanity is an evolving subdivided species with geographically distinct populations. Let us examine some of the genetic consequences of populations occupying different dips in an adaptive landscape. Our widespread species, subdivided into local populations with some genic exchanges between them that disseminate advantageous alleles, has the potential to evolve rapidly when there are ecological changes. Whether the changes are local or worldwide, some of the populations might be able to respond to them easily because each has a somewhat different adaptive potential. A species like ours has a better chance of meeting environmental challenges than it would if all our populations were pretty much the same.

The differences between human populations are, in part, due to differing adaptations. But other evolutionary forces play roles, especially genetic drift. Fluctuating population sizes provide the opportunity for drift and create reoccurring local bottlenecks across much of the human species, limiting the amount of genetic variation (a bottleneck is when there is a brief period of intense selection or of very small population size through which only certain genes survive, and come to characterize the population). But this pattern does not apply to humanity uniformly, because for most of the time since people colonized lands outside of Africa, the majority of all people lived in Africa.

African populations, large and spread over a wide range of habitats between almost 40° north and 40° south of the equator, have been more variable than populations of other regions and have very much less drift and bottlenecking in their histories.

The main reason why geneticists developed the Eve theory is to explain two things: why human mtDNA and some nuclear DNA variation is so small, and why mtDNA and other genetic variation is higher in Africa than elsewhere. The two theories, Eve and Multiregional evolution, supply alternative explanations of the low level of human mtDNA variation and the pattern variation in certain other genetic systems.[69] When there is only a small amount of variation in a gene, when it appears in only one form or in just a few varieties, it could be the gene has a recent origin in the bottleneck of new species formation. During bottlenecks, variations are accidentally lost, and a population founded from just a few individuals needs time to reevolve genetic differences. Thus, low levels of genetic variation could mean a species founded from a small, isolated population just hasn't been around long enough for other variations to evolve as the result of mutations. But this is only one possible explanation, and is not accepted by all.

Ironically, the most widely accepted evidence against the Eve theory has not come from the fossil record, but from the ongoing genetic analysis and interpretations of population geneticists.[70] Moreover, genetic studies have provided considerable support for Multiregional evolution, verifying the validity of some of the assumptions the model requires and, perhaps surprisingly, confirming some of its predictions. These are reviewed in several publications; those by Francisco Ayala, John Relethford, and Alan Templeton[71] are particularly definitive and clear.

Analysis of genetic variation, especially in mtDNA, has established or implied the validity of three key elements of the Multiregional model:

1. Intermixture is a long-standing aspect of the human condition.
2. Prehistoric population sizes were dramatically larger in Africa (the center) than in the rest of the world, possibly by as much as an order of magnitude.
3. The history of mitochondrial DNA (or any other gene in the nuclear DNA) does *not* reflect population history; there were bottlenecks for many genetic systems, mitochondrial and nuclear, but different systems bottlenecked at different times.

LONG-STANDING INTERMIXTURE

While subdivision into local demes with some genic exchange is a good description of recent humanity, what evidence do we have to show it characterized our species in the past as the Multiregional model asserts? Or, to the contrary, does modern human genetic variation all derive from a single recent source, passing through a bottleneck? The key issue is whether internal subdivisions, with their genic exchanges and differing population size histories, can account for this pattern in which both differences and similarities persist for long periods.

There actually is genetic evidence reflecting the presence, and consequences, of ancient subdivisions and persistent intermixture between them. The same nuclear and mitochondrial variations are found over and over again, in the most widely separated populations. Eve theorists interpret this to mean that all of these genetic variations were present in a highly variable population that was the last common genealogical ancestor for humanity. But such an interpretation requires us to assume that once they divided, descendant populations were no longer in contact to cause these similarities. An alternative explanation is that the divided populations were not really separated, and the distribution of these genetic variations could be the result of persistent genic exchanges in the past.

In such a species, contact and interbreeding between populations allows parts of the genome to have an ancestry more recent than the first population of the species. The problem for scientists is to examine the

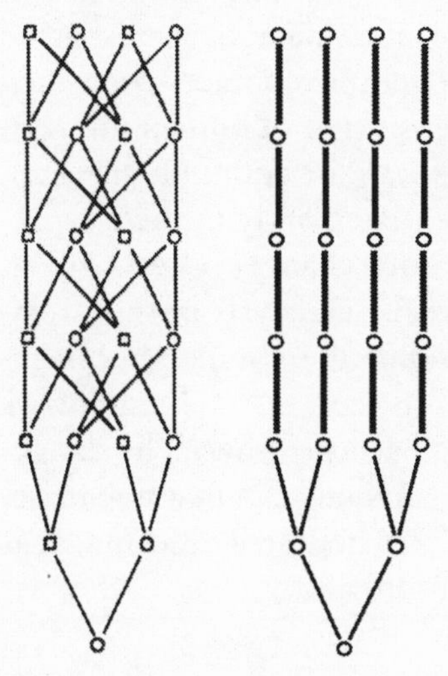

Two models of population evolution are shown in this comparison[72] of evolution within a subdivided species. To the left are populations within a species that diverge but then exchange genes and continue to remerge and divide. Common ancestry among them is a fuzzy, perhaps irrelevant concept. To the right are populations (in a different model these could be species) that subdivide and thereafter remain separated. For them the last common genetic ancestor is also the last common genealogical one.

question of whether variation in the last common ancestral population or interbreeding accounts for widespread genetic variations *without assuming the conclusions.*

Mitochondrial studies are especially useful for this because the evolution of different lineages can be traced. These lineages can never mix together and their individual histories could be independent of population history, just as the history of, say, a single Scottish name might be different from the history of the Scottish people. But because many lineages are found in the same population, their different histories may reveal facts about population history.

Alan Templeton[73] analyzed data on mtDNA distributions and showed that the worldwide distribution of genetic lineages could only be attributed to early genetic divergence between regions *and* continued subsequent genetic contacts between populations. The magnitudes of gene exchanges were controlled only by the distances between the populations, according to his analysis. Templeton showed the pattern of dispersion failed to reveal *either* a single recent source *or* an "Out-of-Africa" population expansion, and he concluded: "The geographical associations of mtDNA are statistically significantly incompatible with the out-of-Africa replacement hypothesis and instead strongly and clearly indicate that all Old World human populations were in genetic contact with one another throughout the entire time period marked by the coalescence of mtDNA."[74] The two elements here are early genetic divergence between regions, and subsequent genetic contacts between populations at magnitudes mitigated only by the distances between the populations. Templeton cautions his readers to remember that genes can spread without population replacement.

Similarly, L. Excoffier showed the more ancient variations in mtDNA were so widely spread that there must have been restricted genic exchanges from an early time. The genetic contacts continued for the entire time period marked by the evolution of human mtDNA variations, according to his studies, from their last common ancestor.

If we assume human populations differentiated early in their evolution and maintained significant genetic interchanges afterward, as these geneticists assert, it follows that a treelike analysis of their relationships is invalid. A tree, with constantly diverging branches, can accurately portray the relationships between species, but it doesn't work when the branches are constantly growing back together. The danger is that in today's world of microcomputers, anybody can fit a tree to genetic or anatomical data. The issue is not whether a tree fits, it is whether this representation of relationships is valid.

AFRICAN ORIGIN OR ANCIENT POPULATION SIZE DIFFERENCES?

The Eve theory depends on genetic evidence indicating an African origin for modern humanity, because as we will see in the following chapter, the fossil evidence is quite equivocal on this issue. But there is no particular reason to suggest Africa was the place of origin for the current mtDNA lineages. The original studies suggesting an African origin were invalid because the computer program used in the analysis was not applied correctly. Then the greater genetic variability in Africans was taken to mean humans evolved there longer: the variation was thought to reflect more mutations and therefore a longer time span for their accumulation. Templeton has argued that no statistical analysis shows that the genetic variation of Africans actually is greater than that of other populations, but even if it is there is another, more compelling explanation. Ancient population sizes expanded first and are larger in Africa than in other regions, which would have the same effect, creating more African variation.

Consider what happens in an expanding population. Average family size is greater than 2, and all variations have a good chance of being passed on; at least, if they are not selected against, they will probably not get lost by accident. But if a population is decreasing, drift can play a very active role since average family size is less than 2. The role of drift is greatly amplified for mtDNA, because for transmission of this molecule the number of *female* offspring is important. Decreasing populations stand an excellent chance of losing mtDNA lineages.

Now, consider a small but stable population, neither increasing nor decreasing. Here, existing variations may each occur in only one individual. If it is a woman, and she has no female offspring, which can happen one fourth of the time in a stable population, her unique variation is lost, her mtDNA lineage terminated. But in a large stable population with the same amount of genetic variability, it is likely each mtDNA variant is shared by many individuals. The odds are the same against one woman having no female offspring, but it is very unlikely all the women with a certain mtDNA variant will lack female offspring. It is much harder for mtDNA lines to end by accident in large populations.

These comparisons show that prehistoric population demography, how large or small populations were in the past and how much they fluctuated, can affect mtDNA evolution. Small, fluctuating populations will lose many mtDNA lines. The last common ancestor for the remaining mtDNA variations will be more recent because there is less remaining variation. Large or increasing populations will retain more variation. For them, the roots will be deeper and the last common ancestor will be farther in the past since more variations are retained.

Thus, ancient population size can dictate how long genetic lineages have existed, and therefore when they arose.[75] Because ancient population size differences are an alternate explanation that unlinks the origin of genetic lineages from the origin of a population, it seems as though genetic analysis cannot help solve the problem of whether today's variation reflects African origin or ancient population size differences.

However, genetic analysis can indicate ancient population expansions. Henry Harpending and colleagues studied the probability distributions of pairwise mtDNA comparisons within populations for evidence of past population structure and size expansions. They conclude: "Our results show human populations are derived from separate ancestral populations that were relatively isolated from each other before 50,000 years ago." These studies clearly reveal there have been a series of recent, very significant, population expansions. Some of these are without question associated with the development and spread of agricultural revolutions. But others are earlier.

This means population size history by itself can explain the pattern of mtDNA variation. If, as we believe and as the archaeological record seems to show,

- there were more people living in Africa for most of human prehistory, and
- human populations outside of Africa were smaller and fluctuated more because of the changing ice-age environments,

we would expect just what we do see—African mtDNA has deeper roots, while in other places the coalescent time is more recent. *But this explanation does not mean the* ***populations*** *living out of Africa have a more recent origin, or that they originated in Africa.* MtDNA history, in other words, is not population history.

John Relethford and Henry Harpending[76] examined the consequences of the possibility that greater African population size, and not greater time depth for modern humans in Africa, may account for their greater variation.

> Our results support our earlier contention that regional differences in population size can explain the genetic evidence pertaining to modern human origins. Our work thus far has involved examination of the clas-

> sic genetic markers and craniometrics, but it also has implications for mitochondrial DNA. The greater mtDNA diversity in Sub-Saharan African populations could also be a reflection of a larger long-term African population.[77]

Moreover, they write, a unique African ancestry implies there was a bottleneck for the human species, as moderns would be able to trace their ancestry to only a small portion of humanity, as it existed then.

> While this seems at first glance a reasonable notion, it soon becomes apparent that the actual effect of such a [bottleneck] event depends on both the magnitude and duration of a shift in population size. Rogers and Jorde[78] show that given reasonable parameters for our species, the bottleneck would have to be more severe and long-lasting than considered plausible. We have to think of a population of 50 females for 6,000 years, for example.[79]

BOTTLENECKS

It often seems to be assumed that the date of coalescence for a genetic system, the time when all current expressions of the system have a common ancestor, is also the time when the population (or more broadly the species) originated. The idea of coalescence is a consequence of the genealogical relations between genes; all of the extant varieties of a gene must have originated in a single gene. Looking from the present to the past, we can say variants of a gene descend from (or coalesce to) a single form. For gene coalescence to mark the beginning of the species the gene is found in, at the time of speciation the species must have passed through a bottleneck of small population size. But the widespread distribution of a number of rare variations in nuclear DNA shows there cannot have been a severe bottleneck for those genes or, as Xiong Weijun and colleagues suggest, even a significant population reduction. If there had been, these rare variations would have disappeared.

Xiong Weijun and colleagues[80] examined the distribution of several very rare alleles in the nuclear DNA. These were found in two individuals, one of Japanese and the other of Euro-Venezuelan origin. Assuming the rare variations were derived from a single mutant ancestor, their complexity indicates a time of origin of at least a half million years ago, but their persistence since then at a very low frequency shows there was no population bottleneck over that period; in other words, no single, more recent origin for both populations.

According to Francisco Ayala,[81] some human genes have structures so similar to chimpanzees' that the human and chimpanzee forms must be descendants of little-changed genes in the common ancestor. They diverged, when the species separated, some 6 million years ago. For instance, humans and chimpanzees share many common genetic variations for the major histocompatibility complex genes, an important part of the immune system. Ayala calculates that with divergence times at the age of chimpanzee-human divergence, certain histocompatibility complex genes such as the human leukocyte antigen could not have passed through bottlenecks of less than 100,000 copies—not much of a bottleneck, and certainly not the kind envisioned by the Eve theorists.

In a related study, Li Wenhsiung and L. Sadler[82] compared nucleotide and protein diversity in humans and *Drosophila.* The levels of protein diversity are quite similar, but nucleotide diversity is much lower in humans. They attribute this difference to a small but stable population size through most of human prehistory, rather than to a bottleneck. Their reasoning is because humans and chimpanzees share many common alleles for the major histocompatability complex genes, a severe bottleneck long after the hominid-chimpanzee split would have eliminated most of these shared polymorphisms. A long period of small population sizes is a better explanation.

These genetic studies must be considered along with the mtDNA and Y chromosome analyses indicating low levels of variation that might reflect a recent bottleneck. If these interpretations are correct, it means some genetic systems went through bottlenecks while others didn't. This would be an impossible finding if all human *populations* had gone through a recent bottleneck, but in fact all genetic systems do not have the same history, and therefore the history of the individual genetic systems is not population history—remember, if there *was* a common recent origin for all populations because a new species appeared, all of the genetic systems should reflect this and *there could be none that did not pass through a bottleneck*. But in fact this is what the genetic data, in aggregate, show did not happen. Some gene systems, for instance, mtDNA and certain segments of the Y chromosome, have gone through bottlenecks, but others have not. What this evidence means is that mitochondrial DNA evolution, indeed the evolution of other genetic systems, is not the same as population evolution. The "Eve" of mtDNA is just that, the last common ancestor of mtDNA lines, and *not* the last common ancestor of the humans who carry them.

DATES—THE "RED HERRINGS" OF THE MODERN HUMAN ORIGINS DEBATE

The dates published for the age of mitochondrial Eve are astonishingly variable. Even the Eve theorists Mark Stoneking and Rebecca Cann could be no more specific than an order of magnitude, estimating[83] an age of 50,000 to 500,000 years for when she lived. The Y chromosome analyses[84] have been little better. For instance, in one late 1995 issue of the journal *Nature* there are back-to-back articles on the time of origin of extant Y chromosomes.[85] One of them gives an age of between 37,000 and 49,000 years. The other estimates 188,000 years with a 95 percent confidence interval of 51,000–411,000 years. We are not very confident.

Nevertheless, geneticists like Stoneking have argued that the date when mtDNA coalesces to a last common ancestor, date of mitochondrial Eve, is critical for the Multiregional model, because, he believes, a recent date would disprove it. However, the problem in interpreting the coalescence times as times of origin comes from three false assumptions:

1. Gene coalescence is a regular process of mutation accumulation in neutral systems, and therefore can be timed like a regularly ticking clock with an acceptable range of error.
2. Human populations were isolated from each other after they originated.
3. The history of particular gene systems is the history of the populations they are found in (specifically, Eve's age is humanity's age).

The date of mtDNA coalescence is, in fact, critical for the Eve theory, but as Templeton points out, it is irrelevant for the Multiregional model. *Only if human groups were isolated after Eve's time would her age be of importance.* The finding that human populations were connected by low levels of genic exchanges means any age for Eve could be compatible with Multiregional evolution because her mtDNA type could potentially spread throughout the world at any time. The different bottlenecks in various gene systems, and the absence of any bottlenecks in some, show that while Eve was the mother of all mitochondria, she could hardly have been the mother of humanity.

POPULATION HISTORIES OR GENE HISTORIES?

Multiregional evolution is quite compatible with the notion that mtDNA history is not the same as population history; implying a recent common origin for all living people is not the explanation of genetic variation. Besides the expectation of small population sizes and recurrent bottlenecking in the past, it predicts different genetic systems will have different histories, for if there was a common recent origin for all humanity, each genetic system would have the same history. From the perspective of Multiregional evolution, the key issue is whether internal subdivisions, with their genic exchanges and differing population size histories, can account for the pattern in which both differences and similarities persist for long periods.

Population history is quite distinct from the histories of the genetic systems, but the genetic analyses can and do tell us something about population histories:

1. For a long period of time, geographically disperse human populations have been exchanging genes with each other.
2. There has been no recent bottleneck for all of humanity—modern humans *do not* uniquely descend from a single recent source.

The research implies there were more people in Africa than everywhere else combined, for most of human evolution. Perhaps this was the roughly three times that of any other area Relethford and Harpending suggest, perhaps more, but the relation of Africa, the center, with the more peripheral populations on other continents has turned out to be much as the center and edge hypothesis proposed.

AT WAR?

Multiregionalists are often described as being locked in battle with the molecular biologists[86]—the dirt-covered field paleoanthropologists *versus* the white-coated biologists. This portrays the relationship between these scientists quite incorrectly. In fact they are linked, not separated, by this very problem, of evolutionary pattern and the question of relationships. What links us to our laboratory colleagues is the refutatory nature of science.

When evolutionary models and phylogenetic trees are developed from the study of past and living organisms, these are not facts but hypotheses we have made about the course of evolution. We paleoanthropologists studying human evolution turn to a plethora of sources to develop these hypotheses, beginning with the fossils themselves but also including their descendants and the relationships their study suggests. When dealing with fossils, the hypothesis of relationship is based on morphology—the physical character of the bones and teeth themselves, their size and shape, and the specific details of their construction. Several different techniques are used to construct phylogenetic hypotheses from morphology. Some techniques look at all characters when assessing relationships; others are more selective, only using those that are unique for a group of species. Some techniques weigh characters by giving them more importance in the analysis; some give special credence to functional relationships between characters, so a foot adapted to bipedal walking is not considered as different evidence from a pelvis adapted to bipedal walking. Detailed discussion of these techniques, that can be broadly classified as either phenetic (using all features) or cladistic (using unique features) is unnecessary here. What is important is the recognition that none of these techniques is foolproof, especially when closely related groups like humans, their ancestors, and human relatives are concerned. That's why they give us hypotheses and not certainties.

Genetic analysis is perhaps the only *independent* way we have to test the hypotheses that come from the anatomical studies. The underlying assumption made in using anatomy to determine relationship brings us back to genetics, because we must assume the variation in the characters we study actually does reflect the relationships between the different populations that have the characteristics. This assumption implies the anatomical characteristics can be used as proxies for genes. Genetic variation can be examined directly in the living but is not a direct source of evidence about past populations. It is an alternative source of information about evolutionary pattern and relationships. The fact is such genetic studies can be used to try to refute hypotheses based on the anatomy of past and living groups, *and vice versa*. As each attempts refutation of hypotheses generated by the other, there certainly *appears* to be a war. Refutations are rarely welcomed, and it is always more desirable to figure out why you are right, in spite of evidence to the contrary, than to admit being wrong.[87] Science is, after all, a social activity. But different sources of information must, in the end, tell the same story because they each reflect the same historic processes. The process of reconciling them through refutations is the road to progress, and at the moment it would appear that the theory of the replacement of all human populations by the descendants of a single one can be refuted. But that doesn't mean there never have been replacements. Ask the Lakota Sioux about this. Ask Aboriginal Indigenous Tasmanians. And ask Neandertals too?

NEANDERTALS AND MODERNITY

The trellis of interconnected populations provides the framework for human evolution. Its details, of course, are another matter. Regional continuity is but a minor aspect of this evolutionary history; the important changes are those with species-wide distribution, the changes that enormously improved humanity's intellectual complexity and adaptive potential, and resulted in the human populations of today. These changes almost certainly took place just as Dobzhansky[88] suggested, as advantageous genes or gene combinations could spread widely through the complex mesh of interconnected populations because selection promoted them. At some point, perhaps quite early in *Homo sapiens'* evolution, common elements of selection following from the spread of ideas and behavioral innovations became important as well. For sure, exchanges of ideas also mean exchanges of genes, but the ideas, and their implications for the magnitude and direction of evolutionary change, have the potential of spreading much more quickly. Sharing these important, advantageous genetic variations creates a common vector for evolutionary change worldwide, without disrupting the pattern of population differences.

The specific aspect of the Multiregional model that has proven to be most contended, and perhaps most contentious, involves the issue of modern human origins. It provides a good example for illustrating how we believe the evolutionary process works in our species. Multiregional evolution could mean that modern humans do not have a single point or place of origin. Instead, they may have appeared because of the coalescence of rapidly spreading adaptive traits that originated at different times and in different places, each conferring advantage on those who had them. Modernization, in other words, might have been an ongoing process in widespread populations, and not a single dramatic event in a single population. Of course, some populations may have spread widely because of anatomical or behavioral advantages they accrued, mixing with some of the populations they encountered and replacing others. This is part of any model of genetic interconnections, and must be a normal part of any model of human evolution that uses the present to interpret the past. It is only the extreme case of complete, universal replacement without mixture that the Multiregional model denies took place. Modern humans did not all descend uniquely from one population that replaced the others. As Weidenreich said, there was no single Adam or Eve.

So if there is anything to be gleaned about modernity from the Multiregional model it is that modern humans do not have a single source. Modernity is not a "thing," a set of features that linked together because they were spread by a single population. But it's not that modern humans have many sources, rather that modern *features* do. What links modern features together is how they contribute to successful adaptations, and this, with the differing genetic histories of populations from place to place, explains why modernity is not the same everywhere.

This is why it has proved impossible to provide an acceptable definition of modernity. Repeated attempts at a definition based on skeletal variation[89] have failed because when they were applied to skeletal samples, it was found they did not include all recent or living people.[90] How could this happen? It comes back to the importance of Neandertals, because these anatomical definitions of modernity cast them in the role of "other." These definitions are based on the assumption Neandertals were *not* modern humans. Indeed, they were constructed to exclude Neandertals. However, when the definitions were applied to populations around the world, it was quickly discovered that significant numbers of Holocene and recently living Aboriginal Indigenous Australians were not "modern." This problem, of course, is not with the Aboriginal Indigenous Australians who are each and every bit as modern as the authors of the definitions, but with the definitions themselves and their focus on Neandertals.

And so the Neandertals have been key players in many respects. Our evolutionary biologist from outer space could return home with a host of reasons why they have been important in understanding human evolution, far beyond their numbers or their role in grandmothering Europeans. In fact they have played many roles, biological, intellectual, and popular, as Europe's contribution to the Multiregional pattern of human evolution.

11

MODERN HUMANS, MODERN RACES?

NOTHING IS WORSE for the self-esteem of an evolutionary biologist than to be considered a typologist. Typological thinking is the opposite of the populational approach and therefore anti-Darwinian; to be a typologist is to be a counterrevolutionary! The implications of typological thinking are even worse for *human* biologists, since not only is it anti-Darwinian, it is socially irresponsible. The commonly held concept of race is a typological construct. And the study of human variation (read: race) has for most of the history of anthropology been an exercise in typology. Nowadays, even though most anthropologists have abandoned the type concept of race, the legacy of racism in our field lies only just below the skin; prick it and you get a reaction!

One afternoon I was sitting with our son, Benjamin, then just a few weeks old, in the audience of one of Milford's talks in which he was demonstrating regional continuity in the Australasian fossil record. We were in Capetown, South Africa, in the summer of 1989, to study the Klasies River Mouth Cave specimens, several handfuls of fragments that some paleoanthropologists believe are the earliest modern humans known. It was just days before the election of DeKlerk, and racial issues were on everyone's mind. In the middle of the talk a friend and colleague leaned over to me and said, "This really sounds like nineteenth-century typology!" He meant it. I was worried and even a little hurt. How could Milford be perceived as a typologist, and worse yet, by extension, a racist? I struggled to rationalize my friend's perception; it *had* to be that his own preoccupation with the racial politics of his country was coloring his view of Milford's presentation—that's all. Surely an unbiased audience would be able to see that Milford was merely showing the evolutionary continuity over time of features within a geographic region, not claiming that these features were how races were defined, and certainly not that they corresponded to any kind of long-term racial reality!

Deep inside we know there really is no unbiased audience. If some well-trained scientists see demonstrations of regional continuity as manifestations of the old type concept of race, how will other members of the public and scientific communities react?[1] No one on the planet is removed from the sociopolitical contexts of race—and what about those who actually *want* to hear voices from science to support their ideologies? Yet the task is far more daunting than recognizing and dealing with racial prejudices. Intricately connected to racial constructs are even more basic scientific ideas about the meaning of variation, and these are what directly underlie different views on modern human origins.

RACE AND TYPE

Race is a factor in the recent modern human origins debates but not, perhaps, for the obvious reasons. Yes, sociopolitical contexts affect scientists, consciously and unconsciously,[2] and influence the construction and acceptance of theories. But the primary cause of differences in human origins theories would seem to lie in science, not politics. We're quite sure that none of the major protagonists in the human origins debates has a racist agenda, although, as we've seen in previous chapters, this could not be said in the past. But they do have different views about how science should be done, different views of how past species or species extending over time should be defined and interpreted, and ultimately and most important, different views of the meaning of variation, all seemingly unrelated to racial politics. Yet, inadvertently, these scientific concepts bring us back full circle to race, because different human origins theories result from different underlying concepts of variation—the same ones that help to form scientific perceptions of race. Therefore different human origins theories are arrived at using the same conceptions of variation that help define race. This then ties back to race in its sociopolitical context: modern human origins theories, once constructed, directly address ideas of racial histories—how different geographic groups of people were related, and how they interrelated over the years. Origins theories clearly have sociopolitical implications, and it would be naive to think their discussions have been taking place in a sociopolitical vacuum.

All biologists deal with variation—it is at the heart of Darwinian thinking—and to some, understanding the importance of continuous variation within populations comes very easily. We consider some of our friends "intuitive Darwinians": these people are usually sloppy. Their offices are a mess, with desks stacked high with piles of papers that seem to have no order; pens, pencils, dissecting tools, calipers, toothbrushes, and candy canes are crammed indiscriminately into old coffee cups; and in the corner on the coat rack, tripods, sample bags, and pterodactyl models are hanging with the lab coats. Overcoats are tossed over available chairs. There is nowhere to sit. If you are lucky enough to find the biologist you are looking for in all the mess, she might apologize for the rubble, and say in a sheepish way how she really should be cleaning her office presently. But she really doesn't care deep down inside. She knows where everything is, more or less.

We also know excellent biologists who are neat and tidy and extremely well organized, but we think they have to work harder. They have to overcome the urge to organize and classify their subjects and thoughts about their subjects in the same way that they organize the rest of their lives. They are not intuitive Darwinians. The point has been made by biological historians such as Ernst Mayr that the largest impediment to an understanding of Darwin was a world view based on Platonic essentialist thinking that is diametrically opposed to the sort of probabilistic *populational thinking* based on the importance of *individual variation* that underlies natural selection.[3] Selection (whether it is acting as a "pruner" or as a "creative" source of variation, developing new phenotypes and coadapted genomes as new combinations emerge from shifting gene frequencies) can only act on variation present in the population, and the amount and distribution of that variation is both a major influence on, and a product of, the evolutionary process. It is this limitation that links all evolutionary change to history.

The essentialist legacy often prevents us from accepting this variation as important. It stems from the world view, espoused by Plato and intrinsic to Western thinking since then, that the world is made up of distinctly different things that are reflections of pure, fixed ideals, or essences. Variation is deviation, and deviations from the ideal type are imperfections or flaws. Dog breeding reflects this philosophy. There is, for example, a breed standard for the Golden Retriever, although many well-bred dogs will deviate from it in some way. The "imperfections" of these members of the breed are considered unimportant in describing what the breed is like; i.e., the range of variation is unimportant in depicting the breed: only the ideal type is described. The typologist or essentialist (or Golden Retriever breeder) focuses on the essence of a category or population and ignores the deviants from that essence as unimportant to the character of the category. In contrast, a populational thinker looks at all the variation, all the deviance from the norm or average, as intrinsic to the character of the category. The populational focus is on diversity within groups, while the typologist seeks homogeneity and ways to unify and standardize categories. Now the dog breeder with her typological focus still recognizes variation and its implications and may prevent her "imperfect" pups from breeding, seeking to "purify" the line. She recognizes a need to limit deviations to keep her focus on the type. It's not that she doesn't know variation is important, but she doesn't think it's important to understanding the category itself.

Typological thinking is a part of our cultural heritage, a part of our mind-set; it is the way most of us organize the world. And even if we "know" not to apply it to biology, it seeps in anyway. Therefore, *intuitive* understandings of populations and the random nature of selection are foreign to many of us who must consciously avoid the classifiers' tendency to ignore the fuzzy edges or further divide the edges into separate categories. If biological variation is "overpackaged" in this way, the reality of the subject is obfuscated, since it is variation in and of itself that is of the utmost importance—it is not just there to be parceled this way and that. The amount of variation and its distribution in populations, races, and higher taxa are the key to evolutionary biology. Variation is the product of evolutionary histories, the result of natural selection and other evolutionary forces, and is the material on which current and future evolutionary processes act. When variation is classified away, an understanding of evolutionary processes and histories can never be reached. Therefore it is not surprising that different perceptions of the meaning of variation and of what constitutes its packaging in race and species determine how the fossil record is deciphered by human paleontologists to support or reject various human origins theories.

KLASIES RIVER MOUTH CAVE

In the summer of 1989 we had the opportunity to study the Klasies River Mouth remains, from the southern Cape region of South Africa, one of the most controversial skeletal samples in the human fossil record. Fortune had smiled upon us. Anticipating a new baby in July, we applied for no research funds that year and had no field work commitments. We were model parents, prepared to devote ourselves entirely to the new drain on our resources. Nevertheless, prepared as we were for a quiet domestic summer, we were both very happy to receive Andy's phone call. Our friend and colleague, Andy Sillen of the University of Capetown, was going on leave in the middle of the term to shoot a film on human evolution for the South African Broadcasting System, and wondered if Milford would cover the course he was presenting and teach for him, starting in August. We could go, new baby and all.

Now there were many things about the offer that made it attractive: we would be leaving the heat behind us to enter a late South African winter; we had the use of Andy's most interesting car, a 1960 Volvo; we were offered a wonderful house to live in for our visit; we would get to know the archaeology and anatomy departments at Capetown and all of their exceptional personnel; and, of course, we were visiting one of the most beautiful regions of the world. It was also an extremely exciting time to be visiting South Africa, politically speaking. New elections were to be held in October; P. W. Botha had been behaving mysteriously; there were rumors that he was ill; and the black townships in and around the city were tense and volatile. The desperately needed change was in the air, but no one was sure what the future would bring. All in all, the invitation to South Africa was a godsend. But we were most excited because it would give us the opportunity to study the Klasies River Mouth remains, one of the most controversial skeletal samples in the human fossil record.

The Klasies remains are important because they constitute perhaps the strongest fossil evidence for a *recent* African origin for modern humans and controversial because they are very fragmentary and incomplete and therefore difficult to interpret. The recent African origin theory has received a lot of attention in recent years and is in many ways the antithesis of Multiregional explanations for modernity.[4] Versions of a single recent origin idea have been a part of paleoanthropological debate for a long time but took on new life as it became better understood that new species virtually always arise through a process of allopatric speciation.

A small, usually peripheral population may be prevented from interbreeding with other populations of a species when it becomes isolated. If the isolation is long enough for behavioral or biological obstacles to interfertility to emerge, a new species is formed. Many single origin theories for modern humans, therefore, posit that modern humans are a new species that began as a small geographically isolated population. Accordingly only a small portion of the human fossil record represents the progenitors of modern humanity, the others having not contributed to modern folk because they or their descendants were replaced by them. Today's recent African origin theory was dubbed the Eve theory[5] because the last common mtDNA ancestor was carried by a woman, mtDNA descent being along matrilines, and it was assumed that the mitochondrial Eve was a populational Eve. Eve was African because the greatest mtDNA variation was found in Africa, and this was taken to mean modern humans lived there longest.[6]

Many older approaches to modern human origins also recognized modern humans as a new species, but most of their proponents understood speciation very differently from the modern evolutionist. They described fossil species as samples that are as different from living species as living species are from each other, and recognized anagenesis (the transformation of one species into another without splitting, within a single clade or lineage) as a valid form of speciation. For the most part these workers looked at species transformations in the fossil record as part of worldwide phases, so that as *Homo erectus* became *Homo neanderthalensis* became *Homo sapiens*, each transformation was direct, without branching. Some paleoanthropologists of the past century, like Schwalbe, recognized a fundamental similarity between the stages idea and some branching models. But for him branching was seen as another way of achieving stages, not as a source of diversity. In the stages of human evolution approach, the vertical component of evolution, change through time, was emphasized, while diversification in human evolution was overlooked. At some arbitrary place in a temporal continuum, often defined by gaps in the fossil record that were assumed to be the result of bad luck for the fossil hunters, a new name was coined for the fossils and a new species was defined.

In contrast, some earlier paleoanthropologists regarded modern humans as a new species arising from a single source and replacing other populations. This assessment, as discussed in the last chapter, was treated as worldwide, but it actually focused largely on Neandertals and reflected the conviction that Neandertals could play no part in European ancestry. It was not necessarily a statement against anagenesis.

These views still have a profound influence on the human origins issue.

After the modern synthesis, the biological species concept, emphasizing reproductive isolation and variation, was gradually incorporated into the thinking of students of human evolution, and slowly some workers started to look at the evolution of modern humans as a speciation event, the result of branching evolution, and one that required the separation of one fossil population from the rest. These ideas gained momentum in the mid-1970s, with the initial discussions of punctuated equilibrium[7] and the ensuing debates. But the biological species concept was meant for living populations, and its application to species with time depth or to fossil samples was always problematic because interbreeding could only be inferred, and indirectly at that. For most paleontologists the focus shifted to questions of how biological species concepts could be applied to their data sets. The development of species definitions based on genealogy[8] provided answers, and in the past decade several different approaches to the species concept acceptable to paleontologists have developed and come into use.[9] They have even spread to paleoanthropological studies.[10] Here, as always, the question was first and foremost about the Neandertal issue; the new species, modern humans, evolved elsewhere, entered Europe and replaced Neandertals as the residents of Europe. It was assumed that the same process occurred to "Neandertals" everywhere, Neandertals having become synonymous with "archaic humans."

However, in spite of the success of punctuated equilibrium theory and the genealogical species concepts, the idea that modern humans are a new species was accepted by only a minority of paleoanthropologists, even among those who embraced a replacement explanation for their spread. This is because when modern humans are interpreted as a subspecies[11] rather than as a species, they could have replaced indigenous populations by interbreeding with them and genetically swamping them out.[12] Explanations similar to this were far more common—an obvious application of European colonial history to the ancient past. In this interpretation modern humans were an entity and had a single source, but their distinction was not that of a species but of a highly successful population. Until the first substantial mtDNA analyses inspired the Eve theory, this was the most common replacement view. However, in order to incorporate the mtDNA data as evidence reflecting *populational* history,[13] there had to be a single recent origin of modern humans, and Eve was at its beginning. Thus, it was argued, mitochondrial Eve was at the origin of, or directly ancestral to, a new species.[14] So after the significance of the mtDNA publications began to seep into paleoanthropology, the theory that considered modern humans as a new branch of the family tree reappeared, and later human evolution was described within a punctuated evolutionary framework.

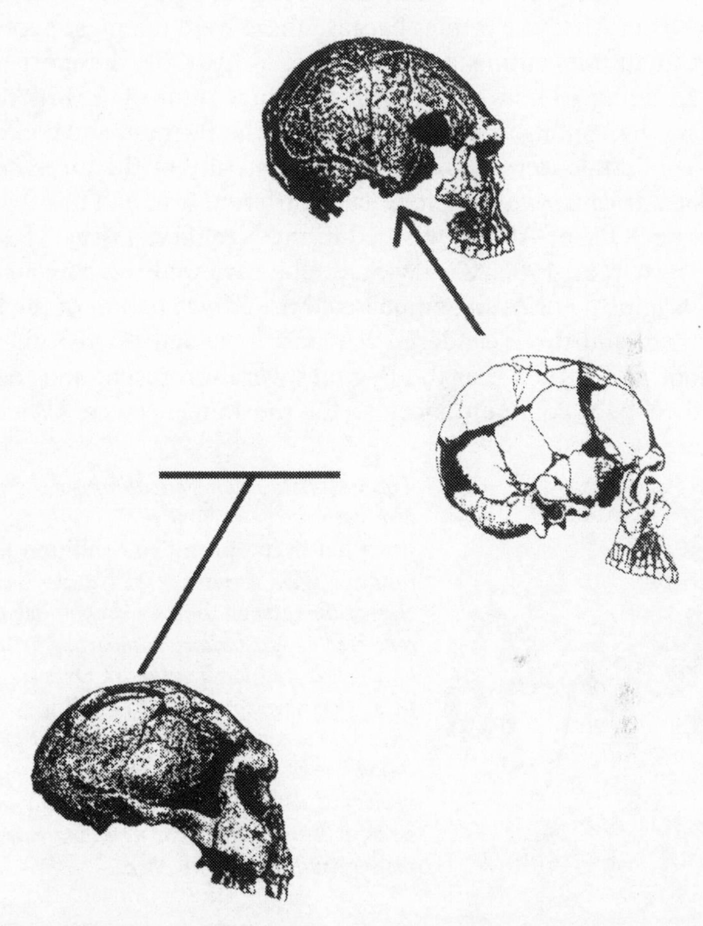

The "Eve" theory means that no archaic populations except for African ones were ancestral to modern ones. It means that recent Australian natives such as Kow Swamp 1, above, *must be the unique descendants of the earliest moderns such as Qafzeh 9, and cannot be related to earlier Australasians such as Sangiran 17* (below) *despite their similarities, as this figure[15] shows. Multiregional evolution posits both Qafzeh 9 and Sangiran 17 as potential ancestors of Kow Swamp 1, arguing that elements of special continuity between the two Australasians show that complete replacements by moderns could not have occurred in the region.*

OUT OF AFRICA

The Eve theory, of course, is not the only "Out-of-Africa" contention—it is just the most extreme. The idea that modern humans have a recent origin in Africa is an old one, occurring well before the modern synthesis. As we noted in Chapter 8, in his 1905 essay on the origin of humanity, Julien Kollmann argued that the human races were derived from African Pygmies because these were the most neotenous of all human populations and therefore, in his view, the most primitive.[16] In doing so he was proposing the first "Out-of-Africa" theory. Kollmann had published in some detail on the Pygmies and contended that Neandertals were more divergent than any of the races derived from the Africans, and therefore had a different origin. Thus from the beginning "Out-of-Africa" was tied to the Neandertal issue.

The next year, 1906, Gustav Schwalbe reviewed the current competing origins theories and emphasized the *human* nature of the fossils from Trinil and the Neandertal remains.[17] He addressed Kollmann's assertions and argued that the Pygmies were too recent and too specialized to be a common ancestor for the human races. Whether a

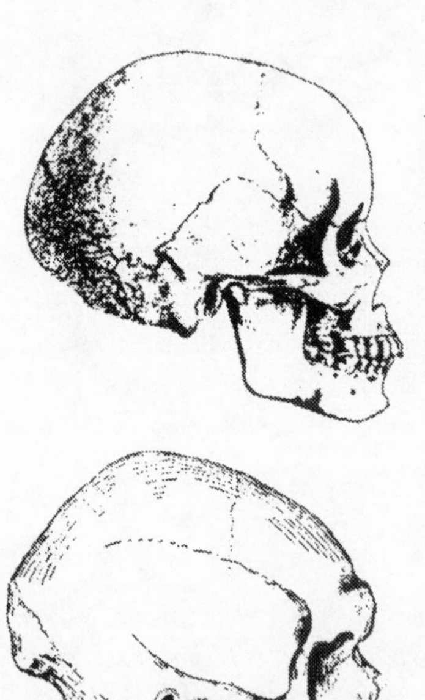

The issue European paleoanthropologists grappled with at the turn of the century and before was what is the primitive condition for humanity? *Were primitives the Native Aboriginal Australians,* (seen below),[18] *and (what were conceived of as) similar Neandertals, with large browridges and low foreheads? Or were primitives the neotenous Pygmies* (above)? *It all seemed to depend on your theory about how evolution works. Of course, whether we use the genealogical definition of species, or just common sense, it is now clear that no living race can be more primitive than another.*

stages or a branching interpretation of modern human origins was accepted, Schwalbe contended that fossil hominids were either the direct precursors of modern humanity or their closest relatives.

The debate continued to evolve through this century, incorporating ideas developed from the modern synthesis, but a specific Out-of-Africa theory did not come back into focus until recently. When William W. Howells summarized the competing modern human origins theories 70 years after Schwalbe,[19] the same fundamental question remained—did modern humans derive (in his words) "from more archaic hominids, already present in the same regions," or "from a single origin"? In his 1976 paper he presented human origins models and their supporting evidence as the dichotomy between single point of origin and parallelism, two models that ignored gene flow. He realized "the two hypotheses are irreconcilable." Although clearly favoring the recent origin model, Howells concluded "present evidence does not disprove either."[20] At the time he felt that the time and place of origin could not be settled, although just a year earlier (and too late to be referenced in Howells' paper), Reiner Protsch published the first of the current Out-of-Africa models, ironically in a different issue of the same journal.[21]

Protsch developed the model that is fundamental to all Out-of-Africa theories. If all modern humans are uniquely descended from Africans, we can expect three predictions to hold up:

1. The earliest modern humans should be found in Africa.
2. Only in Africa should there be transitional fossils leading to modern humans.
3. The earliest modern humans outside of Africa should share unique, presumably African, features that reflect their common origin.

Protsch examined evidence for the first of these and argued that the earliest modern humans were African. In his scheme, only an infant from the South African site of Border Cave was directly dated[22] to an age as old as Neandertals, but this was enough. Because he regarded Border Cave as representing a more modern population, Protsch concluded Africans evolved first and all living races descended from them.

Peter Beaumont and colleagues developed additional details for the African origin theory, also based on Border Cave, arguing that the spread of modern humans could be linked to their development of modern technology.[23] They accepted early date estimates for other

African specimens said to be modern, such as Omo Kibbish from Ethiopia and Klasies River Mouth Cave. Günter Bräuer derived his African origin theory from Protsch's work,[24] adding to the list of ancient "moderns" from south of the Sahara and addressing other predictions of the Out-of-Africa theory—again in terms of the Neandertal issue.

In his first presentations of the idea, Ngaloba (Laetoli 18), Omo Kibbish, and Florisbad were discussed as a transitional group, borderline between archaic and modern anatomy (over the years the composition of this group has changed). Bräuer envisioned the origin of modern humans as a process with two elements: a single African source, and the hybridization and replacement of archaic populations with the expanding moderns. He gathered support for all three predictions of the theory, not only finding evidence of moderns and near-moderns (i.e., transitional specimens) in Africa but claiming that African features can be found in the earliest moderns elsewhere. He claimed, "The ancestors of the inhabitants of Europe and western Asia of some 30,000 years B.P. consisted of modern Africans with some admixture of Neandertals."[25]

However, in both prior and subsequent analyses of the same post-Neandertal European material, many other scholars failed to find convincing evidence to support such a proposition.[26] In fact, these Europeans[27] and other early moderns such as the Australasian and Chinese remains lack most of the cranial and facial features used to characterize living or fossil Africans.[28] Limb proportions, especially leg length, in the descendants of these post-Neandertal folk, are said to suggest a warm climate origin according to one recent paper and a humid climate origin according to another,[29] but as we discussed in the last chapter, we think they may just as reasonably reflect a pattern of cultural mobility that came to differ from the Neandertals.

In their development of the paleoanthropological side of the Eve theory, Chris Stringer and Peter Andrews took the strongest Out-of-Africa position. Their theory proposes

> Africa as the probable continent of origin of *Homo sapiens* [meaning modern *Homo sapiens*], with an origin for the species during the early part of the Late Pleistocene, followed by an initiation of African regional differentiation, subsequent radiation from Africa, and final establishment of modern regional characteristics outside of Africa.[30]

The main elements of this Out-of-Africa theory are

1. The African origin
2. African regional divergence
3. Subsequent spread and the founding of other regional features

The order is important because of the clear prediction that the earliest moderns in other regions should have some shared African features if the theory is true. And whether or not it is true remains the topic of an ongoing lively scientific and public debate.

ARCHAEOLOGICAL ARGUMENTS

The fundamental issue addressed to the African archaeological record is whether modern behaviors begin there: can they be found in Africa first? This is an important implication for any Out-of-Africa theory because one would expect a highly successful group of Africans spreading around the inhabited world and replacing successful indigenous peoples everywhere to leave a record of the elements in Africans' culture (technology, organization, etc.) that reflected the advantages that let them do so.[31]

But archaeological data used to support Out-of-Africa have proved to be more inconclusive than not.[32] Several archaeologists have provided worldwide summaries of archaeological change that support the notion that important technological developments happen first in Africa. The Berkeley archaeologist J. D. Clark argues[33] that the earliest modern humans can be associated with ecologically specialized tool kits first appearing in Africa some 200,000 years ago. These tool kits are characterized as Middle Stone Age, the Middle Paleolithic of Africa. The Middle Stone Age industry is similar to the Mousterian tools[34] that are associated with archaic humans from Europe (exclusively Neandertals) and western Asia (Neandertals and others). However, Clark admits that "the fact that the technology and behavior of the first modern humans in Africa do not appear to be all that different from those of the Neanderthals, poses a major problem" to his theorizations.

The problem is that few human remains are actually found directly associated with the Middle Stone Age's *earlier* manifestations. The oldest of the Middle Stone Age associated "moderns" are the much later Klasies specimens, which are only half the age of the earlier Middle Stone Age. In fact the "modern" or "near-modern" people from the

African Middle Stone Age are not particularly earlier, or more modern, than other similarly "modern" or "near-modern" people described from Indonesia and both east and west Asia. Furthermore Africans dated earlier than the roughly 100,000-year-old Klasies "moderns" such as Ngaloba and Florisbad[35] are notably more archaic. While we do not know who made the *earliest* Middle Stone Age materials in Africa, it is almost certainly populations similar to Ngaloba and Florisbad.

The African Middle Stone Age is as complex as Middle Paleolithic variations in other places, if not more so. Some of its distinct qualities include

1. Grindstones for the preparation of plant foods
2. Use of marine resources such as shellfish and seashells transported over 100 km
3. Use of bone including barbed bone points
4. The hafting of the spear and other projectile points

Each of these elements is short-lived and narrowly distributed. They did not prevail, as did similar attributes that persisted and spread much later in time. They reflect the marked regional differences within Africa: early variants from the southernmost part feature long blades and woodworking tools such as burins, central African sites combine Levallois-based tools and more traditional Acheulean bifaces, north Africa is more dominated by Levallois-based technologies. In east Africa the Kenyan site of Baringo has a Middle Stone Age layer just be-

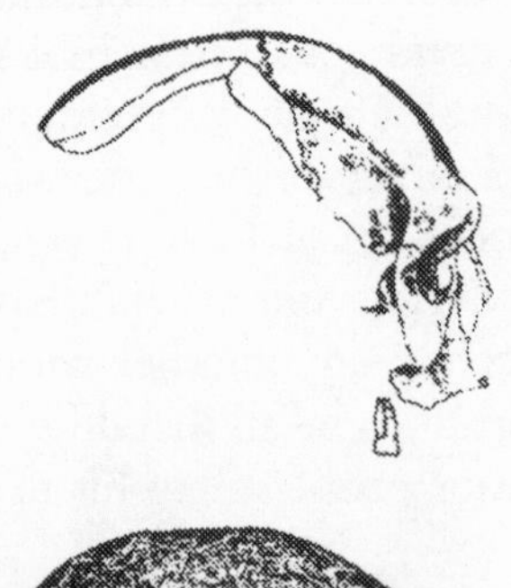

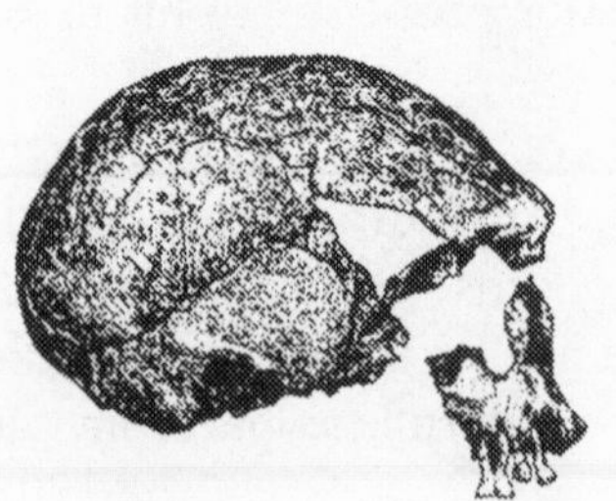

The earlier Middle Stone Age African cranium from Florisbad (above) *and the Ngaloba cranium from Laetoli.*[36] *These remains are notably more archaic than the so-called earliest moderns from the region; while Bräuer*[37] *regards them as "transitional," other paleoanthropologists such as Rightmire*[38] *argue that transitions to modernity are difficult to establish in the African Middle Stone Age skeletal material.*

low a volcanic tuff dated to 250,000 years ago, with long, thin blades that were struck from a preshaped core (similar blades are reported from the lowest Mousterian in the Hayonim Cave in Israel). Blades like these have historically been taken as a marker of modern human behavior when they are found much more recently, in the Late Pleistocene, although in principle they are not different from other flakes taken off of prepared cores, the hallmark of Middle Paleolithic industries everywhere.

Other examples are the bone artifacts and some articles of personal adornment, such as ostrich eggshell beads, appearing in the African Middle Stone Age; perhaps the best evidence of these is from several Zaire localities at Katanda, along the Semliki River near where it flows into Lake Albert.[39] There, barbed bone points and grindstones were found, dated to about 90,000 years (although with a probable error range of almost 25 percent). If the dates are verified and actually relate to the artifacts, these grindstones and bone points are much earlier than anywhere else,[40] which could imply important technological developments that perhaps mark the road to early modernity. However, these materials will probably remain controversial as long as they are unique, and until the question of whether the artifacts have the same age as the deposits is resolved. Ironically it is an Eve theorist, Richard Klein, who may be their most vocal critic, quipping, "They stand out like a big, sore thumb."[41]

There could well be a dating or association problem, but these discoveries may also mean that there is marked regionalization and little cultural contact among the Middle Stone Age Africans, according to John Yellen. Another African archaeologist, Larry Robbins, adds: "This is nothing compared to the sophistication seen in Europe at 35,000 to 40,000 years ago. . . . What you have at Katanda is one part of that suite of things that indicates modern human behavior—not necessarily the whole package."[42] Not to confuse the story, we would just like to add that there is every reason to believe those 35,000- to 40,000-year-old Europeans that Robbins mentions are Neandertals.

Africa may differ from other areas, but if it does so it is in the extent of its marked regionalization. The aforementioned technologies are local—none spread throughout the entire Middle Stone Age range—and on the whole they do not seem to reflect particularly more progressive behaviors. These and other similarities to much later industries and technologies are short-lived and disappear, hardly the pattern we would expect if they were heralding a new, superior, pattern of behavior. As Steven Kuhn puts it, "If the blades really do provide early glimmerings of the modern mental condition . . . it would be more comforting if they had spread, instead of popping up and disappearing."[43] The fact is that it is difficult to find any continuities between these artifacts and the European ones appearing much later, in the Upper Paleolithic where indirect percussion is used to strike blades from prismatic-shaped cores. These blades are produced in a more sophisticated way than those at Baringo.

The increasing regionalism is indeed important, but we believe for a rather different reason. We consider it a direct consequence of the increasing population densities that mismatch analysis of mtDNA (see Chapter 10) reveals to have been earliest in Africa. With more population interactions there are more isolating mechanisms and a far greater potential for regionalization than other more marginal, lower-density areas developed.

Richard Klein developed an Out-of-Africa model[44] to account for some considerably later changes, the so-called human revolution of post-Neandertal Europe.[45] Yet although it has nothing directly to do with the Eve theory, it addresses the behavioral advantages of modern humans and the question of whether there are fundamental distinctions for them. He argues that only in Late Pleistocene Africa can the appearance of modern behavior be linked to modern form. He believes the underlying cause for all of the important Late Pleistocene developments—the major transformation in human behavior that he proposes occurred some 40,000 years ago—"reflects the last in a long series of biologically based advances in human mental and cognitive abilities." Fossil and genetic evidence, he continues, "suggest that the critical neural change occurred in Africa." So now we are back to biology as an explanation, but a seemingly untestable one.

A similar archaeological model for modern human origins proposed by another Eve theorist, Philip Allsworth-Jones,[46] is narrowly confined to explaining the African evidence. He argues that modernity of the fossils from Klasies and Border Cave show that the transition happened in Africa first. The variable mix of modern and archaic features, especially at Klasies, is explained as the natural consequence of finding a transitional population, but so certain is Allsworth-Jones of their modernity that the main issue he raises is not whether the hominids from these sites are modern but whether they are more similar to Negroes or to San. Citing Klein, he attributes the success of these moderns to "neurological change leading to syntactic language." But evidence of syntactic language is not found in their archaeological leavings, and in fact the dissociation of modern anatomy and behavior has plagued all Out-of-Africa theories, including these archaeological expressions. From the southern Cape to the Levant (often a part of

Africa, at least as far as its fauna are concerned), true behavioral modernity of the sort addressed by the Klein and Allsworth-Jones models is much later than what is regarded as anatomical modernity.

THE EARLIEST MODERNS?

The question of whether the earliest moderns are found in sub-Saharan Africa is beset by two issues: the very problematic dating of the two allegedly early sites with more complete specimens, Border Cave and Omo Kibish, and the question of what it actually means to be modern, exacerbated by the fragmentary nature of specimens from the third site, Klasies River Mouth Cave.

At Border Cave, well-preserved remains have been found that are very modern in appearance.[47] However, despite the best efforts of archaeologists such as Beaumont, they are unprovenienced,[48] and chemical analysis suggests they are probably quite recent. Perhaps the situation at Omo-Kibbish is marginally better, but if so not by much. From their condition and fossilization it seems quite likely that the Omo specimens are Pleistocene remains, but as one Out-of-Africa supporter describes Omo: "It is unfortunate that the situation at Omo leaves quite a lot to be desired in terms of the contextual associations of the three hominid specimens found there."[49] In fact, the three Omo specimens were found at different locations and may not even be roughly contemporary. The positions of the specimens cannot be clearly related to the single date that was determined for site,[50] and the dating technique used was recognized as unacceptable, even by the scientists who applied it in 1967. There has never been a reanalysis or redating because the site can no longer be located. To regard these specimens as "dated" in any sense is to make a travesty of the serious attempts under way to determine dates for the human fossil record,[51] but without dates their importance for understanding modern human origins is muted.[52]

The Klasies sample is the exception to the provenience and dating problems surrounding most Late Pleistocene African material, but unfortunately the specimens from this site are far more fragmentary and their morphological interpretation is more problematic than others. Their morphology has often been interpreted in the context of the more complete specimens from Border Cave and Omo, as Fred Smith notes.[53] The critical parts not represented in Klasies fragments, after all, are there to be seen at Omo or Border Cave, so traces that *might* reflect modernity at Klasies are interpreted to *actually* indicate it because they are found in the other early African moderns. The difficulties with this reasoning are evident.

The Klasies River Mouth site consists of a number of caves and shelters along the southern Cape coast of South Africa, on the Indian Ocean.[54] The main site at Klasies consists of five caves and shelters, identified as 1, 1a, 1b, 1c, and 2, that have provided evidence of Middle Stone Age occupation. Of these all but shelter 1b represent the same depositional sequences—the same layers are represented at these places and they can be stratigraphically connected to each other. Most of the human fragments are from cave 1 and shelter 1a, though the most famous and modern-appearing mandible is from shelter 1b. Two stratigraphic levels have yielded human remains in the appropriate time period.[55] In summer, 1989, we came to Capetown at a very exciting time. The archeometry laboratory at the University of Capetown, part of the archaeology department, had just published a series of papers that established the antiquity of Klasies using a variety of different methods, all providing the same results. Louis Binford, a respected archaeologist and paleoanthropologist, had been very critical of the initial dates and associations from the site,[56] but this new information showed that one level with humans yielded dates in excess of 90,000 years. These dates were independently confirmed by oxygen isotope studies, amino acid dates, ESR dates, and several microfaunal analyses. The close temporal relationship of KRM 41815, the shelter 1b mandible, to the rest of the Middle Stone Age hominids, an initial concern of ours, was also well established on archaeological and faunal grounds, and ESR dates confirmed it. Therefore the evidence was very strong that all of the Klasies main site hominids, including the mandible from shelter 1b, were old, dating to about 100,000 years ago.

It certainly seemed to us as though the fossils from Klasies River Mouth were the best dated of the specimens said to represent the earliest "anatomically modern" *Homo sapiens*, and they are therefore a critical sample in the evaluation of the Out-of-Africa model.[57] However, while the Klasies material is often referred to in the literature as modern,[58] we knew from experience that modernity means different things in different places. There had been no systematic comparisons with Holocene South African skeletal material, and thus its modernity was, in our minds, unconfirmed. Andy's fortuitous invitation allowed us the opportunity to conduct such an analysis.

On arriving at the National Museum in Capetown, our first reaction to the Klasies collection was surprise, mostly because it was obvious that the specimens were extremely variable and also because there were so few of them. It was easy to see how this fragmentary material had been interpreted in the context of Border Cave (when there was still a possibility that Border Cave could be ancient). There was very little of it there, and the temptation was strong to relate the fragments to the anatomy of the more complete specimen.

Virtually everyone working on these remains has claimed that the Klasies people were anatomically modern. These assessments were made on the basis of features considered diagnostic of modern humans. In spite of the difficulties and contradictions involved in the definition of modern humans discussed in the last chapter and below, there is a widespread feeling that certain anatomical features are characteristic of modern humans. Brain size is no longer an issue (archaic humans had brains that were every bit as large as our own) but cranial shape is important (the possession of a "high, rounded" forehead is usually considered a modern trait[59]). Other traits considered modern include small browridges divided into central and outer portions, generally small facial size and reduced muscularity, small teeth, the presence of a chin, and gracile limb bones.[60] For years we had seen pictures and casts of selected Klasies head parts: one with tiny dentition, another without any browridge, and mandibles of various sizes, at least one with a chin; the best preserved mandible, from cave 1b, has a well-defined chin that makes it look modern. Milford had worked on this mandible and several of the other specimens some years before when Ronald Singer at the University of Chicago lent him his Klasies collection just before it was returned to South Africa. Others had seen the material or casts of it as well, and Alan Mann warned us that from his examination of KRM 41815, its chin may be more apparent than real. This mandible had lost its front teeth during life. When this happens, the bone surrounding the missing tooth roots will begin to dissolve slowly (this is called resorption, and happens to people with periodontal disease as well: it's why their teeth fall out if the disease is untreated). The resorption here was quite evident, he told us, and as the top of the symphysis (the front of the jaw) receded, it made the bottom appear more prominent and chinlike.

Of equal importance, experience has shown us, again and again, that features regarded as diagnostic of modern humans are best considered regionally, since geographic variation can confound assessments of temporal trends. For example, Neandertals have much larger browridges than living Europeans, and they are always continuously developed across the forehead. A significant number of recent and living Indigenous Aboriginal Australians have large, continuously developed browridges. Does this make them more primitive than

Europeans? Does this make the Neandertals modern? Or does it show that modernity must be considered regionally? We think the latter.

Many of the features considered diagnostic of modern humans everywhere reflect increases in gracilization (reduced size and muscularity). Specimens in one area of the world that exhibit more gracile regional features because of a diminution of body size will appear more modern than their contemporaries elsewhere. They, of course, are not any more modern than gracile present-day populations are more modern than robust ones, but it is easy to confuse regional and temporal characteristics. When comparisons ignore these regional signals, they can produce bizarre results. For instance, the cheekbones of Neandertals are more gracile than those of some Eskimos, and their browridges are thinner and less developed than those of some recent Australians. Are Neandertals therefore modern humans? To avoid similar problematic conclusions we felt it important to compare the Klasies sample with a Holocene sample from the *same* region, just as French Neandertals are continually compared with modern French. After all, the archaeologist excavating at Klasies, Hilary Deacon, had suggested that the recent Khoisan populations from southernmost Africa were their direct descendants.[61]

Finding an appropriate sample was easier said than done, however. First, the region in question may be quite circumscribed. There is some evidence that the populations indigenous to the southern Cape are morphologically different from other African populations, but Deacon's suggestion that the ancestors of the Khoisan have great antiquity in the area and may even be represented by the Klasies remains

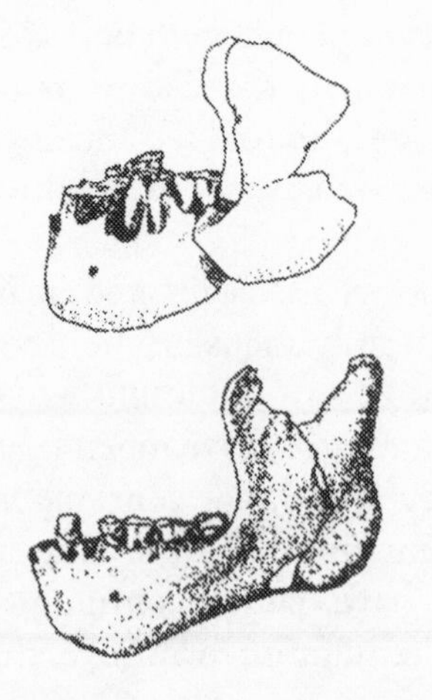

KRM 41815, the Klasies shelter 1b specimen showing the maximum chin development (above), *compared with another mandible attributed to an early modern African, the approximately 35,000-year-old Dar es Soltane 2 Moroccan.*[62] *This much variation seems to have been normal in Late Pleistocene "early modern" specimens and makes it critical to identify modernity from the sample and not just the most modern elements in it.*

is based on archaeological and not anatomical evidence. Workers such as our friend Alan Morris have been quite disparaging about the possibility of finding the necessary anatomical signs.[63] Yet, Deacon's evidence includes data that suggest the coastal and nearby populations of southern Africa remained localized, with very little migration into and out of the region until quite recently. Therefore, to err on the side of caution we decided to use only local specimens, from the southern Cape.

But we quickly realized that modern, present-day or historical people from the southern Cape would be a poor comparative sample, most obviously since there has been so much admixture in the past few centuries with other Africans, Europeans, and Asians (especially Indonesians, many of whom were brought to the Cape by Dutch settlers, leaving a dramatic mark not only on gene pools but on South African cuisine). Furthermore, even if one makes the precarious assumptions that the modern Khoisan are the descendants of the Pleistocene South Africans and that there has been little admixture between Khoisan and Bantu speakers,[64] there are potential difficulties in using a recent San comparative sample. Many anatomical and archaeological collections are problematic since they are typologically based.[65] In the past, South African San and Bantu skeletons have usually been identified on the basis of morphology (i.e., what the classifier thought that the morphology represented, which of course was based on preconceived ideas), leading to continued circularity. Milford came to realize this early in his African studies, a decade before. He was once working on the Florisbad cranium, housed in the museum at Bloemfontein. He noticed shelves of modern crania and was interested in using them for comparisons, but there was something funny about their labels. One cabinet was labeled "Bushmen," another "Bantu," and a third "mixed." He asked how the "mixed" were identified, expecting to hear something about their gravegoods or burial position, or even information on gravestones. But no, what he heard was that all came from a single graveyard. The physical anthropologist working in the museum then saw that some were "obviously" Bantu, others Bushmen. Others, though, were confusing and difficult to identify as one or the other. These were the mixed.

Skeletal remains of the local ethnic groups have not been consistently identified with any reliability, and their actual ranges of morphological variation are far from clear. Fortunately, however, many of the collections at the archaeology department had been radiometrically dated as part of a recently completed Ph.D. dissertation by Judy Seeley. Judy very kindly gave us her unpublished data, and using her dates, we were able to isolate an appropriate mixed sex sample. The other source of information we used was the dated material from the nearby anatomy department, then being published by Alan Morris.[66] We only used specimens for our comparisons that were reliably dated to prehistoric periods and included all the specimens we could find with such dates. Very few specimens in the university's collections fell into the appropriate temporal range. It took us a couple of weeks to isolate the comparative sample. The sample included the oldest dated skeletal material from the southern Cape, other than Klasies itself. Most of the later remains have been radiocarbon-dated to between 1,000 and 9,000 years ago.

The Klasies remains are very fragmentary, and there is no compelling evidence that any of the remains belong to the same individual: each fragment probably represents a different person. Yet, in spite of its fragmentary nature we can go beyond individual pieces of isolated anatomy and address the issue of its sample characteristics. The fragments most frequently discussed as modern are one or two of the mandibles (there are five of them, four with the chin region preserved) and the frontal bone. The rest of the tiny sample is usually ignored in discussions of modernity, but we felt it should not be. After preliminary analysis, our initial observations were confirmed and quantified, and we reached the primary conclusions that the Klasies sample was surprisingly variable, particularly in terms of size, but the *sample* was not modern in the sense that it was not much like our comparative sample.

The problem was that while some individuals appeared modern, others looked archaic. For instance, while the supraorbital region of the sole frontal lacks browridges and fell well within the modern range, a facial size of archaic proportions is indicated by the distance between its eye sockets and verified by the size and shape of an isolated zygomatic bone. These elements are as large or larger than most Middle Pleistocene Africans, some more than twice as old.[67] Anatomical details are archaic and unlike our recent Cape sample, such as the thickness and triangular cross section of the outer orbital border[68] and two of the four mandibular symphyses lack chins and a third has only a weak mental trigone.[70] Some individual specimens fell into the mod-

Mandibles from Klasies River Mouth Cave;[69] some are reversed. From the left these are KRM 41815 (the specimen with a chin), 13400, 21776, and 14695.

ern range for anywhere between a few and all of the features that we analyzed. But when compared *as a sample* to our prehistoric modern Cape inhabitants, it was clear to us that the Klasies sample on the whole was not modern. A few modern features do not make a modern *people*, because if they did, the earliest appearance of modern features anywhere would mark the origin of modern humanity.

WHAT DOES IT MEAN TO BE MODERN?

So, what does the variability of the Klasies sample tell us about modern human origins? If a couple of individuals within the sample have some of what are considered modern features, does this provide evidence that the sample is modern? The proponents of the Out-of-Africa theory think so, as they must, because for them modernity happened all at once and in only one way. We don't, and the question is why do we interpret the same variation so differently? In a sense, this is the crux of the matter—and it is linked to how people view variation and preconceived ideas of whether modern humans represent a new species. Do we focus on individuals or on the sample as a whole? We believe it can only be validly argued that the first appearance of modern features indicates the first appearance of modern humans *if it is assumed that modern humans represent a distinct entity*, a new species with a unique set of features. If so, one would expect that the earliest appearance of the modern features shows that the new species is there: only a new species, with its reproductive isolation, could have truly unique features.

But if modernity did not arise because of a new species, if its origins are more disperse and gradual, we might envisage a sort of cone of modernity, with its apex stuck into the past at the point of appearance of the first modern feature. The base of the cone is in the present, encompassing everyone, and the more ancient time slices produce smaller and smaller samples of modern features.[71] Is the first feature that appears the point of origin of modern humans? We think not, for the same reason that people with more gracile traits living today are not more modern than people with more robust ones. Such a cone makes it seem as if modernity has a single origin, and our perspective from the present makes it appear first somewhere. But this is an artifact of the view, not a reflection of the evolutionary process. It is best to view evolution from the past, to understand the origins of the present, and not *vice versa*.

We can hardly make this point strongly enough. Our colleague Erik Trinkaus wrote that modernity need not be restricted to the features of recent humanity, since

> the earliest representatives of modern humans will necessarily not possess all of the features which can be used to distinguish recent human samples from archaic groups, since there has been significant evolution in (at least) skeletal robusticity since early modern humans emerged in the late Pleistocene.[72]

His contention can only be valid when there is an expectation that modernity emerges gradually. However, if modern humans were a new species that could spread and successfully replace aboriginal natives *because of the advantages its modernity conferred*, Trinkaus would be wrong. Modernity would appear, away from its place of origin, as a package of features, linked together, because it was not the advantages of the individual features that allowed them to spread but the superiority of the (modern human) population they were in. But, on the other hand, if single origin is the wrong theory, we have no particular reason to call the Klasies sample modern.

We can't have it both ways. If the Klasies sample is interpreted to be modern because of a few modern features, it would have to mean moderns were evolving there, at the southern Cape, and their even more modern descendants would subsequently spread around the world, bringing a package of modern features with them. We should not expect to find transitional samples with modern features in other regions if the single-source interpretation is correct. We should especially not expect to find other samples in which the few modern features are *different* from the Klasies ones. In other words, treating the variation at Klasies as indicating modernity because of a few modern features *presupposes an answer to the very question we are examining*, the issue of whether modern humans represent a new distinct entity. We cannot begin with the assumption that modern humans are a new species when that is the question we wish to answer.

The few so-called modern features we identified at Klasies are continuous ones[73] that have variable expressions in both modern and fossil humans; there was considerable past variation in so-called modern features throughout the world, just as there is today, and like today most features appeared just about everywhere, though at greatly different frequencies. The question is, under these circumstances do we have any particular reason to call Klasies modern, at least any more modern than its contemporaries in other regions of the world? The meaning of the variation at Klasies must be assessed in the context of variation in contemporary populations from other regions.

THE WORLD 100,000 YEARS AGO

Looking around the world as it was at about 100,000 years ago, variable "transitional" samples are not at all unique to South Africa. In western Asia the Skhul and Qafzeh remains may be dated to about the

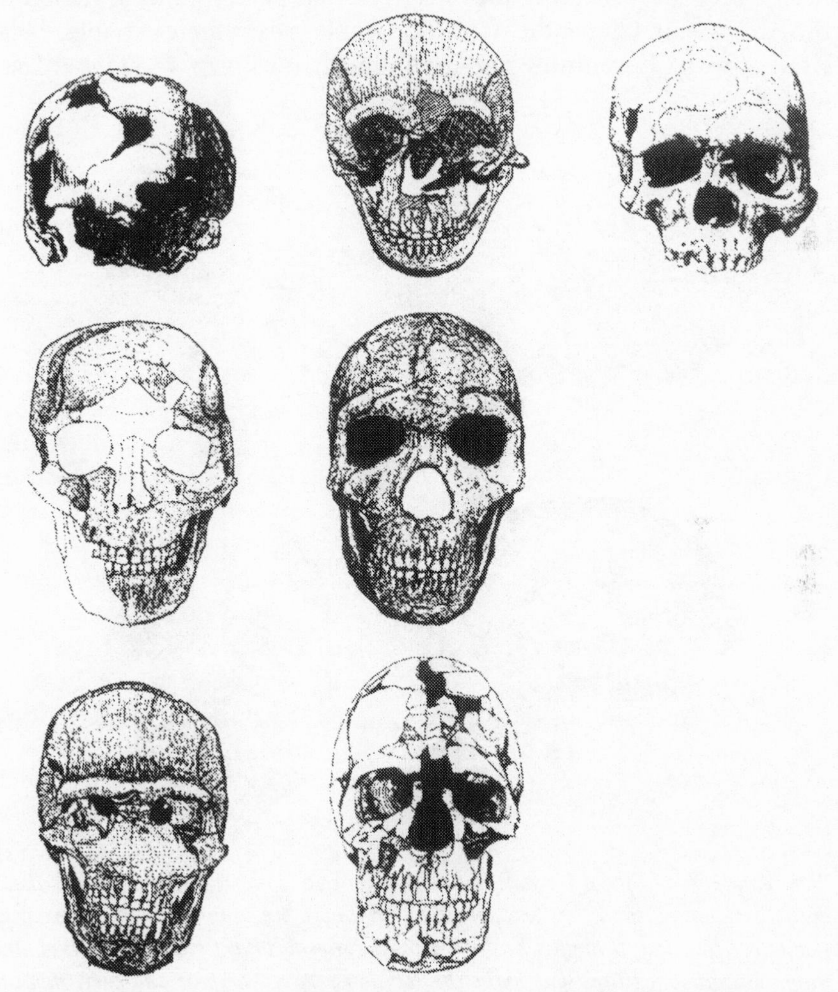

Facial views of Late Pleistocene crania from western Asia.[74] From the top row, from left to right, *these are Qafzeh 3, Tabun 1, Qafzeh 6*, second row, *Skhul 4, Amud 1*, third row, *Skhul 5, and Qafzeh 9. Two of these are considered Neandertals.*

same age as Klasies and are thought by many to represent modern humans. Some of these Levant specimens have high rounded foreheads, reduced browridges, and more linear builds postcranially. Other specimens, however, have crania that appear more archaic, even specifically more Neandertal-like (these are not necessarily the same) in these elements, and some of the postcranial skeletons are astonishingly robust. Like Klasies, there is a mix of archaic and modern features in a single sample.[75]

Moreover, there are early moderns found Out-of-Africa at the same age as or possibly even earlier than Klasies, and not just in western Asia where they are (probably incorrectly) often perceived as a branch of the Africans. In China, the Xujiayao sample is extremely variable,[76] like Klasies or the Levantines; and the Jinniushan woman, thought at least

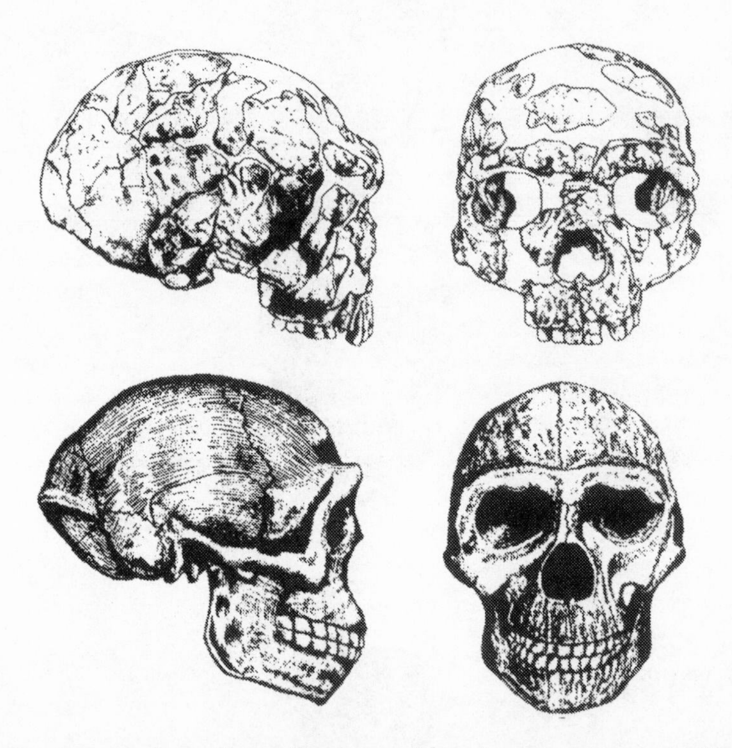

Two female crania from China, Jinniushan (above) *compared with the Zhoukoudian 11 reconstruction,*[77] *its probable ancestor. There are similarities in the orientation of the face and its angle, the nasal root, the supraorbital form, sagittal keel, and numerous other features that make the relation between archaic and early modern crania in China easy to perceive. Jinniushan differs in its much larger cranial capacity, dramatically thinner cranial bone, much reduced supraorbital region (and the absence of other cranial tori).*

by some workers to be about this age[78] and considered even older by others, is clearly transitional between earlier samples such as Zhoukoudian and recent or modern Chinese.[79]

Jinniushan is no less modern human than approximately contemporary Levantine samples.[80] It is the relatively complete cranium and partial skeleton of a woman who demonstrates the earliest appearance of a very thin, large and globular cranial vault, vertically short, broad face, sculpted zygomatic bones, and maxillary notch—all aspects of modern craniofacial morphology.

Similarly, the Late Pleistocene Australian cranium WLH 50, a modern human on behavioral grounds if no other,[81] shows marked and detailed similarities to the Indonesians from Ngandong. These links across an ocean barrier imply that the Ngandong folk may be transitional as well, between the earlier Sangiran folk and the later Aboriginal Native Australians. This early Australian is a modern human,[82] of course, despite features that appear "archaic." Modernity in Australia involves long sloping foreheads and large browridges, along with the significant brain size expansions and other changes that occur in all areas. What these comparisons clearly demonstrate is that the criteria for modernity are different in different regions of the world. The earliest Australians, while they don't look much like Africans *or* modern Europeans, crossed the ocean in seagoing bamboo vessels, clear behavioral evidence of modernity not associated with any of the "modern-looking" Middle Paleolithic specimens discussed above.[83] This emphasizes something we have long suspected. The focus on west Asians and the Middle Stone Age Africans as the earliest moderns reflects the fact that they have some modern *European* features, anatomical and perhaps behavioral, but ignores the early appearance of *different* modern features that characterize other regions.

Finally even Neandertals, while quite different from the people who follow them, may be considered transitional when considering the sample as a whole.[84] These robust, archaic-looking early Europeans were more variable than people think.[85] There are considerable differences in forehead shape, browridge development, mandibular shape, mastoid process, and other features. A single Neandertal site with a large sample, such as Krapina, can express the entire normal range,[86] and it is big! Moreover, some Neandertal specimens are quite modern, closely resembling some of the "early moderns" from central Europe in terms of browridge development and other aspects of cranial shape. A comparison of individuals such as Spy 2 (Neandertal) and Mladeč 5 (early modern) seen in Chapter 9 shows this clearly. A number of Neandertals have chins, presumably a modern trait—certainly one at

Klasies. And there is a persistence of Neandertal discrete traits in post-Neandertal local populations,[87] suggesting to us that Neandertals contributed significantly to the gene pools of modern European populations, and therefore that some of their genes are found among moderns.

So the human world of 100,000 years ago encompasses many variable populations. Virtually all of these have members with so-called modern features. If one focuses on carefully selected features on a few individual specimens, each, perhaps all, of these groups can lay claim to modernity because there are *some* individuals with *some* modern features. But the basis for this claim varies from region to region, and the "modern" characteristics are variations within populations that otherwise appear to be archaic when compared with late Pleistocene or Holocene remains from the same regions. Therefore, in our view, none of the samples are anatomically modern.

DEFINITIONS

However we define modernity, then, it must be a populational distinction. We contend that it is misleading and incorrect to use a single feature, a particular piece of modernity, to justify describing the entire sample as modern or to

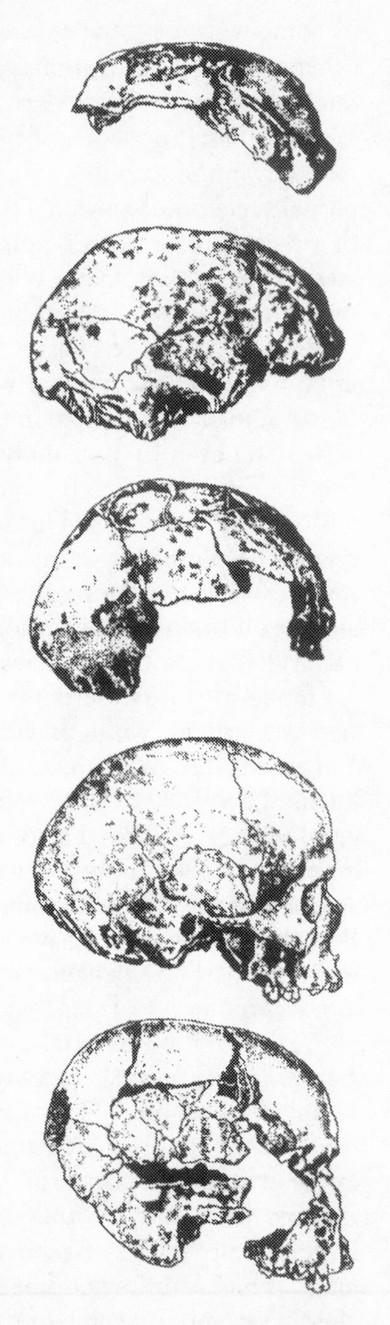

Adult crania from Mladeč (from top, *pictured are Mladeč 4, 5, 6, 1, and 2). Mladeč 4, 5, and 6 are males; Mladeč 1 and 2, females. This early post-Neandertal sample shows considerable variation. Some individuals retain a number of Neandertal features; others appear more "modern."*[88]

argue that the first appearance of a modern feature marks the origin of modern humanity. The latter would assume that there was a single origin for modern humanity—the very issue that at the moment is most contentious. Anyway, where would we start? Perhaps with the chin of the Swartkrans specimen SK 74 or the flexed cranial base of ER 17400 from the region east of Lake Turkana, two much earlier African specimens that are australopithecines. Perhaps with the sharply curved lower cheek surface (maxillary notch) of ZKT 11, a middle Pleistocene Chinese cranium from Zhoukoudian, or the combination of a large and thin vault of Jinniushan?

As we interpret the sample, the Klasies specimens have elements of modernity largely because of their gracility.[89] Many of the Klasies specimens are quite small; one of the mandibles has the smallest molar teeth Milford has ever recorded, and he has measured many teeth. Postcranial remains are diminutive as well, among the earliest in *Homo sapiens* to be Khoisan-sized.[90] This may contribute to the impression of modernity, because smallness usually creates skeletal gracility.

Size is important to this whole issue because it varies around the world, as do its consequences. But reduced size and muscularity don't make gracile people more modern than their contemporaries, either in the past or today. It is true that over time, people have become more gracile in a number of features, but it is equally true that at any given time some populations are more gracile than others. If this variation doesn't suggest different evolutionary grades today, why should it represent different evolutionary grades 100,000 years ago? There is nothing to suggest that Klasies' contemporaries around the world were either morphologically or behaviorally more archaic than the southern Africans themselves, and still less that Klasies is the vanguard of a new species.

Proponents of the Out-of-Africa theory largely differ with this philosophy and the approach to modernity it suggests, as some of the recent work on Klasies demonstrates. Most of the specimens from Klasies were found in excavations during 1967–68. However, new material has been discovered in excavations conducted between 1984 and 1989 by Hilary Deacon from the University of Stellenbosch. This consists of two small maxillary fragments and a proximal ulna. They have recently been described by Out-of-Africa supporters, who see this material as modern,[91] but actually seem to show the same variable pattern as the larger Capetown collection. One of the maxillary fragments is quite small, while the other is robust, consistent with the variation we recorded in other elements of the collection. The ulna was also first described as modern, based on its diminutive size and reduced robusticity;[92] however, in another more comparative study the morphology of the ulna was described as archaic,[93] although with elements that appear in some modern samples. Based on these accounts, the new material seemed consistent with our interpretation of the original remains and was subject to the same varying interpretations by different workers with different philosophies. We wanted to take systematic comparative observations on this smaller sample as well to be sure.

We knew about the new material when we arrived in South Africa and hoped to include the fragments in our comparative study. Unlike the material from the earlier excavation, these specimens were housed at the University of Stellenbosch in Hilary Deacon's laboratory rooms. After we had been in Capetown for several weeks, we were invited by Deacon to visit Stellenbosch, just about an hour's drive from the city. We accepted with some trepidation: we were secretly a little concerned because we were having difficulty with the car Andy lent us. This was a 1960 Volvo, and from what we heard it was Andy's pride and joy. When we arrived in Capetown, Andy had already left and John Parkington, an archaeologist and department head, picked us up and took us to the department where we were given the key to Andy's car. It was a very nice car—a period piece, and we treated it gingerly. Nevertheless, we began to experience some minor difficulty with the vehicle soon after we arrived in South Africa. Every now and then it wouldn't start. There was nothing wrong with the motor; the key simply wouldn't turn in the ignition. It often took an hour of jiggling and swearing to get the key to budge. We had already taken the car to the mechanic two times, and nothing was wrong with the ignition. There was nothing to be done. We had not yet taken Andy's car on anything like a trip, so we were understandably a little worried.

Arriving in Stellenbosch without mishap, we found a beautiful rural university town in the middle of terrific wine country. We made our way to Deacon's rooms and Anne Thackeray, his assistant met us. Anne had recently published on the archaeology of Klasies, and we spent some time discussing the various cultural horizons of this complicated site, until Deacon arrived. He was a charming man with interesting stories, and he kept us amused for several hours, showing us slides of the site and its stratigraphy. "We aren't going to see the bones," Milford whispered to me, midway through the slide show. "How do you know?" I hissed back. "Wait and see," he said.

Deacon took us out for a magnificent pub lunch, and we returned to the university. I could see that he was becoming more uncomfortable, as he was running out of things to do with us. He showed us some more slides and finally invited us to look at the fish bones from the site.

At this point Milford, who is well known for his "directness" (sometimes called "lack of manners" by some of his unenlightened detractors), asked him directly about the hominid material. Deacon apologized and said we couldn't see it since it was being studied by Phil Rightmire. Milford said he thought that Phil, whom he knows quite well, wouldn't mind (which Phil, of course, verified later), and that we weren't going to describe it, just use it as part of a sample, and that we wouldn't publish until after Phil did—all to no avail. Deacon still politely refused. It is not an uncommon practice to limit access to collections for a variety of reasons, but it has happened very rarely to us. Milford has studied virtually the entire human fossil record, in all parts of the world, and has only run into this difficulty one or two times. But he knew when to give up; besides, any disappointment over the bones was dwarfed, by a much more immediate concern that loomed large in the parking lot: would the car start?

On the way home, we discussed once again what it means to be modern. We remember this discussion clearly because out of it came a short publication.[94] From my Neandertal perspective I could see it was the Neandertal comparison that made Klasies seem to stand out so well. We had come to realize that the widespread characterization of non-European skeletal remains such as those from Klasies, Skhul, and Qafzeh as "modern" reflects more than anything the fact that they are *not* Neandertals. We realized there is much to be said for the idea that human populations are modern when they behave in recognizably modern ways, no matter what they look like.[95]

But when does modern behavior appear? The last decade's surprise is that there is actually no indication of modern behavior in any of the early populations generally considered to be "anatomically modern." The idea that a change in behavior causes a response in biological evolution as populations adapt to meet their new circumstances is fundamental to evolutionary thinking, but it wasn't working here. Instead, there is no archaeological evidence that any of these "early modern" populations enjoyed the kinds of advantages from the behavioral capacities that have been suggested would characterize anatomically modern humans and account for their subsequent dramatic success. All of these fossils are associated with Middle Paleolithic industries (the Middle Stone Age is, more or less, the African variety of the Middle Paleolithic). How does Multiregional evolution address this situation?

Pleistocene human evolution was much more than merely the irregularly paced worldwide accumulation of advantageous changes. Becoming anatomically modern was a complex process. It involved numerous aspects of variation, often differing from place to place. And anatomical modernity is clearly independent of behavioral modernity, which came much later (we expand on this point below). What this means is that we can rule out explanations of the appearance of modern humans by either:

1. A modern population expanding and sweeping around the world and being successful because of superior behaviors (whether it replaced native peoples, mixed with them, or both)
2. A set of modern behaviors quickly spreading around the world, changing the magnitude and direction of selection acting on different populations in the same way

If the first were true, it would mean that modern anatomy and behavior would have appeared at the same time, and they don't. If the second holds, modernity would mean the same thing everywhere, and it doesn't.

We know that in spite of its incessant use, "modern human" has proven to be an elusive and slippery term to define. There is no consensus on definitions of modernity, tied as they are to views of variation and of species. Yet this problem lies at the heart of the entire controversy over modern human origins. We don't know how "modern human" is defined, because not only are different definitions contradictory, but when definitions of modernity have been proposed for skeletal remains, the inadvertent consequence has been that the definitions successfully excluding archaic groups also exclude some members of contemporary populations, and this, of course, will not do. We must begin with the precept that all living humans are modern!

Few scientists have actually attempted to define modern humans morphologically, and the brave souls who have, smashed into a wall of difficulties. Michael Day and Chris Stringer[96] proposed an anatomical definition that could be used on fossil remains—a combination of features that were given specific metric or morphological descriptions, like high foreheads, rounded crania, reduced or absent browridges. Their predicament probably arose because the major purpose of their definition was exclusionary; that is, these paleoanthropologists focused on finding criteria that segregated Neandertals from modern humans. They didn't focus on unique characters all humans share and did not take into account the gamut of modern human variation to arrive at some criteria for the species as a whole. Their attempt, in other words, did not begin with the precept of what modern *is*, but rather with what it is *not*. Their definition showed that European Neandertals were definitely not modern, as it was designed to do, but in doing so provided a definition that also excluded substantial numbers of recent[97] and living[98] Aboriginal Indigenous Australians. The definition, of course, was withdrawn but two problems remain. Not only do we lack an operational definition for diagnosing modernity in the fossil record, but we must question whether such a definition is even possible.

If modern humans are a new species with a recent, unique ancestry, reproductively isolated from their contemporaries, we should not be confronting these embarrassing difficulties—a unique definition of this group should be possible. However, it may be close to impossible if the Multiregional model is correct, because regional evolution means that the range of modern variation for all of the features deemed taxonomically relevant for modern humans actually encompasses the way these traits are expressed by many fossils. For example, take the Klasies frontal, which lacks a browridge. This makes it seem modern because the ancestral condition in Africa is to have a large browridge and this would be an important, early change to a condition that is common in humanity today. But many contemporaries of Klasies have large browridges, just as some contemporaries of living Africans have large browridges, because modern human populations range from those with many large-browed people to those whose people have none at all . If we apply the principle that the present is a valid guide for interpreting the past, the criteria we apply to interpreting human variation today should be used to interpret human variation 100,000 years ago. This makes it impossible to arrive at a definition that simultaneously includes the variation of all living people and excludes all members of archaic groups in a Multiregional world.

There are other problems with defining modernity if we think "modern" features did not evolve simultaneously. A populational approach to understanding the place of modern features in archaic populations suggests that even as they increased in number and frequency, the modern features were only part of the normal variation of populations. Were the people that possessed them more modern than their siblings that did not? Of course not. Any definition of moderns, therefore, must include many ancients and make it seem as though for long periods of time archaic and modern people were coexisting not just on the same continent or in the same region, but in many cases within the same family.

More important than the issue of how to define modernity, we think, is understanding the evolutionary processes that produced it. Modernity, the way we look at it, was not the appearance of a set of anatomical details, but a process and a pattern of change. We liken the process to throwing stones into a small pond. The stones are of different sizes and are thrown sporadically, corresponding to the appearance of features at different times and in different populations. From these strikes a series of ripples (advantageous features) spread over the pond surface, the ripples from each strike finally coming to interact with the others. The modern features, the ripples, do not create modernity, but rather, we envision the interference patterns formed by these overlapping ripples as defining modern humanity, a definition based on the anatomical variations in different regions. Each set of ripples spreads over the entire pond and contributes to the evolution of the species, but the angles and intensities of the ripples differ from place to place, and so their interactions create different patterns. Each region is different, and yet it is a singular process that unites them, diffused through the network of interconnections between the populations.

BACK TO BEHAVIOR

It is possible to show quite directly that a worldwide network of interconnected populations through which both genes and ideas could spread was a reality in the Late Pleistocene. Ideas, like the genes, can be promoted when their acceptance leads to advantageous changes in behavior, and they can leave traces behind. Evidence of these dispersal networks is found in the distribution of the Middle Paleolithic stone tool industries, an example of which is associated with the human remains at Klasies, and in this industry are insights into how modern behavior evolved. The Middle Paleolithic developed out of the invention of an important technology, the prepared core technique for making preformed flakes that could be used as tools. The invention of the prepared core or "Levallois" technique was truly insightful and revolutionary. It came from an increasing knowledge of raw material, an understanding that stone can be shaped by softer materials than stone (such as bone or wood), improvement of manipulative skill, and improvement in the neurologically based ability to preconceive form and translate the mental image accurately into performance. To a significant extent, the Levallois technique allowed artifacts to be manufactured that were largely independent of the type, quality, or distribution of raw material (the naturally occurring stones, or other materials, the tools are made of). Local raw materials were brought to the site where tools were made without any prior modification, while more desirable materials from distant sources were imported after being partially or fully processed.

Then came the reduction of the core, the block of material that would be the source of the flake tool.[99] The flake is the finished tool and usually needs no additional retouch on its extremely sharp edges, at least until there is use wear. In the recurrent technique the core is prepared so that a series of sequential flakes (or, if they are long, blades) can be removed from each prepared surface. As the core becomes smaller and smaller, the flakes or blades diminish in size. As far as we can tell, this reflects behavioral changes that were related to the evolution of parts of the brain involved in mental mapping. It was not

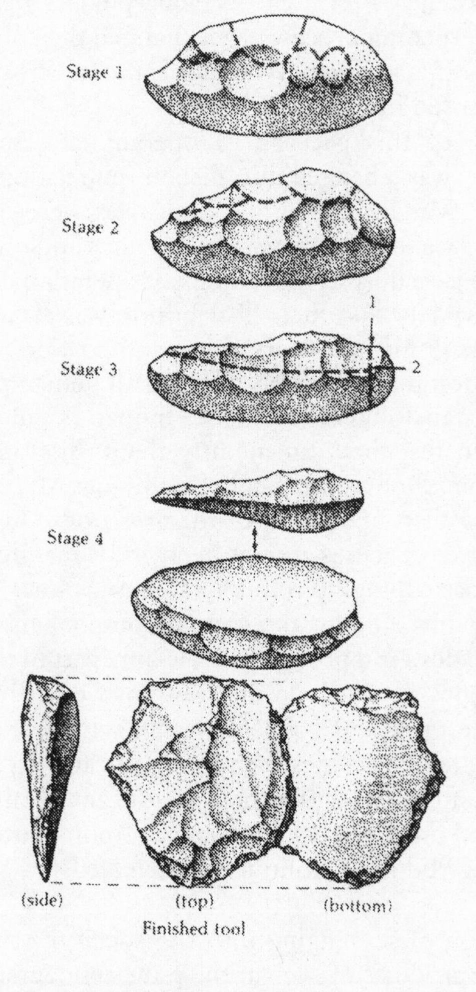

The prepared core (or Levallois) Middle Paleolithic stone toolmaking technique.[100]

until people could make complex mental maps that integrated the process by which seemingly arbitrary operational steps could produce a desired final form in a way that was independent of the materials the form was imposed on, that a complex manufacturing technique such as the Levallois could arise. It is widely believed that the ability to make complex mental maps evolved as an adaptation for interacting in a complex social space and as part of language evolution.[101] The idea that toolmaking reflects these intellectual achievements,[102] however dim the reflection, is suggested by several current reconstructions of cultural evolution.[103]

These tools were used, resharpened, and used again. Some of the flakes were further retouched to achieve a desired shape, but many tools were resharpened after their once-sharp edges were worn down by use. The resharpening process often changed their shape, and tools of one "type" became another. Worn tools modified by resharpening were carefully chosen for continued use.

As knowledge of this package of operational sequences spread widely, selection was changed for distant populations around the inhabited world. Middle Paleolithic industries were everywhere, in Europe, Africa, Asia, and Australasia. Not all Middle Paleolithic industries used the Levallois technique or this operational sequence, but these spread so widely that their distribution was clearly worldwide. The Middle Paleolithic was everywhere and so were the important changes in selection that ultimately affected all human populations.

It is true that "anatomically modern" human populations, according to those who feel they can identify them, first appeared in the Middle Paleolithic. Modern human behavior, says Mary Stiner,[104] also emerged in more than one place. In the arena of food procurement, modern behavioral capacities are not likely to be manifest in the addition or loss of a basic foraging component such as scavenging or hunting,[105] but rather involve how the various components are integrated in adaptive strategies. And procurement is only part of the problem, as successful strategies also include the transport, dismemberment, and processing of the prey, and cooperative behaviors ranging from the hunt itself to related necessities such as infant care. To provide an example, Stiner contends that the most significant changes leading to modernity in food procurement strategies in south-central Europe occurred late in the Middle Paleolithic and involved

- Incorporation of scavenging into the social matrix for hunting, pooling the products of both at the same residential places
- Spatial and temporal overlap of hunting and scavenging activities

- Increased focus on hunting prime adult mammals in an intermediate body size range
- The beginnings of the use of stone-lined hearths to contain heat and intensify its effects

Stiner envisages the behavioral transition to modernity as multistaged and not as the sweep of a single set of "advanced behaviors." It is now clear that in different regions, the changes to modernity came in differing combinations and sequences. For instance, the evidence of modernity Stiner discusses are in an industry that in Europe is invariably associated with Neandertals. Another aspect of Neandertal behavioral modernity is in the hearths found at the Levantine Neandertal site of Kebara (but only rarely in the European Mousterian). Stringer and Gamble,[106] who generally think of Neandertals as anything but modern, contend that hearths are a clear sign of modern behavior because they show an understanding of how to maximize a fire's heat and the forethought of putting that knowledge into practice. Yet the association of hearths with Neandertals is unequivocal. Hearths are consistently found in the French Châtelperronian, a European Upper Paleolithic stone tool industry associated *only* with Neandertals. In fact, Neandertals of the Châtelperronian also built huts, wore body decorations, and made stone tools, all behaviors once thought to reflect the invasion of modern humans, before the true makers of the Châtelperronian were discovered.

Yet at the same time, populations regarded as anatomically modern are found in earlier Middle Paleolithic sites, which do *not* show most of the evidences of modern behaviors discussed above. At Levant sites such as Skhul and Qafzeh, the idea that there is a link between modern behavior and anatomy, whether it is causal or a consequence of modern people with better ideas sweeping around the world, is smashed. We are forced to conclude that the idea that culture and biology can be linked this way is just as wrong in the past as it is today and wonder if the unlinking of these two isn't one of the better indicators of modernity.

We predict we will never find a cradle of modern humanity, because a single source for modern humans does not exist. Modern features or their increased frequencies appear at different times in different places. In many samples from this and other time periods, there are characters whose genetic basis is still found in modern human gene pools. Modern behaviors, too, seem to have multiple origins and combine different elements in different places. Rather than a focus on origins, *per se*, we think the truly interesting issues when it comes to modern humans have to do with evolutionary processes—how gene flow, selection, and stochastic factors work together, and most important, how the cultural factors we think *really* represent modernity arose, diffused, and influenced biology and behavior.

WHY DO WE SEE VARIATION SO DIFFERENTLY?

The differences between Multiregionalism and many of the recent single source theories of modern human origins devolve to whether or not modern humans represent a new species and therefore what taxa the fossils represent. We see all the Late Pleistocene fossils and modern humans as members of a single variable species. Paleontologists who view variation as we do are often called "lumpers." On the other hand, Eve theorists recognize these same fossils as members of different species; they are "splitters." We need to answer the often-asked question, How can different people see the same data so many different ways?

The answer, for us, is found in the insight that we see things not as they are, but as we are. As Geoff Clark and Cathy Willermet note: "We select among alternative sets of research conclusions in accordance with our biases and preconceptions—a process that is, at once, both political and subjective, and, to some extent, inevitable in any scientific endeavour."[107]

The central difficulty, the place where biases and preconceptions have their maximum effect, we believe, is over the basic problem of all biology—variation. The problem is easy to describe because it reflects a fundamental issue in dealing with paleontological data. Different views of variation are often linked to whether the viewer is primarily inductive or more deductive, and if refutational approaches to data analysis are important to the scientific philosophy. Different views of variation and just what it is we think science is supposed to do are, in turn, linked to methodology.

If you ask different people "what is science," you'll get a lot of different answers. Many people equate science with technology, some think of science as the finding of "facts," others, such as our courts, consider science to be what scientists do. But for us science is fundamentally a way of asking questions; it is determining what questions are important and framing them in such a way that their answers in some way increase our understanding of the world. An older consensus holds that the best way to do this is to follow the "scientific method"—that is, a series of steps that include at least three criteria:

1. Creation of a testable hypothesis (or explanatory statement about the world)
2. Testing of the hypothesis through experiment or observation
3. Repetition of the test to make sure the outcome is consistent

Within this methodology there are varying approaches to science that, in effect, differ in their emphasis—either on hypothesis formation or on hypothesis testing. Scientists who think inductively are mostly interested in hypothesis formation. They see the hypothesis as the end result. Historically, and even in the present day, they often don't view the hypothesis as a hypothesis at all, but rather believe it to be some objectively sought "truth" that the inductive analysis of the data led them to. Inductive thinkers start the scientific process with specific observations and proceed to a general theory that fits these observations.

Purely inductive approaches are problematic,[108] though; they do not focus on testing hypotheses, but rather impart the idea that if one has the data, logical ordering of these facts will lead the intelligent scientist to the truth about the operation of the world—the famous idea that "data speak for themselves." We believe this sort of emphasis does not tell you much about the world, and we are not alone in this belief.[109] First, the data do not speak for themselves; they need to be interpreted in a framework that is influenced by a vast number of things, from the historical period in question[110] to socioeconomic factors[111] to particulars of the scientist's personality and training.[112] While philosophers and historians of science may debate which of these factors is most important, it seems clear that all exert an influence over the collection and interpretation of data. Furthermore, it is generally impossible for data to fit only one model. There can be many explanations for a single set of observations, many theories compatible with the same data set; there is no such thing as logically derived truth.

Alan Templeton,[113] in a critique of the Eve theorists, makes this point quite clearly. He distinguishes hypothesis *compatibility* and hypothesis *testing* and argues that this distinction describes methodological differences that are "the central issue" for the Eve debate.[114] This is not the first time the question of inductive approaches to paleoanthropological questions has come up, and surely it will not be the last. From a time much earlier in his career, when he focused on australopithecine questions, Milford fondly remembers many hours spent arguing this point with the late Glynn Isaac, one of the preeminent African archaeologists of the century. They would sit at the tables in front of the Ainsworth Hotel, the favored spot of paleoanthropologists working at the nearby National Museums of Kenya in the 1970s. Isaac thought of himself as the classic inductivist, but unlike many of our colleagues, he realized his acceptance of this approach was an explicit methodological decision that he had taken, and he was interested in defending it, especially against an outspoken Popperian deductivist like Milford. These were much needed breaks from the boredom of repetitive analysis taking place across the street, at what was then a very pleasant spot.

"Why limit yourself to disproving *one* hypothesis?" Isaac would argue. "And how do you know which is the simplest to disprove first?" Isaac's position was that the best approach is to "line up all of the possible theories, gather evidence, and find the best supported one."

Inductive science reached its heyday with the logical positivists who believed, as many workers still do, that probabilities can be assigned to competing hypotheses and that the best hypothesis is the most probable one. Although it is now often recognized that inductive approaches can be flawed, the preoccupation of many scientists with "objective understanding from data" speaks of still prevalent strong inductive tendencies, and also of the predominance in modern research of digital computers and the programs available to run on them. Technological advances have inadvertently promoted inductive science, as they opened the doors to the analysis of vast amounts of new data. With exciting new techniques to arrange and analyze data, more and more paleoanthropologists have focused on this aspect of the scientific process. We worry that modern science will become increasingly characterized as "technology looking for a problem."

We are concerned that some of the Out-of-Africa research is an example of this. Its paleoanthropological supporters are virtually all from a positivist tradition. They are convinced this approach is the only one that objectively derives facts from bones. We believe this scientific philosophy is both a reflection of and a source of their views of variation, and that this, even more than the differences in species definitions, is perhaps the most important reason these views differ from ours.

When Chris Stringer entered doctoral research, his thinking was quite different than it is now.[115] He thought, as was popular then, that Neandertals represented a stage of human evolution and that they were ancestral to modern humans, but he changed his mind during the course of his doctoral studies. Stringer was frustrated with the descriptive and subjective methods of analysis of fossils and wanted to find some way of objectively analyzing bones to determine relationships. His primary goal as a student was methodological. He wanted to develop objective methods of skeletal analysis.

As luck would have it, just as he was embarking upon his graduate studies, W. W. Howells published "Cranial Variation in Man,"[116] introducing and popularizing the idea of using a multivariate metric approach to understanding human variation. Stringer decided to use this statistical approach on the relationship of specimens from Jebel Irhoud, a Late Pleistocene Moroccan site, to the Neandertals of Europe. To his surprise, his multivariate analysis showed Jebel Irhoud skulls were different from the penecontemporaneous Neandertals—and similar to modern humans. This changed his thinking about Neandertals and the pattern of modern human origins.

Most multivariate analyses pertinent to the modern human origins issue are statistical methods that allow the simultaneous treatment of many measurements (variables) to assess relationships.[117] The statistics are supposed to produce measures of "biological distance" in one way or another, since it is assumed that metric similarity = morphological similarity = degree of biological relationship. Since they involve a large number of variables that can be "objectively" obtained through measurement, it is felt that the conclusions are "objective" as well. "Objectivity in, objectivity out."

Multivariate techniques are attractive because they seem to give the data an opportunity to speak for themselves. However, there are many problems with the incautious use of these techniques that stem from a variety of sources.[118] Our literature is now full of misapplied multivariate studies. One problem is that the fossil record consists of many incomplete specimens that don't yield large batteries of measurements or traits. There are techniques for filling in missing data, and some analyses use them in order to incorporate incomplete specimens, although there is no reason to believe they are valid. Avoiding this, other multivariate studies are limited to the most complete specimens and are therefore based on small samples.[119] Therefore these analyses may inadvertently be typological, or based on unwarranted assumptions. Milford, one of the first to use digital computing to analyze very large data sets,[120] was writing about these problems long before there was a Multiregional issue: first, over the question of whether multivariate analysis of measurements that reflect function provides a functional analysis or distinguishes functional groupings,[121] and more recently in debate over the issue of how to compare variation in fragmentary and incomplete specimens.[122]

Paleoanthropologists can easily forget that what their results really indicate is mathematical, not anatomical, and may have little to do with actual morphology. A good example of this can be found in the controversy that has surrounded the analysis of the 30,000-year-old Zhoukoudian Upper Cave, or Shandingdong, crania. Weidenreich considered the three crania as Mongoloids, whose variability showed there were never "pure races" in the past. Other anatomically leaning paleoanthropologists have concurred over the years[123]—if anything, even more specifically linking them to living Chinese populations than Weidenreich felt he could do. With so long a pedigree of anatomical studies, the complete male skull UC 101 should be a good test case for multivariate analysis. However, three multivariate analyses have contradicted these anatomical studies and concluded the specimen is not Mongoloid,[124] one study even suggesting that the heretofore uncontroversial female cranium UC 103 was *African*.[125]

Something went seriously wrong here,[126] and we think it is in the assumptions of the techniques used. The multivariate analysis of measurements is not an anatomical analysis. The multivariate clustering of a Shandingdong cranium with modern Australian, Melanesian, and African nearest neighbors based on 33 measurements does not necessarily mean the "close morphological resemblance" that has been asserted.[127] The argument is sometimes used that "the numbers don't lie," and what computer analysis yields is more reliable than what the eye tells us or what one gathers through morphological analysis. We have difficulty with this idea, as it flies in the face of our experience as paleoanthropologists. The bottom line is that just because it is mathematical doesn't mean it's "true" or objective or has any more validity than other forms of analysis.

The danger of using multivariate analyses to address the human origins issue is that the analysis presupposes the solution. When you plug your data into a statistical program, you will get an answer, whether you are using the appropriate statistics or not. Its like adding up the diameters of apples and oranges and taking the average. There *is* an average, but what is it an average of? Most analyses that try to assess relationships between fossil and living groups assume there was a branching pattern of population divergences that created the variation. If the populations are species this makes sense, and this certainly would be the case if the Eve theory were correct. But we believe all of human history is of a single species. Branching analysis of humans will always depict variation as patterns of divergence; it will always give branching patterns that *could* account for the variation. But if human evolution was modeled like Weidenreich's network, it proceeded like Moore's stream with channels diverging and merging, then the pattern a branching analysis determines is invalid because branching didn't cause the variation. The very question we wish to resolve is which model underlies the variation. By using a branching analysis, one presupposes the conclusions. We cannot learn whether branching analysis is valid by doing a branching analysis.

We certainly don't mean to imply that all use of multivariate analysis is questionable. There is increasing use of multivariate techniques for refutation, and such approaches have become the core of anthropological research, from forensic identification to phylogenetic analysis. Multivariates are at their best when the research is circumscribed and problem-oriented, and data are analyzed to answer specific questions.[128] Many of the younger scientists in our profession have grown up with technologies and algorithms beyond the wildest dreams of their professors. They see it quite differently. We simply want to raise the point that multivariate analyses do not actually provide objectivity, and just like any other analysis, special care needs to be taken to recognize intrinsic biases.

POLYGENISM RAISES ITS HEAD

One consequence of the Eve and Out-of-Africa debates has been a real shift in the focus of publications, and the conversion of Multiregional evolution from a theory about process into one about origin. The attitude taken has been that a Multiregional theory of multiple origins is being contrasted with an Eve theory positing a single origin. This, quite erroneously, seems to make the debates analogous to the polygenist/monogenist debates that have plagued the history of anthropology—a point not lost on several of our antagonists who continue to liken Multiregional evolution to Coon's polygenism.

But the intellectual alignments are quite different. It is ironic that the Out-of-Africa and Coon's theories both treat race the same way. They view modern racial variation as *branches* ascending from a common ancestor, and treat racial differences as the consequence of divergences. They differ in when that ancestor lived; unlike polygenists, Out-of-Africa theorists posit an ancestor so recent as to make racial differences negligible.[129] But they both begin with the same mental image of what races are: clades or branches to be treated the same way as species. Both theories pay lip service to genic exchanges, but it is not significant in their models of how races evolve, or present in the computer programs they use to depict relationships. It may as well not exist.

Coon's polygenic vision of races as branches of humanity extending into the distant past remained a real intellectual tradition, even in the last half of this century. However, the situation is more complicated because essentialism is everywhere in paleoanthropological thought, even in the assertion that there *is* such a thing as "modern humans"[130] and the precept that modern humans are an entity so distinct that their appearance is a problem to be solved. Intrinsic typological thinking, which may be even stronger in anthropologists than in other evolutionists, prevented any other explanation but parallelism to account for long-term differences. This explains the continued haunting by polygenism: Multiregionalism seems to give temporal depth to "race."[131] The only way we have been taught to deal with races as historic phenomena is typologically. The accepted alternative to race in the study of human variation, clines, are looked at entirely two-dimensionally without time depth. Studies of clinal variation are either very shallow temporally (as Birdsell's analysis[132] of clines in Aboriginal Indigenous Australians) or they are in the present, on the cusp of a single point in time. So when anthropologists look at race as something historical, they immediately revert to typological thinking—to the way race as a historical entity has always been treated. It follows that when anthropologists rejected the type concept of race, they also inadvertently rejected the idea that races, or large geographic groups, had any temporal depth. These seemed to be polar opposites.

The polygenism legacy is strong, and we have knee-jerk reactions against it that prevent us from looking further to realize that adding a temporal dimension to geographic variation does not have to be racist or typological at all. We have to consciously think of races as dynamic, changing entities with temporal depth because they are not the diverging branches on an evolutionary bush but the constantly separating and merging channels in a stream. But this is easier said than done, because the complex, intertwined relationships between race and human evolution make it difficult to pull any single thread away from the knot. One solution is to get rid of race. While defining race is critical for anthropologists, many avoid the issue[133] because it is complicated and uncomfortable and increasingly not politically correct. But by avoiding the question of race (that is, by erasing race as an aspect of human variation) applied to historical process, anthropologists trade one bias for another.

And, as far as the modern human origins debate is concerned, it is much too late to avoid these questions about race. Several of our colleagues have now fabricated "predictions" about how closely related the races are supposed to be, according to the competing origins theories,[134] with clear implications that Multiregional evolution promotes racial typology by assuming greater racial difference. The Multiregionalists actually believe races are transitory and very closely related because of the persistence and magnitude of intermixture, especially historically. But this may be irrelevant, because it is the perception that becomes the reality. What is obvious is that entrenched ideas about what races are strongly influence perceptions, especially over the creation of and reactions to theories of human evolution.

In any event, Multiregional evolution is not a theory about the origins of modern races. While we view some characters differing between today's races as long-standing, we do not mean to imply that the races themselves are static, unchanging entities. It's not the races, but changing combinations of regional features that can be traced into the past. These features neither describe nor define races—their combinations simply exist in high enough frequencies, in certain regions, to support hypotheses about ancestry. Some, such as the pattern of incisor shoveling in Asians, are virtually ubiquitous. Others exist at lower frequencies, or are only important when they appear in combinations.[135] We do not believe the races are like branches on a tree of humanity that have been around for a million years, and it is Coon, the polygenist, *not* Weidenreich, who wrote "*Sinanthropus* and the Mongoloids, *Pithecanthropus* and the Australoids."[136] A time traveler studying the past would find that some, probably most, biological features that characterize certain regions change and disappear, and new ones originate. Geographic groups are very dynamic and are constantly changing as complex populations band together as a species due to the evolutionary forces that underlie Multiregional evolution. We don't think you can find a single, unique origin for modern human races. Modern humans have no exclusive definition except as all those alive recently and today. Modern races are a consequence of tremendous population expansions and contractions of the Holocene era and before; they so overlap that social rather than biological criteria define them.[137]

But races are also a part of human variation: they reflect evolutionary processes and affect the current course of human evolution. Understanding race and all its complexity is critical to perceiving ourselves as an evolving species. Branching models are simple to comprehend but they ignore the complexity of the evolutionary process in our species. In our species most genetic variation exists within populations, not between them, because of how ubiquitous intermixture has been throughout human history and prehistory. The present gives us our best model for interpreting the past, and that present is not one of constantly diverging trees, but of ethnogenesis, a river broken into channels, some large and some small, constantly dividing and merging in a Multiregional pattern of evolution. Our species has far greater antiquity than so-called modern behavior, and during its history, in spite of many expansions and extinctions of populations, modernity had many sources as populations all around the world contributed to it.

LEAVING SOUTH AFRICA

By October, our semester at home had already begun and we had to go back to teach. We would be returning to a Michigan autumn, perhaps our nicest season, though a long and cold one that usually brought snow before it merged into winter. But we would be experiencing no culture shock. We were leaving with very vivid images of an apartheid society that were all the more haunting because they were not foreign. No matter how often, or enthusiastically, anthropologists write about the nonexistence of human races, on the ground races are a social reality.

The scariest thing about apartheid is that it works. It is an institution meant not only to suppress but to shield members of the elite from the horrors of poverty and despair facing so many people. As we walked through Capetown on our first day, our initial impression was that we were in any beautiful American city, and we could have been. I remarked that of all the places I'd been, this was most like a city in the States. The malls were the same, the stores were the same, the people were very similar in many ways. There were multiracial faces on the street, as in any American city. Most of them were white, as in many American cities. It began to sink in that even though 80–90 percent of the population is black,[138] we would never know it from daily experience. This was, after all, the point of the system. It was so easy for fundamentally good people to close their eyes to injustice. Indeed, one had to seek it out. The newspapers and election campaigns made it seem as if we were in a real democracy—we were, if you happened to be white. Blacks were not citizens, and were confined to "townships," a stone's throw from the center of Capetown, but nominally another city. It was against the law for blacks to live in Capetown although the townships on the cape flats could easily be seen from the beautiful city in the hills. If we needed any further confirmation of the social character of "race," the apartheid laws and classifications provided that testimony, where the arbitrary nature of racial categories was blatant. Until very recently, on every birth certificate, one's race was recorded. Categories included, white (for Europeans, mostly), black (for native Africans, mostly) and colored (for mostly everyone else). This last category included people of racially mixed parentage, Indians, and other non-Europeans. There were clear economic and social agendas behind such biological parceling. The Japanese, South African trading partners, were legally white, while Chinese people were colored. On visiting South Africa, Jesse Jackson was considered white, since if he were black, he would be subject to all of the discriminatory laws facing all other black South Africans.

Now, of course, the laws have changed dramatically. Apartheid as a legal policy is no more, but many of its institutions continue. This should surprise none of us. How many visitors to Washington, D.C., venture the few feet from the Capitol that it would take to see how many Washingtonians live? Our "townships" are more permeable than those of the old South Africa, but are we any more aware of the realities within them? They are reported in our papers, but they remain a world away, if we so choose—and most of white America does. People from the suburbs may drive through their townships on their way to work, looking neither right nor left, to keep the images out. Their borders may be more permeable, but we are looking at matters of degree, not kind. The legacies of apartheid continue long after legal barriers are lifted.

We were leaving South Africa on the eve of the national elections. As they approached, social tensions were on the rise. Virtually every night we could see the fires burning in the townships, evidence of that paradoxical, inwardly directed violence born of frustration, inequity, and poverty that we see every day in our own inner cities. We could see it from the university, but it was never reported in the paper. It officially didn't exist. Nevertheless, everyone knew about it and talked about it at the integrated University of Capetown, where 25 percent of the student body was nonwhite. This violence was starting to take an outward direction, we heard, mainly along the major highway that went past the townships. Increasingly, every day, rocks were being thrown from bridges and along the sides of this major thoroughfare, causing accidents and major damage to vehicles and people. This was not reported in the paper either, but it was common knowledge. This was the road to the airport that we would have to travel on the very eve of the election. Andy was returning and planning to take us in his car.

This, of course, made us a little nervous. We were now afraid to turn it off. Several days earlier, we decided to visit the Cape Point, where the Atlantic and Indian oceans meet. The parking lot was some distance from the point, and we had decided it would be better to let the car run, risking theft of the collector's item, rather than turn it off. Surprisingly, no one stole it. We lived in fear that it would stall on our way to the university or another appointment and that we would never get it started again.

We saw Andy the day before we were to leave. He was going to interview Milford for the Multiregional evolution section of his film, and we would have dinner with him and his fiancée at their house. It was time to relinquish the car; we would be chauffeured home and to the airport the next day. When we pulled up in front of their house, we breathed a sigh of relief, but had residual worries about how we were going to get to the airport, and dreaded telling Andy about the malfunction of his pride and joy.

We waited for the appropriate moment. After a candlelight dinner and several glasses of wine, Milford broached the subject.

"Well, Andy," said Milford slowly, seeking just the right words. "There is something you should know about your car."

Andy grew pale. "What's wrong with it?" he asked in a shaky voice. "You did drive it here, didn't you?"

Then we told Andy and Tina all the stories; the one about getting stuck after dark at the South African Museum; the one about getting stuck on a wine-tasting tour; the one about getting stuck at the university while making a turn into the road behind the department. That time we had to push the car out of the way of rush-hour traffic. There were many more, and it took over an hour to regale him with all the details.

Andy had been laughing loudly throughout the entire story.

"Unsympathetic so and so," I thought to myself. "But he's taking the bad news quite well." In the end, we suggested he see his own mechanic forthwith, one more familiar with his car, and that we would be happy to help him with the repair costs.

"That won't be necessary," he said. "There's nothing wrong with the car."

"Oh, but there is," we said knowingly and in unison.

"Oh, no," said Andy. "There always was a key that didn't work well. I marked it with a red spot and only kept it as a spare in case of emergencies."

Disgusted, Milford threw the key onto the table, its faded red dot glowing irreverently in the candlelight.

EPILOGUE: AFTER EVE

THERE WAS NEW EXCITEMENT in the spring of 1995. Robert Dorit, a molecular biologist from Yale, and some colleagues published the results of their analysis of the Y chromosome.[1] This so-called sex chromosome excites interest because it is, in a way, the male counterpart of the female's mitochondrial DNA. The Y chromosome is found only in males (who are XY, with an X chromosome received from the mother and a Y from the father), never females (who are XX), and therefore must be passed down along male lines. In other words Y chromosomes form male lineages, just as mtDNA chromosomes form female lineages, or for that matter pretty much as family names traditionally form male lineages[2] in our society. There was hope that a lineage based on genetic variations on the Y chromosome could be established and that it would be an independent test for the mtDNA lineage analysis. Could all Y chromosome variations be traced back to an "Adam" who was, more or less, from the same population as the mitochondrial "Eve"? That would be a stunning confirmation for a theory that seemed much more dead than alive.

The paper was to come out in the journal *Science*, and as is often the case in these days of public science, advance copies were faxed to various interested scientists for comment, with the word "EMBARGOED" stamped boldly on them, meaning that press silence was required until the *Science* publication was out. *Science News* sent Milford a copy and soon *The New York Times* and *Time* called him for comment. This publication was to be a well-publicized event—a clear reflection of the heightened public interest in modern human origins, as if any additional evidence for this interest was needed!

The researchers studied a small segment of the Y chromosome, some 729 base pair units long, that forms a single gene the authors say is "involved in sperm or testes maturation." They chose a segment of the chromosome that does not recombine, so base pairs on the Y chromosome could not mix with base pairs from the X chromosome when the cells reproduce, and the researchers determined the actual sequence of DNA bases in the section. The surprise was that in the 38 men they examined, "chosen to represent a cross section of geographic origins," there was no variation at all in the gene. Every one of the 729 base pairs was the same, again and again. Why?

Selection is discounted as an explanation for the absence of variation, the authors reason, because the Y chromosomes of other primates that they examined show variation in this section of the chromosome. They report differences between chimpanzee, gorilla, and orangutan that they take to mean no selection is working on the region, reasoning if there *was* selection, deviations from the optimal genome would have been eliminated. Discounting selection in the apes, they assume random mutations account for the variation. (We find this reasoning questionable and simply wonder if sperm or testes maturation might be under different selection in a colonizing species that has been suffering population reductions, i.e., bottlenecks, and responding with rapid expansions for the last million years of its history, when compared with ape species that have not been able to respond to population decreases and are virtually extinct.)

The *Science* paper posed the question: if mutations cause the variation in apes, why is there no variation in humans? These researchers wrote in their abstract:

> The invariance likely results from either a recent selective sweep, a recent origin for modern *Homo sapiens*, recurrent male population bottlenecks, or historically small effective male population sizes.[3]

But they have an opinion about which explanation is correct, since they then calculate the coalescence time for this gene, ***a calculation that assumes common recent origin for all modern humans and historically large male population size***. This is an unusual calculation anyway, as Jon Marks, also of Yale University, pointed out in an electronic mail message to us on the same day—how can there be a coalescence time for something that never diverged to begin with?

The article got to press: all gloves were off. In the *Times* report Dorit was cited as saying:

> The simplest explanation for what we find is a recent single origin for modern *Homo sapiens*. It's significant that a completely different part of

> the human genome appears to be telling a similar story of relatively recent origins.[4]

Relatively recent? Really? On NPR that same day Dorit admitted his calculations could mean an origin more than 800,000 years ago.[5]

In the *Times*, on NPR, and elsewhere, Milford wondered out loud whether common recent origin was the simplest explanation. Mitochondrial researchers had mostly given it up[6] and turned to examining the evidence they had for different waves of population expansion at different times, bottlenecks, and varying population sizes. It seemed as though a rather different story was emerging. True, the Y chromosome evidence *could* be interpreted to show a recent bottleneck for all humanity, but if that was the correct explanation for the Y chromosome segment, publications showed that other genetic systems could not have passed through this bottleneck.[7] All these genetic studies must be considered together, because if modern humans had a common recent origin this should be reflected as a bottleneck *all* genetic systems passed through. Recent origin demands that the histories of these genetic systems are the same as the history of the population they are in. No analysis of a genetic system should show a bottleneck was impossible, but in fact several analyses did. This means that the histories of the different genetic systems are not the same as population history. Each tells its own story but not the story of the species. When Wilford of the *Times* asked what this meant for Multiregional evolution, Milford responded that

> the new findings and the African Eve research reflected the history of different genetic systems and not necessarily the evolution of early human populations as a whole. "This is not bad news for our theory."[8]

In fact, it was not any news at all.

The next day Clara Nii Ska, the anthropologist and wife of Wub-e-ke-niew, an Ahnishinahbæótjibway, called Milford from Minnesota. She called to tell Milford she agreed with him, or at least with how he had been quoted by the wire services (citing the *Times* article). Clara explained why it is that those patrilineages (male lines of descent) being discussed have nothing to do with population history. The lack of variation in the Y chromosome study was because they were probably sampling the same patrilineage over and over again. Whose? Why Europeans, of course.

Wub-e-ke-niew wanted to learn about his own line of descent, and to do so he worked out the genealogy of the Ahnishinahbæótjibway.

He entered some 60,000 names and relationships into his computer. What he learned is that the vast majority of patrilineages could be traced, not to Aboriginal Indigenous Americans, but to European sources. He wrote a book on this topic.[9] Clara, who had done much of the analysis, estimates that some 99 percent of the people who identify themselves as Ahnishinahbæó^tjibway have European patrilines. Reasons for this are complex and purposeful. In a second phone call, Clara told us she believes the high level of European patrilines in the descendants of indigenous peoples is not unusual in areas of European colonization. It reflects the colonization process, and in some cases subsequent government policy.[10]

So there is yet another interpretation for the Y chromosome studies, if not the one that satisfies, perhaps the one that is true.[11] The lack of variation in male-transmitted Y chromosomes from 38 men sampled around the world may well be the best reflection we will have of what happened during the centuries of European conquest, colonization, and displacement of native peoples.

Populations' contacts result in several different historic patterns. They may mix freely, each taking mates of both sexes from the other. In this case the descendants will have mixed nuclear DNA from all sources, and also mitochondrial lines from populations. But at the other extreme, for instance, at the time of contact for numerous Pacific islands by ship crews of male Europeans (one thinks of the fate of the *Bounty* mutineers), the men may take wives from the Aboriginal Natives. In this case the nuclear DNA of the descendants will be equally mixed from both sources, but the mitochondrial lineages, only passed on from mothers, will be from the Natives, and not the sailors.

The *Bounty*, of course, was an extreme case, but there are similar patterns throughout the history of colonizations. It seems mostly to be the indigenous women that contributed to the next generation as Europeans contacted native groups, and ironically the successful invaders could well multiply the number of native matrilines,[12] even as their cultures were driven to extinction. Now we were beginning to see the other side of the genetic coin, the male story, and it made sense to us in the context of Multiregional evolution. The contrasting lack of variation in Y chromosomes suggests that there are only a limited number of sources for the patrilines in the populations of today. Only passed from father to son, patrilines would *decrease* in colonizations because many potential native fathers were marginalized, or killed, and did not pass on their Y chromosomes.

This pattern, in which today's large populations are descended from only a limited number of past small ones, with descent in many cases largely incorporating only native women, could help explain the oft-noted similarities of modern people, and the distribution of human nuclear DNA in which so many gene variants seem universal, differing only in frequency from population to population. The European expansion involved more than pillage and plunder. The rest of the story may be revealed in our genes, a recent episode in the chronicle of race and human evolution.

NOTES

Introduction

1. From *Wernher von Braun*, recorded by Tom Lehrer on "That was the Year that Was," Reprise Records (R 6179), Warner Brothers, New York.
2. But, of course, no sciences can be separated from human society. As Gould (1981, pp. 21–22) says, ". . . science must be understood as a social phenomenon, a gutsy human enterprise [that] progresses by hunch, vision and intuition."
3. Eugenics is the selective breeding of human beings, with the goal of improving the human species. In the 1920s the eugenics movement in England was considered a progressive, positive movement to improve humanity. Less than a decade later, its negative side prevailed, as the "improvements" were increasingly accomplished by preventing "undesirables" from having offspring. Sterilization and other social programs to limit the breeding of "genetic defectives" reflected the unavoidable fact that the definition of what is desirable and what is not is social and political.
4. We are making a distinction between hominid and human: while hominids have existed for 6 million years, hominids that we would recognize as humans are relative newcomers. Physically "modern" people are only recognized within the last 100,000 years or so.
5. "Conscious" is the key here, as scientists clearly are part of society and their disagreements must be seen in this context. As Sapp (1987, p. xiv) puts it, "Once we fully realize that scientists are constantly negotiating what science is, then it is necessary to consider social interests on both sides of scientific controversy." There *are* agendas.
6. Popper (1962).
7. Mulkay and Gilbert (1982) note that whenever there are conflicts in science, they are played out in a social context in which the competitors' errors are explained by social rather than scientific reasons such as "dislike," "prejudice," or "incorrect training."
8. The publication of *The Bell Curve* (Herrnstein and Murray, 1994) is but one recent unfortunate example of biological determinism in public science. In it (for example, in the discussion beginning on p. 535) the authors actually advocate alterations in public policy to accommodate the biological capacities of different socioeconomic/racial segments of society.

Chapter 1: Multiregional Evolution and Eve: Science and Politics

1. After Wood (1992).
2. After Walker *et al.* (1982).
3. Wolpoff, Wu, and Thorne (1984).
4. Thorne (1981).
5. Koobi Fora is a famous site on the eastern shore of Lake Turkana in Kenya where the fossilized remains of australopithecines and other early humans were found.
6. Africa can be considered the center of the human species' geographic range.
7. Kow Swamp was the focus of Alan's Ph.D. dissertation.
8. The Pleistocene is the epoch of the Ice Ages, beginning some 1.75 million years ago.
9. Hiorns and Harrison (1977).
10. Clinal variation is the continuous, gradual variation of a trait over geographic space. In the context of geographic patterns of a trait's variation, clines are the directions in space that follow its unchanging expression, and gradients are the directions that are perpendicular to the clines and therefore mark the path of the trait's maximum change.
11. Made up from figures in Larsen, Matter, and Gebo (1991).
12. Thorne and Wolpoff (1981).
13. We discuss Weidenreich at length in Chapters 7 and 8.
14. Adapted from Howells (1993).
15. Cann, Stoneking, and Wilson (1987).
16. Stringer and Andrews (1988).
17. Wolpoff *et al.* (1988).
18. Relethford (1995) discusses this point accurately and in good detail.
19. One especially useful discussion of this is found in Gamble (1994), a book we find very insightful in spite of its support of the Eve theory.
20. The name "Eve" was used to describe the last common mitochondrial ancestor (the "lucky mother" as Cann put it in 1987) in a 1983 *Nature* paper reviewing mtDNA research to date, written by N. Barton and S. Jones; Wilson gave a plenary address before the 1987 AAAS meetings called "The Search for Eve," the tape of which encouraged Milford and colleagues to make their position public. The events surrounding the naming of the total replacement theory of the 1980s the "Eve theory" are summarized in M. Brown (1990, especially pp. 107–110).
21. Kimura (1979).
22. After Ridley (1993, Figure 6.3).
23. After Tattersall (1993).
24. Avise (1989) draws this analogy quite well.
25. To be specific, the Eve theory assumes that the last common ancestor for all existing mitochondrial DNA is the same as the last common ancestor for all existing nuclear DNA.
26. Gribbin and Cherfas (1982).
27. Similar situations are discussed by Rousseau (1992).
28. Of course, all science is sooner or later political (Bell, 1992).
29. This is not the first time this contrast has been made. Confronted with the geochemists' high-tech-derived evidence for an asteroid collision at the time of (presumably causing) dinosaur extinction, Robert Bakker expressed similar sentiments: "The arrogance of those people is simply unbelievable. They know next to

nothing about how real animals evolve, live and become extinct. But despite their ignorance, the geochemists feel that all you have to do is crank up some fancy machine and you've revolutionized science. . . . In effect, they're saying this: 'We high-tech people have all the answers, and you paleontologists are just primitive rock hounds.'" (quoted in Raup, 1986, pp. 104–105).

30. Wilson and Cann (1992, p. 68).
31. Cited in M. Brown (1990, pp. 217–218).
32. Wilson and Cann (1992, p. 68).
33. Kotter (1974).
34. And this is far from the first time that political consciousness raised its head. When Franz Weidenreich first came to China, in 1936, and was in the process of thinking through his Polycentric theory, the predecessor of our own, he was very reluctant to publish on it. Weidenreich was worried about the way the evolutionary model he was developing might be misconstrued in the charged racial atmosphere of the time.
35. Gould (1987, p. 19).
36. Gould (1988, p. 21).
37. Jackson (1994).
38. This was later revised and published as Lieberman and Jackson (1995).
39. Lieberman and Jackson (1995, p. 239).

Chapter 2: A First Lesson in the Politics of Paleoanthropology

1. Becker (1932).
2. And promoted that same year in an earlier "This View of Life" (Gould, 1987).
3. That is to say, "developed" by Gould and colleagues, for Darwin understood it perfectly well, writing in the 4th and later editions of *The Origin* "Many species once formed never undergo any further change . . . and the periods during which species have undergone modification, though long as measured by years, have probably been short in comparison with the periods during which they retain the same form." See Dennett (1995, pp. 282–299) for further details.
4. For instance Gould and Eldredge (1977).
5. As Dennett (1995) points out, however, they have to some extent confused gradualism with a kind of constant speedism.
6. Gould (1988).
7. Actually, the "new" 100,000-year date was actually not all that new. When the first of these sites with the "moderns" was announced in the January 26, 1935, issue of *The Illustrated London News* in a full-page spread, the headline was "Palestine Man 100,000 Years Old." Jakov was just finishing up a biography of Gorjanović and his place in the history of science, which he knew would be published for the forthcoming anthropological congress. To tease the team who had just published the "new" dates, many of whom were coming to the congress, he reproduced the full page from the 1935 newspaper in his book (Radovčić, 1988, p. 107), proclaiming the same ancient date for the site, more than a half century earlier. We can still remember his broad smile as he handed out copies of his book to them, waiting for the page to be discovered.
8. Gould (1988, p. 20).
9. Gould (1988, p. 21).
10. Gould (1988, p. 21).

Chapter 3: Polygenism, Racism, and the Rise of Anthropology

1. Vallois (1952).
2. McCown and Kennedy (1972, p. 32).
3. Home, cited in Stocking (1968, p.45).
4. Stocking (1968, p.45).
5. "Racial thinking" is a phrase used by Stocking (1968) to describe the 19th-century European view that human differences (social, behavioral, and physical) are primarily racial—that is, deep, intrinsic, and largely immutable. We use it this way throughout the chapter.
6. Mayr (1982a).
7. Blumenbach (1806, *Contributions to Natural History*, reprinted in Bendyshe, ed., 1865b, pp.310–312).
8. Flourens (1847, Memoir of Blumenbach, reprinted in Bendyshe, ed., 1865b, p. 57).
9. Burkhardt (1977).
10. Gould (1994).
11. Adapted from Greene (1959, Figure 8.2).
12. In Bendyshe (1865b, p. 235).
13. Gould (1991b).
14. Becker (1932, p. 47).
15. Stocking (1968).
16. Moorehead (1969).
17. Morell (1993).
18. Burkhardt (1994).
19. Mayr (1972).
20. Burkhardt (1977, p. 134).
21. Burkhardt (1977, p. 135).
22. Mayr (1982a).
23. Mayr (1972).
24. Mayr (1988).
25. Mayr (1988).
26. Burkhardt (1977, pp. 166–174).
27. Lamarck (1809, p. 351), translation by Burkhardt.
28. Mayr (1982b).
29. Coleman (1964).
30. Stocking (1968, p. 39).
31. Burkhardt (1977, p. 195).
32. Stocking (1971).
33. For in-depth, academic treatment of this whole topic see Stocking (1968, Chapters 1–3).
34. Mayr (1982a).

Chapter 4: Slavery and Its Reverberations

1. The historian George Stocking (especially 1968 and 1987) discusses this point at some length.
2. For example, Dr. Charles Cauldwell (1772–1853), an outspoken polygenist discussed in Stanton (1960). Moreover, the relationship between polygenism and American politics has been treated in detail by Stanton (1960) and also discussed in Gould (1981). In this chapter we include some of their more salient points.

3. Spencer (1983).
4. Morgan (1877).
5. Stocking (1968, p. 39).
6. Gould (1978).
7. But compare Gould (1978) and Michael (1988) on the issue of whether Morton altered his conclusions by using compressible seeds to measure cranial capacities.
8. Gould (1981, p. 54).
9. Gould (1981, pp. 54–69).
10. These are Morton's terms, not ours, and we do not vouch for the accuracy of his averages or, for that matter, for the reality of the racial groups they are said to describe (also see Michael, 1988).
11. Like Cuvier before him (Coleman, 1964) he differed from progressionists, including many paleontologists, who believed that history revealed an irresistible "march of progress" as humanity and its world moved toward perfection, and who argued that the small brains of some races were a sign of their primitiveness or relation to the apes. Morton (as Cuvier) did not believe in a progression, as one race evolved into another, nor did he think that races represented different evolutionary stages (neither Morton or Cuvier believed in descent with modification).
12. Stanton (1960).
13. Stanton (1960).
14. Coleman (1964, p. 182).
15. Stanton (1960, p. 101).
16. Cited in Mayr (1988, p. 185).
17. Stanton (1960, p. 100).
18. In this, of course, he was quite different from Cuvier.
19. From a letter from Agassiz to his mother, translated by Gould (1981, pp. 44–45).
20. Lurie (1960).
21. Examples of the complexity of the relationship among slavery, politics, and polygenism can also be found in the work of earlier scholars, many of whom strongly disavowed links between science and society. Charles White (1728–1813), an English surgeon, was an early polygenist who agreed with Lord Kames but drew different conclusions. He had strong views of European superiority, and was convinced that there were a number of different human species that conformed to a natural principle of gradation in accord with the Great Chain of Being. However, as convinced as he was of European superiority, he disavowed any political (or religious) importance to his views and in fact supported abolition. He condemned the slave trade saying that "the Negroes are, at least, equal to thousands of Europeans, in capacity and responsibility; and ought, therefore, to be equally entitled to freedom and protection. Laws ought not to allow greater freedom to a Shakespeare or a Milton, a Locke or a Newton, than to men of inferior capacities; nor show more respect to a general Johnstone, or a Duchess of Argyle, than the most unshapely and ill formed." (Stanton, 1960, pp.17–18).
22. One of the hallmarks of these debates was the dichotomy of the nature/nurture issue and its link to the polygenism/monogenism controversy. Samuel Stanhope Smith (1750–1819), a great and vocal monogenist, painted those who weren't complete "nurturists" as polygenists and infidels. He contended *all* racial features were plastic and changeable with climate. Those who argued for any nondevelopmental differences between races he tarred with the polygenist (un-Christian) brush. Given the platonic view of variation prevalent at the time, where "fixed"

features were considered those from creation, and there was no understanding of genetics or evolution, this stark dichotomy may be more understandable than it now appears. Nevertheless, to be branded unpious had a poor effect on the scientific community; Smith was not well liked. (Stanton, 1960, Chapter 2.)

23. Agassiz (1850) cited in Gould (1981, p. 45).
24. Agassiz (1850) cited in Gould (1981, p. 47).
25. 10 August, 1863, in Gould (1981, p. 50).
26. Lest the reader think reasoning like this is restricted to the past century and the question of slavery and race, a book just published even as we write these words (Shreeve, 1996) uses physical repugnance as the basis for explaining how Neandertals and modern people, both equally "human" by archaeological standards, could live side-by-side but supposedly never mate with each other; it contends they just weren't interested, unable to recognize each other as potential mates!
27. August 9, 1863, in Gould (1981, p. 49).
28. Gould (1981, p. 49).
29. August 11, 1863, in Gould (1981, p. 50).
30. Gould (1981, pp. 49–50).
31. Stanton (1960, p. 62).
32. Stanton (1960, p. 159).
33. Spencer (1986, p. 147); we further discuss Gobineau below.
34. Cosans (1994).
35. As Desmond (1982) suggests.
36. Desmond (1989).
37. Burkhardt (1977, p. 38).
38. Dobzhansky (1980), Medvedev (1971).
39. Schiller (1979).
40. Spencer (1979).
41. Gould (1977, p. 84).

Chapter 5: Polygenism After Darwin

1. Apart from the few large, fairly complete pieces, the materials were stored in three locked glass cabinets, in boxes sorted more or less by type. Left temporal bones were put together, right parietal bones, humeri, and so on. The bones were piled atop each other, and with so good a potential for breaking, they rarely were moved.
2. Courtesy of Jakov Radovčić.
3. His major monograph was published in 1906, and there were additional detailed publications throughout his active career, as Radovčić (1988) reviews.
4. Drawing by Karen Harvey.
5. Mann (1975).
6. Radovčić (1988).
7. Dubois (1894), and see Howell (1994b) and Theunissen (1988).
8. According to Trinkaus and Shipman (1993, p. 44) it was not remembered until the early 20th century.
9. Many of these were still thought to be quite primitive, if not specifically anthropoid.
10. Gorjanović wrote on this in the popular Zagreb magazine *Priroda* 20 years later, with the optimistic view that his Krapina discoveries and French Neandertal fossils had settled the issue for good. He was wrong (Radovčić 1988, pp. 210–221).

11. Hrdlička (1930), Trinkaus and Shipman (1993), Wendt (1956).
12. Bowler (1983).
13. After Weinert (1947, Figures 8 and 25).
14. More substantial descriptions and comparisons were later published by the better trained professionals Schwalbe (1899) and Weinert (1928).
15. Theunissen (1988).
16. According to Howell (1994b).
17. In the same preliminary reports section of the *Archiv für Anthropologie* that Gorjanović published in.
18. Ranke (1896).
19. As Gould (1977) points out, this idea was very Newtonian and thereby quite acceptable. It meant that the course of biological evolution was controlled much like the behavior of the physical universe, by laws and conditions set down at the beginning and governing all subsequent changes.
20. Some splitters were attracted to monogenism; these were, by and large, scientists who had a broad vision of what humanity encompassed, including a wide array of variation, from "primitive" to "advanced" races.
21. Radovčić (1988) is a detailed, interesting, and well-illustrated account of Gorjanović's life and position in the intellectual history of paleoanthropology.
22. Rink, Schwarcz, Smith, and Radovčić (1995) place the site within a short period of time, just before 130,000 years ago.
23. After Ullrich (1982).
24. Secondary burial occurs when a body is left to deteriorate, and later the mummified ligaments are cut to disassemble the joints so a small bundle can be made of the corpse, which is then buried. The question at Krapina revolved around the positions of the cut marks: are they like butchered animals or like other cases of bundle burials?
25. The difference in cut-mark location is significant since it indicates that the human bones were treated differently from prey.
26. Many of these are remains of juvenile rhinoceroses which, according to Preston Miracle, reflect cooperative hunting because they were young enough to have been guarded by their very formidable mothers.
27. Courtesy of F. H. Smith.
28. Schwalbe (1899).
29. Dubois (1894).
30. Schwalbe (1906).
31. From the text of a keynote talk given before the First Congress of Serbian Physicians and Natural Scientists held in 1904, translated by Radovčić (1988, p. 81).
32. Gould (1976).
33. Gorjanović kept what he considered to be his more speculative ideas out of his German publications, presenting them in Croatian publications or in lectures.
34. From the text of a talk given before the Yugoslav Academy of Arts and Science in November, 1899, just after his first excavations at Krapina were completed. Translated by Radovčić (1988, p. 34).
35. Heilborn (1923).
36. Courtesy of Kasia Kaszycka.
37. Trinkaus and Shipman (1993, p. 168).
38. Mayr (1982a).
39. Burkhardt (1977).

40. This is an extension of developmental preformation theories, prevalent and popular since Aristotle, where the course of individual development (ontogeny) was thought to be the unfolding of a preprogrammed plan.
41. Bowler (1983).
42. This did not seem to contradict Darwin's thinking, as he was indifferent about whether the human races should be considered subspecies or species.
43. See Keith's review (1915).
44. Klaatsch (1923, p. 103).
45. Klaatsch (1923, pp. 100–102).
46. Keith (1910, 1911).
47. Hauser's reputation was awful, almost without equal. Marcellin Boule, the French paleontologist who just a few years later described the La Chapelle Neandertal skeleton, published vitriolic stories about his escapades in *L'Anthropologie*, his journal. Much of this rubbed off on Klaatsch (Hammond 1982), who did not help matters with his anatomically unlikely reconstruction of the Le Moustier Neandertal child.
48. Unlike those from western Europe, many central European human fossils were destroyed.
49. Gorjanović (1910).
50. The confusion of modernity with gracility constantly plagues paleoanthropology, and is an important element in the support for the Eve theory (Chapter 11). Gracile specimens look modern, although their gracility could result from small size, sexual dimorphism (women are notably more gracile than men), or as in the case of Krapina the youth of the specimens.
51. Wallace (1864).
52. Wallace, however, did not share their racism. He was antislavery and a humanist in many respects.
53. Descent from *any* ape was the issue: he would have been judged equally radical whether he believed it happened once or several times.
54. Wendt (1956, p. 282).
55. Agassiz (1850, p. 114).
56. Nordenskiöld (1929).
57. Gould (1977, p. 77).
58. Topinard (1885).
59. Haeckel (1866, pp. 287–288).
60. Source unknown.
61. Haeckel (1898b).
62. Desmond (1982, p.156).
63. Desmond and Moore (1991).
64. Mayr (1982a, p. 70).
65. Among the paleoanthropologists, most notably Tattersall (1994a, 1995a), but there are others.
66. Haeckel (1883, p. 85).
67. This being the source of the genus "*Pithecanthropus*" that Dubois used to name his Javan find. Haeckel had a dramatic effect on Dubois's thinking, as can be seen in his own attempts to use an ontogenetic process to explain the course of human evolution. We discuss this further in Chapter 8.
68. Bowler (1983).
69. Desmond (1982).

70. Gasman (1971).
71. Lenz (1931, p. 417).
72. Rich (1973).
73. The link of Nazism to Monism, however, is less direct. Certainly through its embodiment of Haeckel's ideas the philosophy of Monism was compatible with Nazism. But many monists were liberal, and in this century the Monist League had a liberal, pacifist bent (Holt, 1975). It disbanded in 1933 rather than becoming "coordinated" into the Nazi state (Richards, 1987).
74. Stein (1988, p. 52).
75. Brücher (1935).
76. Haeckel (1866).
77. Haeckel (1905, p. 390).
78. Haeckel (1883 , p. 85).

Chapter 6: The Last Stand

1. Although never at the same time; we didn't know each other then.
2. A point recently reviewed in a particularly well-written book by Dennett (1995).
3. Mayr and Provine (editors, 1980, especially the Provine essay), and Huxley (1942).
4. *The Origin of Races*, Coon (1962).
5. Haraway (1988).
6. Spencer (1979).
7. Stewart (1981).
8. Courtesy of Harvard Peabody Museum.
9. Spencer (1981).
10. Shapiro (1981) notes this important connection, which reverberated through the generations of physical anthropologists that followed. Pearson was one of the founders of 20th-century biostatistics. Interested in races and their evolution, he developed statistical techniques to deal with large samples that were soon applied far beyond the problems of physical anthropology. The importance of statistical treatment even more strongly influenced Hooton's students such as W. W. Howells, and Howells' students such as E. Giles (Milford's adviser) and G. P. Rightmire.
11. Especially Armelagos, Carlson, and Van Gerven (1982), and Lovejoy, Mensforth, and Armelagos (1982).
12. Spencer (1986, p. 314).
13. Dixon (1923, p. 503).
14. Stocking (1968).
15. Hooton (1931, pp. 572–573).
16. Hooton (1931, p. 395).
17. Hooton 1946, p. 390).
18. Hooton (1930a).
19. Hooton (1937).
20. Hays (1958, pp. 139–142).
21. Piltdown was a fraudulent "fossil," fabricated by combining parts of a recent human skull, an orang jaw, and a filed-down orang canine. Martin Hinton, curator of zoology at the British Museum in 1912, was the perpetrator but before this was known Keith was suspected (Spencer, 1990; Tobias, 1992).
22. Brace (1982).

23. Hooton (1937, p. 97).
24. Stewart (1950, p. 98).
25. Keith (1910, p. 206).
26. Weidenreich is discussed in the next two chapters.
27. Theunissen (1988).
28. Keith (1936, p. 18).
29. Keith (1948, p. 388).
30. Keith (1948, p. 387).
31. Keith (1948, p. 388).
32. Keith (1948, p. 388).
33. Keith (1950, p. 696).
34. "We are fairly safe in assuming that the Australian is far less intelligent than the Englishman" (Hooton 1946, p. 158).
35. Kühl (1994, p. 5).
36. Hooton (1936).
37. Hooton (1939a, p. 248).
38. Hooton (1939a, p. 246).
39. Herrnstein and Murray (1994).
40. Hooton (1939b, p. 11).
41. Hooton (1939b, p. 12).
42. Hooton (1939b, p. 12).
43. Boas (1902), Stocking (1976).
44. Barkan (1988).
45. Barkan (1988, p. 183).
46. Huxley and Haddon (1935).
47. Stocking (1968).
48. Harris (1968). It was widely believed at that time that culture *was* biology in the sense that people of different races could be expected to have different cultures because of their biological distinctions, that solutions to sociological problems were biological. Boas's word was far from the final one, as Nazi ideology carried the belief well into this century.
49. Boas (1912).
50. Shapiro (1939), Hulse (1962).
51. And, as described above, founder of the American Association of Physical Anthropologists.
52. Hooton (1936).
53. Marks (1995, p. 101).
54. Barkan (1988, p. 203).
55. Hooton (1946, pp. 409–410).
56. We discuss Piltdown in the next chapter.
57. It has been suggested that Coon has been confused with Gates (Shipman, 1994), and we sometimes feel we have been confused with him ourselves.
58. Gates (1948).
59. Hill (1940).
60. Weidenreich (1945b, 1946a).
61. Mayr (1982b, p. 231).
62. Brace (1982).
63. Haraway (1988).
64. Hooton (1946, p. 575).

65. Spencer (1981).
66. Garn (ed., 1964), Lasker (ed., 1960).
67. For instance, see books on culture and human evolution edited by Montagu (1962) and Spuhler (1959).
68. Marks (1995, pp. 56–58).
69. Coon (1962, p. 16).
70. Coon (1962, p. 18).
71. Coon (1962, p. 17).
72. This citation is from the figure title for Figure XXXII.
73. Coon (1962, p. 427).
74. Montagu (1963, p. 362).
75. Coon (1981, p. 21).
76. Coon (1981, p. 89).
77. Coon (1981, p. 85).
78. Hooton (1930b, p. 41).
79. We discuss Weidenreich and his thinking in the next two chapters.
80. These themes are developed throughout the introduction of Coon's 1962 book.
81. A true devotee of Hooton, who systematically collected tremendous amounts of metric data in his study of human variation, Coon collected and analyzed measurements for the human fossil record—in the first systematic attempt to do so for the entire world. Coon, however, occasionally took estimated measurements beyond the limits of acceptability. Just after a very important *Homo habilis* skull was discovered by Richard Leakey, it was pictured, with Leakey, in *National Geographic.* Coon needed to know how big the ER 1470 specimen was: it was important for his theorizing about the evolution of brain size, and this was the first complete *Homo habilis* skull to be found. "Coon calculated it using a rather Rube Goldberg approximation calibrated by the estimated width of the tie worn by Leakey, who was shown holding the skull in a publicity photo" (Shipman, 1994, p. 286f). When Coon saw the photo, he went home and searched through his closet, looking for a tie that seemed similar. But Richard was wearing a narrow tie that day, not one of the older, wider ties that Coon had in his collection. Using a tie for scale that seemed similar but actually was wider than Leakey's, he greatly overestimated 1470's size.
82. Shipman (1994) and others such as Brues, (personal communication).
83. Dobzhansky (1963, p. 172).
84. Coon (1962, p. 337).
85. Coon (1962, p. 369).
86. Coon (1981, p. 366).
87. Coon (1981, p. 354).
88. Montagu (1963).
89. *Scientific American* 267(3), special 1992 issue on "mind and brain."
90. Beals *et al.* (1984) show that brain size tends to be larger in populations adapted to colder climates.
91. Coon (1982, pp. 151 and 149).
92. Coon (1962, p. 664).
93. Coon (1962, pp. ix–x).
94. Shipman (1994, pp. 201–214).
95. *Current Anthropology* 4, number 4.
96. Howell, cited in Shipman (1994, p. 210).

97. Washburn (1963, p. 527).
98. Livingstone (1964).
99. Coon (1968, p. 275).
100. Marks (1995, p. 105).
101. Coon (1981, p. 344).
102. Coon (1981, p. 344).
103. Difficult, but not at all impossible. Tattersall (1995a, p. 215), for instance, asserts "poor Coon . . . was widely and unfairly reviled for propagating a racist doctrine."
104. Stocking (1976).
105. Stocking (1968, p. 140).
106. Boas (1894).
107. Stocking (1968, p. 281).
108. Boas (1894, p. 303).
109. Boas (1894, pp. 303, 306–308).
110. Littlefield et al. (1982).
111. Carleton Coon, courtesy of the Harvard Peabody Museum.
112. Coon (1981, p. vii).
113. Marks (1995, p.105).

Chapter 7: The Straw Man

1. Hooton (1946).
2. Andrews (1945).
3. Howells (1944).
4. Hooton, remember, had graduated from this school.
5. Howells (1942, p. 182).
6. Keith (1936).
7. Especially Keith (1916, 1919), in which he developed the Haeckelian idea that racial prejudice was a major factor in race differentiation, and competition, especially through warfare, a driving force in their evolution. These culminated in his postwar *New Theory of Human Evolution* (1948).
8. Howells (1981, p. 410).
9. Howells (1942, p. 182).
10. Howells (1942, p. 183).
11. Howells (1944), our emphasis.
12. Howells (1944), our emphasis.
13. Howells (1959, p. 244).
14. Howells (1993, p. 124).
15. Weidenreich (1943a).
16. Weidenreich (1943b, p. 256).
17. Weidenreich (1943b, p. 256).
18. Weidenreich (1946c, pp. 417–418).
19. Howells (1950, p. 81).
20. Our emphasis, Howells (1959, p. 234).
21. Howells (1959, p. 235).
22. Howells (1994a, p. 300).
23. Howells (1950, p. 85).
24. Tattersall (1995b, p. 113).
25. Cavalli-Sforza *et al.* (1993), Wood (1994).

26. Lewin (1993, p. 53).
27. Weidenreich (1936a, 1936b, 1937, 1941b, 1943b).
28. For some details of Weidenreich's life there is a short obituary by Gregory (1949), one by Eisley (1949), and we deeply appreciate the additional information provided by Weidenreich's family.
29. Weidenreich (1922b).
30. Weidenreich (1922a, 1923, 1924, 1931).
31. That is, he did not address specific questions of how species were related.
32. Weidenreich (1925a).
33. Weidenreich (1928b).
34. Hrdlička (1927, p. 250).
35. Hrdlička (1927, p. 250).
36. Modified from Trinkaus and Howells (1979) and Santa Luca (1978).
37. Weidenreich (1946a, p. 82).
38. Weidenreich (1946a, p. 91). Weidenreich's choice of the word "hybrid," although he frequently used it in quotations, was unfortunate because it implies once "pure" groups were doing the "hybridizing." Weidenreich did not mean this, but nevertheless used the then very prevalent word, confounding his meaning for those who were unfamiliar with his work as a whole. He had similar problems with other terminology, using literally incorrect, and often typological terms for "convenience." Thus, for example, he continued to use the genus name *Sinanthropus* for the lower cave Zhoukoudian fossils, while actually considering them members of *Homo sapiens.*
39. W. H. Sheldon, a psychologist and physician, later avoided this problem by developing a system in which the three main constitutional types were considered the components of what he called each individual's "somatotype": endomorph, mesomorph, and ectomorph. Each person could be given a grade to represent their degree of endomorphy, mesomorphy, and ectomorphy in Sheldon's system.
40. Proctor (1988, p. 163f).
41. Weidenreich, like many children of German Jews in his generation, was a nonbeliever. Family legend has it that he approached his friend and student Albert Schweitzer with the request to baptize his three daughters. Dr. Schweitzer allowed that he would not do so unless Weidenreich was baptized as well. Weidenreich declined to give up one religion he did not believe in, to enter another he did not believe in.
42. Proctor (1988, pp. 163–164).
43. By definition in the Nazi state, those with at least one Jewish grandparent were considered Jewish. This definition was modeled after an American one, the Bureau of Indian Affairs' definition of a Native American as a person with at least one Native American grandparent.
44. Gieseler and Breitinger (1956).
45. Weinreich (1946, pp. 240–241).
46. Proctor (1988, p. 161).
47. This surprised him—he worried he would never get the position *because* of Keith's influence, since he had so strongly and publicly rejected the validity of one of Keith's most cherished fossils, Piltdown.
48. Weidenreich (1943b, p. 246).
49. Weidenreich (1940a), but bear in mind that he and many others did not consider the much more archaic australopithecine remains to be human.

50. From Wolpoff (1996), after Black (1930) and Weidenreich (1943b).
51. It appears on the cover of Milford's 1997 edition of *Human Evolution*. A second reconstruction is distributed by Krantz casting; it is of unknown origin. A third reconstruction was recently attempted by Ian Tattersall; it has a more African appearance than the others.
52. Weidenreich (1939a).
53. Howell (1994b), Theunissen (1988).
54. von Koenigswald (1956).
55. This is Gould's (1990) suggestion.
56. von Koenigswald and Weidenreich (1939).
57. From Wolpoff (1996).
58. The exact date of Ngandong is uncertain, even to this day. Animal bones from the lowest levels of the Solo River's high terrace, where the specimens were found, were dated by uranium/thorium to 101±10 kyr (kyr means thousand years), but other high terrace dates from a nearby Solo River site are 165 kyr. A potassium/argon date from a volcanic rock near the site approximates 500 kyr. On the other hand, a date of 25–50 kyr determined on other deposits from near the site has been suggested by Henry Schwarcz. Milford often describes their age as 250±200 kyr.
59. Weidenreich (1943b, 1951).
60. Weidenreich (1939b).
61. That is, punctuated equilibrium in today's terminology.
62. Weidenreich (1939b, p. 72).
63. Weidenreich (1940b, p. 380)—this is the paper that Howells cited in 1942, describing Weidenreich as a polygenist.
64. Shapiro (1974).
65. Mann (1981).
66. Weidenreich (1945a).
67. Dodds (1973, p. 98).
68. Gregory (1949) gives a complete Weidenreich bibliography.
69. Weidenreich (1951).
70. Boule and Vallois (1957, p. 521), our emphasis.
71. Trinkaus and Shipman (1993, p. 276).
72. Weidenreich (1949, pp. 149–150).
73. Weidenreich (1939b, p. 70).
74. Weidenreich (1939b, p. 72).
75. Weidenreich (1939b, p. 74).
76. Weidenreich (1943b, p. 246).
77. Weidenreich (1946a, p. 3).
78. Weidenreich (1946a, p. 1).
79. Bowler (1986, p. 144).
80. Gates (1937).
81. Weidenreich (1946a, p. 2).
82. Howells (1950, p. 85).
83. Weidenreich (1925b).
84. Weidenreich (1947b).
85. Weidenreich (1946a, p. 83).
86. Williams (1979, p. 216).

87. Weidenreich (1946a, p. 271).
88. This is a chapter heading in Coon's book, *The Origin of Races.*
89. Weidenreich (1943b, pp. 249–250).
90. Weidenreich (1946a, p. 30).
91. At the time of his death Weidenreich had written most of a book on human evolution. The manuscript is in his notes preserved at the American Museum of Natural History. Left unfinished were several of what may well have been the most interesting parts, particularly on the origin of races, but he did complete much of a chapter called "The Course of Evolution." The citation is from this chapter.
92. Weidenreich (1946a, p. 30).
93. Weidenreich (1946a, p. 86).
94. Weidenreich (1946a, p. 86).
95. Weidenreich (1946a, p. 85).
96 Weidenreich (1943b, p. 254).
97. Tattersall (1995a, pp. 214–215).
98. Like some before him, Tattersall regarded Howells' candelabra as a valid and much-needed simplification of Weidenreich's thinking.
99. Weidenreich (1947a, p. 189).
100. Tobias (1960).
101. Gould (1980); Hammond (1979); Hooton (1954); Spencer (1990); Tobias (1992).
102. Radovčić (1988).
103. Gee (1996).
104. Friedrichs (1932).
105. Weidenreich (1943b, p. 220).
106. Weidenreich (1940b, p. 381).
107. Shipman (1994, p. 202).
108. Weidenreich addresses this in further detail in Weidenreich (1946c and 1947b), an exchange he had with Gates (1944 and 1947) in the *American Journal of Physical Anthropology.* Dobzhansky (1944) also published in the journal on the topic of races and species in human evolution. This was evidently a topic of considerable interest then.
109. Weidenreich (1946a, p. 81).
110. Weidenreich (1946a, p. 87).
111. Weidenreich (1946a, p. 84).
112. Weidenreich (1947a, p. 201).
113. Weidenreich (1943b, p. 254).
114. Weidenreich (1943b, p. 240).
115. Weidenreich (1946a, p. 89).
116. Weidenreich (1946a, p. 91).
117. Weidenreich (1946a, pp. 85–86).
118. Bourdieu (1975).
119. Keith (1950, p. 631f).
120. Coon (1962, p. viii), our emphasis.
121. Howells (1994a).
122. Trinkaus and Shipman (1993).
123. Bowler (1986, p. 109).
124. Trinkaus and Shipman (1993, p. 312), and Tattersall (1995a, p. 214).
125. As noted above, human fossils did not include australopithecines.

Chapter 8: Functional Morphology, Orthogenesis, and the Dubois Syndrome

1. Cann, Stoneking, and Wilson (1987); Cann (1987).
2. Cann, Stoneking, and Wilson (1987).
3. Harwood (1993).
4. Caspari (1933).
5. Harwood (1993, p. 264).
6. The fact that an offspring's cytoplasm is derived from the cytoplasm of the mother's egg provides the opportunity for its contents to be passed on independently of the nuclear DNA; in cytoplasmic inheritance information about a descendant's structure or function is transmitted within this cytoplasm.
7. Meaning nuclear genetics, which was understood according to the Mendelian model by then. The existence of DNA, let alone DNA outside the nucleus, was unknown.
8. Correns, first director of the prestigious Kaiser-Wilhelm Institute for Biology, was one of the most significant geneticists involved in the study of non-Mendelian genetic phenomena.
9. The more orthodox plasmon theorists, such as Correns and von Wettstein, considered the plasmon to be a unitary regulatory system, distinct from the chloroplasts and other particulate elements in the cytoplasm. More "revisionist" plasmon theorists (see Harwood 1993) considered particulate models for the "plasmon." Other workers interested in cytoplasmic inheritance did not recognize the existence of a "plasmon."
10. Harwood (1993), Mayr (1988).
11. Churchill (1968).
12. Ironically, in 1900 Correns "rediscovered" Mendel's writings, like the better-known Hugo de Vries (1848–1935), but independent of him.
13. Weismann (1885).
14. In fact some, like Weismann, were avid selectionists and consciously tried to distance themselves from non-Darwinian hypotheses.
15. Mayr (1988, p. 507).
16. Sapp (1987) notes that virtually all field biologists and paleontologists accepted the notion of dual evolutionary mechanisms, and even the genetics community was divided.
17. Mayr (1982a, p. 526).
18. Mayr (1963).
19. Mayr (1988).
20. Mayr (1988, p. 407).
21. For instance, see the exchange between evolutionary biologists in Brockman (1995).
22. Mayr (1960, 1963).
23. It is now understood that when numerous genes affect a characteristic, their proportions in a population are important for the form the feature will attain. Selection can create the potential for new combinations by changing these proportions (very rare genes will never be found together, common genes will be), and then promote or discourage the results.
24. Gould (1977).
25. Reif (1983).
26. Reif (1986).

27. Desmond (1989).
28. Stein (1988).
29. For instance, see Weidenreich (1930), in his reply to the very critical writings of Curt Stern (1930). Mayr (1980) and Rensch (1980) point to the disdain biological scientists, from embryologists to naturalists to paleontologists, then held for the importance of nuclear inheritance, a term whose meaning was coming to be limited to the transmission of genes from one generation to the next (Sapp, 1987).
30. Mayr (1980, p. 16).
31. Reif (1986).
32. Provine (1980), Pfeifer (1965).
33. Adams (1980b), Dobzhansky (1980).
34. Harwood (1993).
35. Johannsen (1922), Winkler (1924).
36. In German this literally means changes of long-lasting duration.
37. Harwood (1993, p. 110).
38. Harwood (1993, p. 115).
39. Mayr (1982a, p. 549).
40. Rensch (1980), Harwood (1993).
41. In fact, many biologists, especially paleontologists, were sure that traits controlled by nuclear inheritance were trifles, just the variations that occurred within species. They turned to the cytoplasm to account for the inheritance of traits that varied significantly between species (Sapp, 1987).
42. Weidenreich, moreover, was aware of the question of how functional variability can develop and be passed on (1925a).
43. Weidenreich (1913, 1922a, 1924).
44. And, for those that were essentially *non*evolutionary, it was not even necessary.
45. Weidenreich (1922c).
46. For instance Uncle Ernst's adviser and mentor, Kühn.
47. Roesler (1981).
48. Weidenreich (1922b).
49. Rensch (1980, p. 291).
50. Harwood (1993); Reif (1983).
51. Mayr (1982a, pp. 549–550).
52. Speaking later that day, at the Tübingen meeting, the Finnish geneticist Harry Federly gave a "brusque, condescending rejection" of Weidenreich's attempt at compromise. He argued that chromosomal genes were not malleable, and if they were "all of Mendelism would have to be scrapped." Federly rejected the experimental foundation of the *Grundstock* hypothesis and regarded it as an "attempt to sideswipe the problem [that] had no foundation." Federly basically argued that evolution could not be understood through the study of phenotypes (different physical forms), but only through genetics (Harwood 1993, p.120).
53. Harwood (1993, p. 120).
54. Harwood (1993, p. 121).
55. Proctor (1989, pp. 30–38). Of course this need not be the case, the expression of genes could be modified even if acquired characters were inherited, but, according to Proctor (p. 34), neo-Lamarckian inheritance was regularly, and vehemently, rejected in the defense of racial hygiene.
56. Sapp (1987).
57. Wright (1945).

58. Eicher (1987).
59. Caspari (1948).
60. Waddington (1942, 1952).
61. Frisancho (1979).
62. Jablonka and Lamb (1995) and Sapp (1987) discuss the scientific, social, and political aspects of the century-long dispute between nuclear and cytoplasmic inheritance.
63. Weidenreich (1947d, p. 224), although just after Weidenreich died S. L. Washburn made considerable use of Weidenreich's data and ideas in his treatment of the evolution of bipedalism (Washburn, 1950).
64. Weidenreich (1947d, p. 235).
65. Reif (1983).
66. Weidenreich (1947d, p. 222).
67. Mayr (1963, p. 690).
68. Weidenreich (1947d, p. 234).
69. Weidenreich (1947d, p. 222).
70. Weidenreich (1947d, p. 226).
71. Weidenreich (1941a).
72. Weidenreich (1948).
73. Weidenreich (1941a).
74. Gould (1977).
75. Bolk (1929).
76. Kollmann (1905).
77. Schwalbe (1906).
78. Holloway (1972).
79. Gould (1990).
80. Theunissen (1988).
81. Howell (1994b).
82. Dobzhansky (1944, pp. 260–261).
83. For instance as in Mayr (1982a).
84. Harwood (1987).

Chapter 9: Center and Edge

1. Wolpoff, Smith, Malez, Radovčić, and Rukavina (1981).
2. Brose and Wolpoff (1971).
3. Drawing by Karen Harvey.
4. Jelínek (1966, 1969, 1976, 1980a, 1983).
5. Frayer (1993), Wolpoff (1989b).
6. Adapted from Mikić (1981).
7. Drawing courtesy of D.W. Frayer.
8. Weidenreich (1946a).
9. Dobzhansky (1980).
10. Quoted in Trinkaus and Shipman (1993, p. 276).
11. Dobzhansky (1950, pp. 106–107), commenting on a paper presented by T. D. Stewart.
12. Dobzhansky (1963).
13. Coon (1982).
14. Weidenreich (1947a, p. 189).
15. Hemmer (1967, 1969).

16. Hemmer (1969, p. 179).
17. Robinson (1967).
18. Robinson (1967, p. 98).
19. Leakey (1989, p. 55).
20. Leakey (1989, p. 57).
21. Aguirre (1994).
22. Jelínek (1980b, 1982).
23. Jelínek (1978, 1980a, 1981).
24. Jelínek (1978, pp. 427–428).
25. As described in Wolpoff (1986a), Brown (1990).
26. Jelínek (1981, p. 88).
27. Wolpoff, Thorne, Jelínek, and Zhang (1993).
28. Thorne (1981).
29. Mayr (1963).
30. Parsons (1983).
31. Mayr (1963, p. 393).
32. Thorne (1981).
33. Asfaw *et al.* (1992).
34. Gamble (1994).
35. Discussed at length in Wolpoff (1996).
36. Gamble (1994).
37. Wolpoff (1996).
38. Birdsell (1957).
39. Templeton (1982).
40. After Grimaud-Hervé and Jacob (1983), Larsen, Matter, and Gebo (1991), Sartono (1980), Tyler (1995), Weidenreich (1945a).
41. Facial features may also have implications for social/sexual selection, discussed in the next chapter.
42. Templeton (1993).
43. Templeton (1993).
44. Relethford and Harpending (1994, p 251).
45. Coon (1962, p. 589).
46. Rightmire (1976, 1978).
47. Thorne *et al.* (1993).
48. Deacon (1992).
49. White (1987).
50. Hausman (1982).
51. Rightmire and Deacon (1991).
52. Churchill *et al.* (1996).
53. Morris (1992b).
54. Vigilant *et al.* (1991).
55. Brose and Wolpoff (1971).
56. von Koenigswald (1958).
57. Pilbeam (1972).
58. Stringer (1992b, 1994).
59. And there is Tattersall (1994b), and Milford's response to him (Wolpoff, 1994c).

Chapter 10: Multiregional Evolution

1. The history of Neandertals in human evolution is the topic of books by Radovčić (1988), Stringer and Gamble (1993), Trinkaus and Shipman (1993), and Wendt (1956).
2. Haeckel (1866).
3. King (1864).
4. In Boule and Vallois (1957).
5. Boule (1923, translated later in Boule and Vallois 1957, p. 521).
6. Stringer and Gamble (1993).
7. Quennell and Quennell (1945, pp. 66–67).
8. Trinkaus and Shipman (1993).
9. Weidenreich (1943a, p. 256).
10. McCown and Keith (1939).
11. Ironically, this is not too different from the age assessed for them now (see the story about Jakov's book on Gorjanović footnoted in Chapter 2), but for most of the past 60 years they were thought to be half that age, or less.
12. After McCown and Keith (1939), Larsen, Matter, and Gebo (1991), and Vandermeersch (1981, Figures 165, 181, 194, 200).
13. Howell (1951, p. 412).
14. Howell (1952).
15. Howell (1957, p. 337).
16. Howell (1957, p. 339).
17. Stringer (1982, p. 431).
18. Wells (1921, p. 287).
19. Golding (1955, p. 219).
20. Asimov (1958, p. 280).
21. Auel (1980, p. 20).
22. Stringer and Gamble (1993, pp. 31–33).
23. Asimov (1958, p. 312).
24. Auel (1980, p. 20).
25. Unless, of course, Neandertals are the same species as Eve. But with Neandertals representing the epitome of "otherness," no Eve theorist is yet ready to redefine Neandertals as modern humans!
26. Glausiusz (1995).
27. Richard Meindl (1987) suggested a similar argument some years ago but based on an immunity to TB in the heterozygotic genotype; of course, both may be correct.
28. These geneticists identified 50 variants of the gene and calculated how long it took them to evolve from a single original variation. They got their minimum date from the fact that no changes in the gene from parent to offspring were seen in the families they studied. There was, of course, no maximum age estimate.
29. Glausiusz (1995).
30. Shreeve (1996).
31. This is a real piece of Platonic essentialism, as it assumes from the onset that the Neandertals and moderns of the Levant are different things, and then wonders how they stayed that way. But we do not believe it is correct. We agree with McCown and Keith (1939), the only paleoanthropologists to study the entire Mt. Carmel collection before it was split apart. They interpret the variation to be a continuous range, not distinct types.

32. Shreeve (1995, p. 78).
33. Shreeve (1995, p. 81).
34. Frayer 1993, p. 9).
35. The figure is courtesy of D. Frayer.
36. Gamble (1994).
37. Foley (1987).
38. Moore (1994a, 1994b).
39. Owen (1965).
40. Sharrock (1974, p. 96).
41. Wolf (1994).
42. Moore (1994b, p. 937).
43. Morton and Lalouel (1973).
44. Moore (1995, p. 531).
45. Gould and Vrba (1982); Wolpoff (1996).
46. Cartmill (1992).
47. Milford was writing on the question of multiple species at one of the South African australopithecine sites (Wolpoff, 1968b). He argued the null hypothesis must be there was only one species at a time, and it would be difficult to disprove this hypothesis because of the effects of the hominid cultural adaptation. Culture, he believed, was the adaptive mechanism for the hominids, and competing species would have to compete better or change their competitive behaviors by becoming better at using the cultural adaptation. Competition, he reasoned, could only make both compete more and better, and therefore there would never be two hominid species together.
48. Franciscus and Trinkaus (1988).
49. Wolpoff (1968b).
50. Charlesworth, Lande, and Slatkin (1982, p. 476) assert: "The extent of genetic differentiation between two or more local populations is determined by the balance between gene flow and natural selection or random genetic drift."
51. Lewin (1993), Howell (1994a).
52. Brues (1972); Birdsell (1972b); Livingstone (1964).
53. Crummett (1994).
54. After Weidenreich (1937, Plate 1, Figure 3).
55. Drawing by Karen Harvey.
56. This is a point also made by Rogers (1995), who develops matrix models that show continuity is likely to be very difficult to prove from fossils unless it sends a much stronger signal than genic exchanges—a possibility that may occur from time to time, but which we would not expect to be common.
57. Wright (1943, 1986).
58. After Ridley (1993, Figure 8.11).
59. Frayer (1993).
60. After Frayer (1993, Table 7).
61. Stringer (1994).
62. Habgood (1989).
63. Specimens are drawn after Piveteau (1957) and Howells (1959).
64. Halliday (1983).
65. Bateson (1978, 1982).
66. Holloway (1974, 1975).
67. Holloway (1988).

68. For instance, Clive Gamble (1994) believes that information exchange played a critical role in successful colonization of regions outside of Africa, as populations could interchange information about new habitats and the successful strategies for adapting to them. But this requires language and it is here that kinship-based mate choice might be important. The evolution of complex language is as much a matter of understanding complicated messages as creating them, and many linguists believe the same mechanisms are involved in both (complex communication is generated and deciphered according to the same rules). If there is genetic variability in this ability, related individuals are more likely to share it than unrelated ones: the evolution of a complex interaction system such as human language may well depend on successful kin recognition, and choosing mates who are related without being too related.
69. For instance, at the time of this writing there are 1995 publications on the topic by Dorit *et al.*, Hammer, and Whitfield *et al.*
70. These are geneticists who analyze the evolution of gene pools, and many such as Relethford (1995) argue that Eve is a problem of population genetics and not mitochondrial biochemistry.
71. Ayala (1995); Relethford (1995); Templeton (1993, 1994, 1996).
72. Modified from Wright (1969, Figure 7.1).
73. Templeton (1993, 1994).
74. Templeton (1993, p. 65).
75. Relethford (1995); Avise *et al.* (1984, 1988).
76. Relethford and Harpending (1994, 1995).
77. Relethford and Harpending (1995, p. 672).
78. Rogers and Jorde (1994).
79. Relethford and Harpending (1994, p. 251).
80. Xiong Weijun *et al.* (1991).
81. Ayala (1995).
82. Li and Sadler (1991).
83. Stoneking and Cann (1989).
84. The Y chromosome of the nucleus may be useful for an origins determination because it is only inherited by one sex, in this case males. But assuming the point of origin for Y chromosomes is the point of origin for modern humans relies on the same questionable assumptions as the mitochondrial DNA basis for the Eve theory.
85. Hammer (1995); Whitfield, Sulston, and Goodfellow (1995).
86. M. Brown (1990).
87. Kuhn (1962), McEachron (1984).
88. Dobzhansky (1950).
89. Day and Stringer (1982, 1991); Stringer and Andrews (1988).
90. Wolpoff (1986a); P. Brown (1990).

Chapter 11: Modern Humans, Modern Races?

1. Lieberman and Jackson (1995) address just this issue and argue that all of the origins theories not only collide with race, but are unavoidably typological in their treatment of it and "have implications that both magnify and minimize the differences between races" (p. 239). We do not believe either is true.
2. Bordieu (1975), Hammond (1988), Mulkay and Gilbert (1982), Sapp (1987),

Shipmin (1994), and many others detail the implications of the fact that science is a *human* endeavor.

3. Mayr (1988, 1991).
4. There is, of course, more than one explanation, as Multiregional evolution is the framework of how evolution works in our polytypic species, a description of the mechanism of change, not the road map of what actually took place. That map must come from the fossil record.
5. The genesis of this usage is discussed in Chapter 1.
6. However, other explanations for the pattern of variation are discussed in Chapter 10. The differences in mtDNA variation could just as well reflect a much larger prehistoric population size in Africa and much smaller populations with numerous bottlenecks occurring as their sizes fluctuated in most other regions. Differences in variation between central and peripheral populations are a key prediction of the center and edge hypothesis.
7. Gould and Eldredge (1977, 1993).
8. We discuss these definitions in Chapter 9.
9. Wiley (1978, 1981).
10. For instance, in the various essays collected by Kimbel and Martin (1993).
11. Often named *Homo sapiens sapiens.*
12. As described by Howells (1976) and Bräuer (1984).
13. In contrast to the idea that different genetic systems have different histories.
14. Stoneking and Cann (1989).
15. From Vandermeersch (1981) and Larsen, Matter, and Gebo (1991).
16. Kollmann (1903, 1905).
17. Schwalbe (1906).
18. The Australian is drawn from the Klaatsch collection (after Wilder, 1926) and the Pygmy is an Akka (after Flower, 1888).
19. Howells (1976).
20. Parallelism, remember, was Howells' interpretation of Weidenreich's polycentrism.
21. Protsch (1975).
22. As opposed to more indirectly applied dates that are determined from fauna or flora thought to be of the same age as the hominid.
23. Beaumont, de Villiers, and Vogel (1978).
24. Bräuer (1984).
25. Bräuer (1989, p. 139).
26. Frayer *et al.* (1993).
27. No evidence of commonly African cranial features can be found in these Europeans; if anything, their crania uniquely resemble the earlier Neandertals (Frayer, 1986; Jelínek, 1976, 1983; Wolpoff, 1996). The case for an African ancestry is not clear-cut, and the absence of cranial evidence is especially problematic because forensic anthropologists find the most definitive evidence distinguishing today's Africans from peoples from other regions in the cranium, especially the face.
28. Frayer *et al.* (1994).
29. Ruff (1994), Holliday and Falsetti (1995). It is certainly possible that climate is the correct interpretation of the postcranial remains of the post-Neandertals, the people entering Europe as the glaciers retreated had to come from somewhere

but we believe that changes in locomotor behaviors had a significant influence on these proportions as well.

30. Stringer and Andrews (1988, p. 1263).
31. Frayer *et al.* (1993).
32. Wolpoff *et al.* (1994).
33. Clark (1992).
34. Bar-Yosef (1989).
35. Neither of these is clearly associated with any tools, but their estimated dates suggest they are the makers of Middle Stone Age.
36. After Piveteau (1957, Figure 537) and Larsen, Matter, and Gebo (1991).
37. Bräuer (1984).
38. Rightmire (1979, 1984).
39. Yellen, Brooks, Cornelissen, Mehlman, and Stewart (1995).
40. In contrast, barbed bone points first appear in Europe less than 40,000 years ago.
41. Cited in Gibbons (1995).
42. Also cited in Gibbons (1995).
43. Cited in Gutin (1995).
44. Klein (1992).
45. For instance, Mellars and Stringer (1989) use this phrase, but it is widely used in European archaeological circles to summarize changes that include everything from storage to body decoration and cave paintings.
46. Allsworth-Jones (1993).
47. Corruccini (1992), Rightmire (1984).
48. Rightmire (1979).
49. Allsworth-Jones (1993), who as an Eve theorist would be expected to put the Omo situation in the best possible light, as he believes modern humans originated in Africa.
50. Leakey (1969), Butzer *et al.* (1969).
51. For instance, see the discussion of state-of-the-art dating in Aitken, Stringer, and Mellars (1992).
52. Frayer *et al.* (1993, 1994).
53. Smith (1992).
54. Deacon and Shuurman (1993).
55. Deacon and Geleijnse (1988).
56. Binford (1984).
57. Stringer and Bräuer (1994).
58. Singer and Wymer (1982).
59. This is one aspect of a short, high cranium; others include a high, rounded cranial rear and arched parietals (the bones on the sides of the cranial vault).
60. In listing these anatomical features said to mark modernity, we do not mean to imply we believe such a list (or any other list) is valid.
61. Deacon (1988, 1992).
62. After Klein (1989, Figures 6.17 and 6.18).
63. Morris (1992b).
64. Hausman (1982).
65. Morris (1986).
66. Morris (1992a).
67. Wolpoff and Caspari (1996).
68. Smith (1992).

69. Courtesy of D. W. Frayer.
70. The triangular area at the base of the lower jaw, when it is prominent there is a distinct chin. Variation in this feature is discussed by Frayer *et al.* (1993). Our observations on Klasies anatomy have been verified by Lam *et al.* (1996).
71. Wobst (1990).
72. Trinkaus (1993, p. 493).
73. Continuous features are those for which variation is in small gradations, like height or cranial capacity, as opposed to discrete features which are either present or not, like the horizontal/oval form of the mandibular foramen discussed in Chapter 10.
74. After McCown and Keith (1939, Figures 168, 189), Larsen, Matter, and Gebo (1991), and Vandermeersch (1981, Figures 16, 25, 33).
75. This mix of features is found within the large sample at the Skhul Cave alone, according to McCown and Keith (1939), its describers and the only paleoanthropologists who were able to study the whole sample together (it was divided into three parts, now housed on three different continents, after their study).
76. Wu and Wu (1985).
77. Jinniushan is drawn by Karen Harvey, ZKT 11 is from Hublin (1987).
78. Pope (1992a).
79. Chen *et al.* (1994).
80. Pope (1992b), Wolpoff (1996).
81. Davidson and Noble (1992).
82. This is certainly the position taken by the Australian government and courts, who stand behind the reburial of this Pleistocene fossil along with other Native Indigenous Australian ancestors.
83. Davidson and Noble (1992).
84. Jelínek (1969, 1976), Wolpoff (1989b).
85. Brose and Wolpoff (1971).
86. Radovčić *et al.* (1988).
87. Frayer (1993).
88. Drawing by Karen Harvey.
89. And even this is not without its ambiguities; for instance, Smith (1992) suggests that the frontal piece without browridge may be juvenile, accounting for the supraorbital morphology a different way—it wasn't old enough at death to have developed one.
90. Rightmire and Deacon (1991).
91. Bräuer *et al.* (1992).
92. Rightmire and Deacon (1991).
93. Churchill *et al.* (1996).
94. Wolpoff and Caspari (1990).
95. Wolpoff and Caspari (1990, p. 395).
96. Day and Stringer (1982, and again in 1991).
97. Wolpoff (1986a).
98. P. Brown (1990).
99. Bar-Yosef and Meignen (1992).
100. From Wolpoff (1995).
101. R. Wallace (1992).
102. Holloway (1969).
103. Laszlo and Masulli edited (1993).

104. Stiner (1995).
105. It was once thought that there was a simple evolutionary sequence, in which a scavenging adaptation gave rise to a hunting one.
106. Stringer and Gamble (1993).
107. Clark and Willermet (1995, p. 155).
108. Popper (1959).
109. Kuhn (1962); Popper (1962).
110. Adams (1980b), Bell (1992), Gould (1981), Hammond (1982), Harwood (1993), Sapp (1987).
111. Clark (1993), Gasman (1971), Gillispie (1959), Ringer (1969), Stocking (1971).
112. Hull (1994), Mulkay and Gilbert (1982).
113. Templeton (1994).
114. There are good examples of this in two papers addressing multiregionalism, both abstracted from Ph.D. dissertations. Diane Waddle (1994) analyzed a series of theoretically derived mathematical matrices representing population relationships. These were used to pattern graphically genealogical configurations and genic exchanges to compare the two models: Eve and Multiregional evolution. The "expectations" determined from the two models tested were in the form of numbers that were assigned to denote the strength of the various relationships, and the analysis correlates these matrices with matrices of actual anatomical distances between sets of fossil and recent crania from Europe, western Asia, and Africa, seeking to find the highest correlation. Waddle concluded that the single origin model was correct because single origin matrices had higher correlations with the anatomical data. But her data were consistent with both, even though her analysis put Multiregional evolution at the disadvantage by failing to include data from eastern Asia and Australasia that strongly support it, and by the (incorrect) assertion that Multiregional evolution implies that the human races are markedly different (the magnitude of difference between the human races is an observation, not the result of which theory is correct). In fact, according to Lyle Konigsberg and colleagues (1994) a different and more appropriate analysis of her data provides more support for the Multiregional model than for the replacement one. Another attempt to examine the fossil record for evidence of regional continuity is by Marta Lahr (1994), who defines a set of 30 skeletal traits and purports to test the prediction that Multiregional evolution implies there should be higher frequencies of regional features linking fossil and living populations in each of the areas examined. She shows that 37 percent of her features fit this regional patterning, but rejects Multiregional evolution in favor of a single African origin for all modern populations anyway, because the majority of features do not fit it. Both of these studies show that the data are consistent with Multiregional and Eve models. Most important for us, while either had the potential to refute Multiregional evolution, neither was able to do so.
115. Stringer (1994).
116. Howells (1973).
117. For instance see Van Vark and Howells (eds., 1984).
118. Corruccini (1984, 1987); Kowalski (1972).
119. This was just the point Milford had argued with John Robinson when he was at Wisconsin for a year. But others who had taught at one time at this great university—for instance, W.W. Howells and his student G.P. Rightmire—collected large data sets of complete specimens to be used in multivariate analyses.

120. His 1969 dissertation, *Metric Trends in Hominid Dental Evolution*, analyzed all of the tooth measurements of past and present human populations that could be found, or measured at major U.S. museums. He had to write computer programs for all of the univariate statistical analyses and tabular presentations of the summarized data. Life is much simpler today.
121. Wolpoff (1976) .
122. Wolpoff (1992b, 1994d); van Vark and Bilsborough (1991, 1994).
123. G. Neumann (1956) strongly supported Weidenreich's contention that the Shandingdong specimens were generally Mongoloid ancestors of Native Americans, and L. Oschinsky (1964) detailed additional resemblances to "New World Mongoloids." C. Coon (1962, pp. 474–475) wrote that UC 101 "does not conform strictly to a Mongoloid model, but neither do all Chinese alive today. . . . [With the other crania it] bear[s] the same kind of relationship to the modern Chinese that the Upper Paleolithic skulls of Europe do to modern Europeans." In their review of Late Pleistocene human evolution in China, Wu Xinzhi and Wu Maolin of the IVPP confirmed (1985) the Mongoloid affinity of the three crania, and argued that in comparison with modern populations and taking age and sex into account, they were relatively homogeneous. In an earlier, more detailed analysis, Wu Xinzhi (1961) had already concluded "there is no reason to consider the Upper Cave fossils as representing anything other than a Mongoloid population entirely consistent with what is known about the development of modern *H. sapiens sapiens* in North China" (Wu's translation).
124. Howells (1989) and two others based on his data set: Van Vark and Dijkema (1988) and Kamminga and Wright (1988).
125. Wright (1992).
126. Wolpoff (1995).
127. Wright (1992).
128. An example we like is a study by J. Kidder and his colleagues (Kidder *et al.*, 1992). They examine the question of whether a multivariate definition of "modern human" is possible. Using the fossil human remains from Europe and Asia, they concluded that no multidimensional diagnosis of modern human can be both exclusive of archaic populations and inclusive of all humanity. The problem they focused on is not whether modern humans are significantly different from their forebears, but whether the differences can be characterized the same way everywhere or if they are distinct to some degree from one place to another.
129. Of course, racial differences may be negligible for other reasons, such as persistent genic exchanges between populations and similar responses to the same surces of selection. The degree of difference between races does not distinguish between these competing models of their evolution.
130. Clark (1992).
131. Milford may have unwittingly helped promote this interpretation, by asserting that modern human origin is the origin of races on several occasions. If modern humans originated from a single source within the last 200,000 years it would have to be true, and in fact this was what Milford meant. But there are other interpretations of how Multiregional evolution could pertain to race origins, for instance, the polygenic one, and Milford, who regularly writes on top of the tests he gives to undergraduates, "Say what you mean, and mean what you say," finally followed his own advice and became more careful.
132. Birdsell (1972b).

133. Littlefield *et al.*, (1982).
134. Waddle (1994) and Lieberman and Jackson (1995) make such assessments, reaching the same conclusions that (as Lieberman and Jackson [1995, p. 239] put it) "the one-million-year span for the multiregional view implies more time for greater differentiation through adaptation to local ecological pressures." Now granted, this might come from their misunderstanding of how Multiregional evolution is supposed to work, as they believe (p. 238) it suggests "the parallel evolution of human 'races' from ancestral *erectus* types," citing *Milford* for this "insight," and expect that the model relies on a synchronized "simultaneous speciation of *erectus* into *sapiens*" in different parts of the world. But this kind of misconception could never arise in the first place if they understood that races do not evolve by differentiating further and further, like a bush, and that long-term adaptations to local conditions are unlikely in the dynamically changing Pleistocene. Such understandings are improbable, however, because they want to get rid of the race concept, not consider it in an evolutionary context.
135. Habgood (1989); this is to be expected as the vast majority of genetic variations are found within human populations, not between them.
136. These are chapter headings in Coon's book *The Origin of Races.*
137. Barkan (1992); Montagu (1964); Sauer (1992).
138. The correct percentage depends on the definition of black that is used.

Epilogue: After Eve

1. Dorit, Akashi, and Gilbert (1995).
2. Avise (1989).
3. Dorit, Akashi, and Gilbert (1995, p. 1183).
4. Wilford (1995).
5. Soon thereafter Wills reached a similar conclusion for the "date" of mitochondrial "Eve." It is possible these calculations show an older coalescence than once thought, but a more reasonable interpretation is that they are extremely inaccurate as divergence time estimates.
6. For instance as in Harpending, Sherry, Rogers, and Stoneking (1993), and Rogers and Harpending (1992).
7. Excoffier (1990), Excoffier and Langaney (1989), Li and Sadler (1991), and subsequently Ayala (1995), but perhaps of greatest interest two subsequent studies sequencing different Y chromosome segments found coalescent times that were different by almost an order of magnitude (Hammer 1995; Whitfield, Sulston, and Goodfellow 1995).
8. Wilford (1995).
9. Wub-e-ke-niew (1995).
10. In societies where descent is traced along patrilines, political hierarchies can be disrupted when patrilines are prevented from continuing, and this seems to have been a policy, in some cases, meant to force assimilation of Native Aboriginal peoples by destroying their political cohesion.
11. In Isaac Asimov's *Second Foundation* (1953), his labyrinth of inter-nested explanations for where, or who, the *Second Foundation* really is, uses the distinction between the answer that satisfied and the answer that was true. His point, and ours, is that these may not at all be the same.
12. This pattern, of course, it not limited to Europeans. Richards (1995) reports a study of a black community from Viche, Equador, in which the maternal contri-

bution of mtDNA from indigenous Native Americans is estimated as about twice the contribution of nuclear genes. There have been similar studies of the maternal and paternal contributions of Europeans to the African-American gene pool, for instance, Hsieh and Sutton (1992), in which the contributions are found to be about equal. But perhaps not surprisingly, searching for a study paralleling the Ecuadorians is much more difficult. Virtually nothing is known of the African-Americans' maternal and paternal contributions to European-Americans.

GLOSSARY

AAPA. American Association of Physical Anthropologists.
aboriginal. Native, indigenous.
Acheulean. An archaeological industry beginning about 1.5 million years ago and spanning more than a million years, found first in Africa and then across the inhabited world. The forms of Acheulean tools appear to have been preconceived by their makers, the most characteristic artifact being hand axes, bifacially flaked, pear-shaped, pointed general-purpose tools of vastly varying sizes. The most common tools are amorphous and ubiquitous utilized flakes.
acquired characters. Features of an organism that develop because of environmental influence and not from heredity.
adaptation. Changing to fit, or respond to, the requirements of the environment.
AJPA. *American Journal of Physical Anthropology.*
allele. Any of the alternative forms of a gene at a specific locus (position) on a chromosome.
allometry (allometric scaling). Generally, the effect of size on shape. Specifically, any relationship of anatomical variables (Y, a function of the independent variable X) that fits the equation $Y = AX^k$ (A is a constant, k the coefficient of allometry).
allopatric speciation. Species formation when there is geographical isolation.
anagenesis. Anatomical change in a single lineage over time that is thought to be sufficient, by some, to name a new species.
anterior. Front.
apes. The large hominoid arboreal primates, a group whose definition is based on similarity and not common descent (apes are not a monophyletic group).
archaic. Ancient—as in archaic humans, meaning premodern humans.
artifacts. Humanly modified objects.

Aurignacian. An early Upper Paleolithic industry, classically defined in France where it appears about 32,000 years ago, thought by many to be exclusively associated with modern humans.

australopithecine. Referring to members of the genus *Australopithecus,* early bipedal but otherwise apelike human ancestry.

Australopithecus afarensis. An early species of the australopithecines, found between 3 and 4 million years ago in Tanzania, Kenya, and Ethiopia. The species is very small, weighing less than chimpanzees but with gorilla-sized brains and molar teeth. Skeletal evidence and footprints show it was bipedal.

Bauplan. The basic interrelated structural characteristics of a species.

biface. Stone artifact with flakes removed from two intersecting surfaces (flaked on both sides), creating a sharp edge.

biocultural. An approach that emphasizes the reciprocal interactions of human biology and culture.

biogenetic law. The principle that ontogeny recapitulates phylogeny.

biomechanics. Pertaining to the physics of the skeletal system, especially its static and dynamic analyses.

biometric. Studies based on measurements of biological systems; for paleoanthropologists, this means measurements of bones.

bipedal. Two-legged.

bottleneck. A brief period of intense selection or very small population size through which only certain genes survive and come to characterize the population.

bovid. Cattle, oxen, or closely related animals.

B.P. Before present.

braincase. The part of the skull that encloses the brain, i.e., not the face.

browridge. *See* supraorbital torus.

canalization. Adjustments to the developmental process that allow it to bring about one specific end result, regardless of minor variations in the conditions encountered during the growth process.

canine. A conical or spadelike tooth (depending on species) located between the incisors and premolars.

catastrophism. The theory that systematic calamities extinguish species so other species can migrate (according to older thinking) or evolve (according to more recent thinking) to take their place.

character. A feature.

character state. The particular expression of a feature; for instance, the cephalic index might be low, or hair color might be red or brown.

Châtelperronian. An early Upper Paleolithic industry, so far only associated with Neandertal fossils, often preceding, but sometimes contemporary with, the earliest Aurignacian. Also sometimes contemporary with the Mousterian.

clade. A group composed of all the species descended from a single common ancestor; a monophyletic group.

cladistics (cladism). Classification reflecting genealogy (recency of common descent) by means of shared derived characters; also called *phylogenetic systematics.*

cladogenesis. The splitting of one clade (or lineage) into two non-interbreeding entities.

clinal variation. Continuous, gradual variation of a trait over space. In the context of geographic patterns of a trait's variation, *clines* are the directions in space that follow its unchanging expression, and *gradients* are the directions that are perpendicular to the clines and therefore mark the path of the trait's maximum change.

colonizing species. A species with a high rate of reproduction, readily able to take advantage of new habitats because of its genetic or behavioral variation and internal subdivisions.

convergence. The independent evolution of the same or very similar features in two or more species from different features in their last common ancestor.

correlation. The co-relationship of two features. A measure (a number between absolute values of 1 and 0) of how much variation in one corresponds to variation in the other. Correlation is not causation.

craniofacial. Pertaining to the skull and the face.

craniology. Study of the bony skull and mandible.

craniometry. Measurement of the cranium.

cranium (crania). The skull without the mandible.

crest. An elongated bony ridge with a sharp edge. A *simple crest* is created by the directional pull of a single muscle, a *compound crest* by the opposing pulls of two muscles.

cytoplasm. The outer portion of a cell where amino acids are joined into proteins and enzymes and where mitochondria and other cell bodies reside.

cytoplasmic inheritance. The transmission of information through the cytoplasm, such as through the DNA in mitochondria. The fact that an offspring's cytoplasm is derived from the cytoplasm of the mother's egg provides the opportunity for its contents to be passed on independently of the nuclear DNA.

dauermodification. Literally "lasting changes," adaptive changes thought to arise in the cytoplasm after an organism's exposure to environmental stimuli, which with longer exposures become stable and heritable.

deme. A local population, a group of potentially interbreeding individuals at a particular place.

demography. The study of a population's main life-history parameters—its growth, size, composition, and age-specific birth rates and death rates.

dentin. Internal tissue in a tooth crown and tissue of the root, surrounding the pulp cavity and surrounded by the crown enamel. This bonelike substance is softer than the enamel.

Diluvial. Referring to the time of the biblical flood.

dimorphism. Two forms: see sexual dimorphism.

discriminant function. A multivariate statistic that uses combined measurements to distinguish groups that are known to differ.

DNA (deoxyribonucleic acid). The molecule that carries the genetic information (genes) in all organisms except the RNA viruses. It consists of two long polysugar-phosphate strands connected by base pairs and twisted in a double helix. The bases appear in connected congruent sets (like rungs on a ladder): there are only four different bases, and their sequence is the genetic code.

drift (genetic drift). Changes in gene frequencies due to random or stochastic variation and not the result of selection, mutation, or genic exchanges. Drift changes are most prominent in small populations.

dryopithecine. Member of the subfamily Dryopithecinae, a type of fossil ape.

ecology. The interrelationships between organisms or populations and their environment.

ecotone. A boundary region between ecological zones.

embryo. An organism just after conception; for instance, in humans during the first eight weeks of in-utero development.

enamel. The very hard, prismatically structured outer surface of a tooth crown.

Eoanthropus. "Dawn ape," the name given to Piltdown.

ESR. Electron spin resonance, a dating technique that measures the number of electrons trapped in crystals to estimate how long it has been since the crystal formed.

essentialism. The belief that variation is discrete and bordered and falls into fixed classes or essences that each represent perfection. Biological variation away from the "ideal type" is thereby considered the result of imperfection, deviation from the most perfect state.

ethnogenesis. The model of biocultural evolution that emphasizes the impermanent nature of social, ethnic, and biological groups (*ethnos*) in which each population can have multiple ancestors and multiple descendants. Like a river these independently varying aspects of humanity diverge, covary in parallel to each other, and then remerge and realign with other identities, with new divergences following.

eugenics. Selective breeding of humans, individuals or races, with the goal of bettering the human species, often by preventing "socially undesirable" individuals from having offspring.

Eve theory. Sometimes known as "Out-of-Africa" or "Out-of-Africa II" (recognizing that there was an earlier migration from Africa); it is one version of the Out-of-Africa theories stipulating that all modern humans have a common recent origin, in an African population that became a new species and swept around the world replacing native peoples.

evolution. Genetic change, change in a population's gene pool from generation to generation; Darwin's descent with modification.

exaptation. The name given for a character that evolved to fulfill a different function than the one it currently serves.

extinction. The disappearance of a group in part or all of its range.

extranuclear. Outside the cell's nucleus; extranuclear inheritance is based on chromosomes that are not part of the nuclear genome, in this case the mitochondrial chromosome.

fauna. Animals.

femur (femora). Long bone of the thigh or upper leg.

fetalization. The preservation of fetal stages of ancestors in adult descendants.

finalism. The belief that there is an inherent tendency for the world to change in order to attain some preordained goal.

fitness. A measure of fertility and survivorship reflecting genetic variation.

flake (flake tools). A usually sharp-edged stone fragment struck or pressured off of a core (a larger rock or nodule).

flora. Plants.

fossil. Preserved remains of once-living plants or animals in which the replacement of organic or inorganic materials by soil minerals has begun. Naturally occurring casts are also considered fossils.

frontal bone. The cranial bone forming the forehead and the top of the orbits and nose.

frontal trigone. A backward-facing triangular form of the lateralmost part of the supraorbital torus. The apex is created by a prominent temporal ridge, and the torus is thicker at the trigone than it is more centrally.

gene. A unit of inheritance carried on a chromosome, transmitted from generation to generation by the gametes and controlling some aspect of the development of an individual.

gene flow. A form of genic exchange in which genetic material is transferred between populations because of interbreeding or mate exchanges.

gene pool. All of the genes found in the members of a population.

genealogical species. Species that are defined by common ancestry and are treated as distinct individuals with definite beginnings and ends.

genic exchange. The sharing of genetic material because of gene flow or migration.

genome. The entire DNA component of a cell, a structured array consisting of genes and their parts, units of DNA replication, and nonfunctioning regions.

genotype. The genetic makeup of an organism; its total genetic material.

genus (genera). A group of closely related species; a monophyletic category for the taxon above the species level that includes one or more species.

***Gigantopithecus*.** An extinct Asian ape species related to orangutans, known only by their very large, high-crowned teeth and enormous jaws.

glaciation. A period of glacial advance, divided into alternating and increasingly severe colder (stadial) and warmer (interstadial) periods, and separated by longer warm periods (interglacial).

gracile. Slender, delicately built, weak muscle attachments or bony buttresses.

great apes. The four large living apes: bonobos, chimpanzees, gorillas, and orangutans. Great apes are not a monophyletic group.

Grundstock. A hypothesized non-Mendelian form of inheritance for basic anatomical structures, the *Grundstock* (a German name for the basic holdings of a library or museum) was thought to be located in the cytoplasm or perhaps throughout the cell. Changes in the *Grundstock* were described as beyond the ability of mutations and required macroevolutionary processes.

habitat. The normal home or environment of a group.

hand axe. A teardrop- or pear-shaped, bifacially flaked stone implement.

heterochrony. Evolutionary changes caused by variation in the relative time of appearance and rate of development of features.

heterozygosity. The occurrence of two different alleles at a particular locus.

hominid. Extant humans and their unique ancestors and collateral relatives, extending back in time until the split with the line leading to chimpanzees (the closest human relative).

Hominidae. The hominid family.

hominoid. Member of the Hominoidea, the superfamily including humans and apes and their unique ancestors.

Homo. Same; the genus of our species (*Homo sapiens*) and perhaps of others such as *Homo erectus*, *Homo habilis*, etc.

Homo erectus. A species preceding our own that many believe is the first true human. A number of paleoanthropologists subsume it within *Homo sapiens* for this reason and because it is a direct linear ancestor.

homology. A feature in two or more species that is the same because of descent—it evolved from the same feature in the last common ancestor of the species.

homoplasy (parallelism). The separate appearance of a feature with the same character state in two or more species that developed independently from a different character state of the feature in the last common ancestor.

homozygosity. The occurrence of two identical alleles at a locus.

incisor. Broad tooth at the frontmost part of the jaw.

inductivist. One who conducts science by generalizing from specific observations.

industry. A group of archaeological assemblages found over a specific region or time whose artifacts are similar.

IVPP. The Beijing-based Institute for Vertebrate Paleontology and Paleoanthropology.

kinship. Relationships between people that are based on real or imagined descent and (sometimes) marriage.

kyr. Kiloyear, unit of a thousand years (100,000 years is 100 kyr).

Lamarckism. The evolutionary ideas of Jean-Baptiste de Lamarck, often reduced to meaning the inheritance of acquired characteristics, or use inheritance, but including his belief that there is an inner drive to nature, in which simpler forms of organisms are successively transformed into more complex ones.

lateral. Away from the midline of the body.

Levallois. A technique for flake production in which a stone core is shaped like a tortoise shell and a single flake with preformed shape is struck from it.

Levallois flake. A flake struck from a Levallois core.

lineage. A group of ancestral-descendant organisms that are reproductively isolated from other lineages; a line of common descent.

lithic. Of or pertaining to stone.

lumper. One who emphasizes similarities and formalizes variation at higher taxonomic levels (*cf.* splitter).

macroevolution. Evolution above the species level; the evolution of higher taxa and the processes that result from differences in species survivorship or rates of speciation.

mandible. Lower jaw.

mandibular foramen. Opening on the internal surface of the ramus for the mandibular vessels and nerve to pass. There are two distinct anatomies to its rim. In the *common form* the rim is V-shaped, with a groove separating the anterior and posterior parts. In the *horizontal-oval form* there is no groove and the rim is horizontally oriented and oval in shape, the anterior and posterior parts connected.

mandibular groove. A groove extending down from the lower rim of the mandibular foramen or from just below it.

mandibular torus *(torus mandibularis).* Shelflike thickening of bone extending transversely on the inside of the symphysis.

mandibular (mental) trigone. An upward pointing raised triangular form at the base of the symphysis.

marginal ridges. Elevated ledges on the vertical edges of the inner surface of the incisors.

mastication. Chewing.

mastoid process. A pyramid-shaped prominence of cancellous bone on the temporal bone behind the external auditory meatus. Muscles that extend and turn the head attach to it.

mate recognition system. The system of signals (chemical, olfactory, vocal, visual) that bring together potential breeding partners.

matrilineage. A line of descent reckoned through females (mother to daughter).

maxilla. Paired bone of the upper jaw, enclosing the nose and the inner and lower rims of the eye and holding the teeth.

maxillary notch. A distinct angle between the base of the cheek and the outer wall of the palate.

megafauna. Large animals from horse to elephant size.

Meganthropus. A genus name given to remains of the robust, small-brained earliest humans from Java. Most often these remains are classified as *Homo erectus*, or for those who believe that species is invalid, early *Homo sapiens.*

Mendelism (Mendelian). The theory of particulate inheritance; the inheritance of small, indivisible units, one from each parent, that combine in the offspring.

mental eminence. Projecting mandibular trigone, or chin.

microevolution. Evolution of populations over short periods of time, in response to observable causes.

microfauna. Very small animals, such as bats, moles, or mice.

migration (as a cause of genic exchange). The movement of genes caused by individuals moving, including new individuals entering (*immigration*) or leaving (*emigration*) a population, introducing or removing genetic material and thereby changing allele frequencies..

mitochondria. The small extranuclear organelles (bodies) within a cell's cytoplasm that control the cell's production of energy from food through the production of ATP (adenosine triphosphate).

mitochondrial DNA (mtDNA). The single (double-stranded) DNA molecule that controls the development and functioning of the cytoplasm's mitochondrion containing it. Because reproduction is by cloning, mtDNA is usually passed along female lines, as part of the egg's cytoplasm.

molar. A flat posterior tooth.

molecular clock. A means of determining dates of evolutionary divergences using genetic similarities between extant species and assuming that molecular evolution proceeds at a constant rate.

monogamy. A long-lasting exclusive sexual bond (or marriage, in humans) between a single adult male and female; a mating system resulting in social groups based on single pairs and their offspring.

monogenism (monogenic). Single origin.

monomorphic. A species without significant geographic variation; only one form.

monophyletic (group). All the descendants of a last common ancestor (these can be individuals, species, etc.).

morphology. The form, shape, and/or structure of organisms.

morphospecies. A typological species recognized on the basis of morphological differences or discontinuities.

morphotype. A common form defined by anatomical features, thought to characterize a race or species.

Mousterian. Middle Paleolithic industry of Europe and western and central Asia, based on flake tools that are often stuck from prepared cores. Often associated with Neandertals, it is also found with non-Neandertal and mixed populations in the Middle and Late Pleistocene.

Multiregional evolution. The evolutionary model that posits humans evolved as an interconnected polytypic species from a single origin in Africa. The small population effects during initial colonizations outside of Africa and adaptations to local conditions helped establish regional differences, some of which were subsequently maintained through isolation-by-distance and adaptive variation. Advantageous changes spread widely because of genic exchanges and the common background of the evolving cultural system whose elements also could spread. Most modernizing features arose at different times and places and diffused independently according to this model. The key insight of Multiregional evolution is that the causes of diversity and the causes of common evolutionary change were not opposed to each other; genic exchanges were not the opposite of differentiation, they were part of its cause.

multivariate statistics. Statistical procedures that are designed to treat simultaneously (and to assess relationships among) several variables per object.

mutagens. Physical or chemical causes of mutations.

mutation. An error in replication or other alteration of the nucleotide bases creating a change in the sequence of base pairs on a DNA molecule. If the change occurs in the DNA of a somatic cell, the mutation may cause a change in the organism's phenotype (leading, for example, to cancer) but will not affect the organism's offspring; only mutations in the germ cells can cause heritable changes in the offspring.

natural selection. Differences in reproductive success and/or survivorship of individuals that result in the unequal contribution of genotypes to the gene pool of the next generation.

Naturphilosophie. A romantic movement, mainly associated with 18th- and early 19th-century central Europeans, that assumed the unification of all natural phenomena and processes through developmental and transcendental beliefs.

neo-Lamarckism. The theory of use inheritance; the idea that features acquired by the activities of an individual can, if repeated over several generations, become hereditary.

neoteny. Retention of the juvenile features of an ancestral species in the adult form of a descendant species, by slowing down the rate of development.

neural. Pertaining to a nerve or to the nervous system.

niche. The limited portion of the environment, in terms of space, resources, etc., that a species fits and/or which it requires for its survival and reproductive success.

nonmetric trait. Feature whose expression is better or more accurately described as a discrete character state than as a measurement.

NSF. The National Science Foundation of the United States government.

nuchal torus. A thickened bony prominence extending transversely across some or all of the back of the head, on the occipital bone, reflecting the pattern of muscle use as it separates the nuchal plane below from the occipital plane above.

occipital bone. The bone forming the back of the skull and much of its base.

occlusal. The surfaces of the opposing teeth that meet for chewing; in an occlusal view, the grinding or bite surfaces of the teeth are shown.

ontogeny. The developmental history of an individual from egg to adult.

orbit. Bony socket for the eye.

orthogenesis. The theory that evolution is inner-directed and once started in a certain direction cannot deviate from it.
osteology. The study of bones and their variations.
paleo- (palaeo-). Old.
paleoanthropology. The study of human evolution, including the details of prehistory and theories of causation and relationship.
Paleolithic. Literally the old stone age, the period when humans relied on a stone technology to sustain a scavenging/hunting/gathering adaptation.
parallelism. *See* homoplasy.
parietal. Wall. One of the flat paired bones forming part of the lateral side of the skull.
patrilineage. Line of descent traced through males (father to son).
pelvis. The bony structure composed of the sacrum and three paired bones: the ilium, ischium, and pubis, which fuse together in adults as paired innominate bones.
penecontemporary. Living at or almost at the same time.
phenetics. A method of systematics in which relationships are determined by degrees of similarity.
phenotype. Appearance of an individual; the observed set of characteristics, the result of the interaction between genotype and environment.
phyletic. Pertaining to descent (*cf.* phylogeny).
phylogenetic species. A monophyletic group of individuals whose identity can be diagnosed by at least one shared unique feature.
phylogenetics. The study of how genealogical relationships can be determined from morphological similarities that are homologous.
phylogeny. A hypothesis about how fossil and living species are related in a genealogical framework.
plasmon. A cell's cytoplasm and everything in it, as opposed to the contents of the nucleus.
polycentrism. Franz Weidenreich's theory of multiple centers of human evolution connected by a network of genic exchanges.
polygenism (polygenic). Multiple origins.
polygyny. A type of mating system (or marriage, in humans) in which one male mates with more than one female.
polymorphic. Showing a variety of forms; a feature with alternative character states.
polyphyletic. A group in which the last common ancestor is not included.
polytypic. A variable taxon that contains more than one taxon of the next lower category, such as a species with several subspecies or races.
population. A community of potentially interbreeding individuals, usually at a given locality or within a limited geographic region.
postcranial skeleton. All elements of the skeleton behind the skull (below, in humans).
posterior. Back.

preformation. The belief that the egg holds a preformed adult in miniature that is unfolded during development.

prehistorian. A scientist who studies the archaeology and biology of human evolution.

premolar. Tooth lying between the canine and molars, usually smaller than the molars and generally flat except for the most anterior lower tooth in species with a canine cutting complex.

progressionism. The idea that progress is a natural process, intrinsic to the character of natural and social histories.

prognathous. Forward protrusion of the facial region, as a whole or in part.

prosimian. The Prosimii, a primate suborder including lemur, loris, and tarsier species.

provenience. The exact circumstances of how a specimen is related to the deposit in which it is found.

proximal. Closer to the midline of the body, applies to the appendicular skeleton.

Punctuated Equilibrium theory. A model of evolution in which important changes occur when new species are formed; only rarely are changes slowly and gradually accumulated during the stable periods between speciations.

quadrupedalism. Four-footed posture and locomotion.

race. A group of individuals geographically (and for humans, also culturally) determined who share a common gene pool and varying combinations of distinguishing characteristics.

radiometric dating. A process based on nuclear decay, in which the atomic composition changes with time and can be used to determine the age.

ramus. The portion of bone at an angle to the body, as in ascending ramus (mandible), pubic ramus (innominate).

range. Territory normally occupied.

***Rassenkunde*.** Usually translated as "ethnology," it had the meaning of "racial science" in Nazi Germany

rectilinear (rectolinear). Straight-line, as in evolution proceeding in a singular direction.

reductionism. The simplification of a concept in order to understand its basics; the oversimplification of a concept to the point that its essence is lost.

resharpen. Trim the edges of a worn-out stone tool.

resorb. To destroy and remove bone or parts of bone by osteoclasts (bone cells with digestive enzymes).

retromolar space (gap). A space or gap at the rear of a mandible between the back of the last molar and the anterior edge of the ascending ramus where it crosses the alveolar margin.

robust. A large or heavily built body or body part.

Rubicon. A border or barrier.

sagittal keel (torus). A thickening of part or all of the midline of the frontal bone and/or parietal bones where they meet sagittally.

sagittal plane. A vertical plane on the midline that divides the body into a right and left half.
selection. *See* natural selection.
sexual dimorphism. A polymorphic character in which males and females of a species differ in some aspect of their anatomy not directly related to reproduction or birth.
simian. Pertaining to monkeys or apes.
Single Species hypothesis. The theory that only one hominid species at a time could be expected because culture should so broaden hominid niches that competition for limiting resources between species would be inevitable and lead to enhanced cultural abilities and further niche broadening. Only one of the competing species would be expected to persist. The hypothesis rests on the assumption that all manifestations of culture result in effective niche expansion.
skull. The bony skeleton of the head, including the lower jaw.
sociobiology. The study of the biological basis of all social behavior.
sociocultural. Pertaining to society and culture.
somatotype. Sheldon's classification of physical types that were also felt to correspond to personality variables. Three primary somatotypes were recognized: ectomorphs, mesomorphs, and endomorphs.
specialized. (1) Derived, in the sense of differing from the ancestral condition; (2) adapted to a limited range of resources.
speciation. The process whereby species multiply; the acquisition of reproductive isolation between populations, splitting one species into two.
species. In living animals a group of populations (**Biological** species) that can actually or potentially interbreed and have fertile offspring, and are reproductively isolated from other species. *Also see* genealogical, morpho-, and phylogenetic species.
speleology. The study or exploration of caves.
splitter. One who emphasizes differences and formalizes variation at lower taxonomic levels (*cf.* lumper).
stochastic. Random.
stratigraphy. The location or position of fossil or other deposits relative to other buried layers or features.
subspecies. A geographically defined aggregate of local populations which differs, with various degrees of significance (depending on the author), from other such subdivisions of the species.
superciliary arches. Smoothly rounded bulges of bone found on the frontal bone of the skull at its center and extending over the inner portion of the upper orbital border.
supraorbital torus (tori). Browridge: a thickened ridge or shelf of bone above the orbits at the base of the forehead, continuously, although not necessarily evenly, developed from the middle of the cranium to each side.
suture. A joint where two bones interdigitate and are separated by fibrous tissue. The joints between most of the bones of the skull are sutures. Most

sutures join and the bones eventually fuse together as individuals grow older.

symphysis (*pl.*, **symphyses**). A flexible fibrocartilaginous joint found on the midline of the body, such as the mandibular symphysis and the pubic symphysis.

systematics. The science of the diversity of organisms and their relationships and classification.

taxon (*pl.*, **taxa**). A monophyletic group of organisms recognized as a formal unit, at any level of a hierarchic classification.

taxonomy. The theory and practice of classifying organisms.

temporal bone. Complex bone on the side and base of the cranium that includes the ear, mandibular joint, and a portion of the side of the braincase.

tool. An artifact with a functional use.

torus (*pl.*, **tori**). A smooth rounded ridge or protuberance on a bone.

transformationism. A theory of evolutionary change in which it is thought that a species changes from one condition into another, often combined with progressionist thinking.

typological. A single individual or limited variation used to epitomize a sample; *see* essentialism.

uniformitarianism. Originally a concept in geology, the precept that the processes observable in the present can be used to explain the past.

unilinear. A single line.

vertebra (*pl.*, **vertebrae**). One of the bony segments of the vertebral column.

vitalism. The belief in a "life force" that may influence individual development or evolutionary change.

wadi. Gully.

Würm glaciation. The most recent glaciation, spanning from over 100,000 years to about 11,000 years ago.

ZKD. Zhoukoudian.

zygomatic arch. Bony arch on the lateral part of the cheek formed by projections of the zygomatic bone and the temporal bone, enclosing the fibers of the temporalis muscle and for attachment of the masseter muscle.

zygomatic (also **malar**) **bone.** The facial bone that makes up the cheek corner and outer orbital pillar, and encloses the front part of the temporal fossa.

zygote. The fertilized egg that results from the union of two gametes.

REFERENCES AND FURTHER READINGS

Adams, M.B. 1980a Severtsov and Schmalhausen: Russian morphology and the evolutionary synthesis. In E. Mayr and W.B. Provine (eds): *The Evolutionary Synthesis: Perspectives on the Unification of Biology.* Harvard University Press, Cambridge. pp. 193–225.

Adams, M.B. 1980b Sergi Chetverikov, the Kol'stov Institute, and the evolutionary synthesis. In E. Mayr and W.B. Provine (eds): *The Evolutionary Synthesis: Perspectives on the Unification of Biology.* Harvard University Press, Cambridge. pp. 242–278.

Agassiz, L. 1850 Diversity of origin of the human races. *Christian Examiner* 49:110–145.

Aguirre, E. 1994 *Homo erectus* and *Homo sapiens:* one or more species? In J.L. Franzen (ed): *100 years of* Pithecanthropus: *The* Homo erectus *problem.* Courier Forschungsinstitut Senckenberg 171:333–339.

Aitken, M.J., C.B. Stringer, and P.A. Mellars 1992 *The Origin of Modern Humans and the Impact of Chronometric Dating.* Princeton University Press, Princeton, NJ.

Alexeyev, V.P. 1978 *Paleoanthropology of the Globe and the Formation of Human Races.* Nauka, Moscow.

Allsworth-Jones, P. 1993 The archaeology of archaic and early modern *Homo sapiens:* an African perspective. *Cambridge Archaeological Journal* 3(1):21–39.

Andrews, R.C. 1945 *Meet Your Ancestors.* Viking, New York.

Armelagos, G.J., D.S. Carlson, and D.P. Van Gerven 1982 The theoretical foundations and development of skeletal biology. In F. Spencer (ed): *A History of American Physical Anthropology 1930–1980.* Academic Press, New York. pp. 305–328.

Asfaw, B., Y. Beyene, G. Suwa, R.C. Walter, T.D. White, G. WoldeGabriel, and T. Yemane 1992 The earliest Acheulean from Konso-Gardula. *Nature* 360:732–735.

Asimov, I. 1953 *Second Foundation.* Avon Books of Doubleday & Company, New York.

Asimov, I. 1958 The ugly little boy. In I. Asimov (ed): *The Best Fiction of Isaac Asimov.* Grafton, London.

Auel, J. 1980 *The Clan of the Cave Bear.* Bantam, Toronto.

Avise, J.C. 1989 Nature's family archives. *Natural History* 98(3):24–26.

Avise, J.C., R.M. Ball, and J. Arnold 1988 Current versus historical population sizes in vertebrate species with high gene flow: a comparison based on mitochondrial DNA lineages and inbreeding theory for neutral mutations. *Molecular Biology and Evolution* 5(4):331–344.

Avise, J.C., J.E. Neigel, and J. Arnold 1984 Demographic influences on mitochondrial DNA lineage survivorship in animal populations. *Journal of Molecular Evolution* 20(2):99–105.

Ayala, F.J. 1995 The myth of Eve: molecular biology and human origins. *Science* 270:1930–1936.

Baker, J.R. 1974 *Race.* Oxford University Press, New York.

Banton, M. 1987 *Racial Theories.* New York: Cambridge University Press.

Banton, M., and Harwood, J. 1975 *The Race Concept.* David and Charles, London.

Bar-Yosef, O. 1989 Mousterian adaptations—a global view. In *The Fossil Man of Monte Circeo: Fifty Years of Studies on the Neandertals of Latium. Quaternaria Nova* 1:575–591.

Bar-Yosef, O., and L. Meignen 1992 Insights into Levantine Middle Paleolithic cultural variability. In H.L. Dibble and P. Mellars (eds): *The Middle Paleolithic: Adaptation, Behavior, and Variability.* University of Pennsylvania Museum Monograph 78, University Museum Symposium Series IV:1–163–182.

Bar-Yosef, O., and B. Vandermeersch 1993 Modern humans in the Levant. *Scientific American* 268(4):94–100.

Barkan, E. 1988 Mobilizing scientists against Nazi racism, 1933–1939. In G.W. Stocking Jr. (ed): *Bones, Bodies, Behavior. Essays on Biological Anthropology.* University of Wisconsin Press, Madison. pp. 180–205.

Barkan, E. 1992 *The Retreat of Scientific Racism: Changing Concepts of Race in Britain and the United States Between the World Wars.* Cambridge University Press, New York.

Barton, N.H., and B. Charlesworth 1984 Genetic revolutions, founder effects, and speciation. *Annual Review of Ecology and Systematics* 15:133–164.

Bateson, P. 1978 Sexual imprinting and optimal outbreeding. *Nature* 273:659–660.

Bateson, P. 1982 Preferences for cousins in Japanese quail. *Nature,* 295:236–237.

Beadle, G., and E.L. Ephrussi 1935 Transplantation in *Drosophila. Proceedings of the National Academy of Sciences USA* 21:642–646.

Beadle, G., and E.L. Tatum 1941 Genetic control of biochemical reactions in Neurospora. *Proceedings of the National Academy of Sciences USA* 27:499–506.

Beals, K.L., C.L. Smith, and S.M. Dodd 1984 Brain size, cranial morphology, climate, and time machines. *Current Anthropology* 25:301–330.

Beatty, J. 1994 The proximate/ultimate distinction in the multiple careers of Ernst Mayr. *Biology and Philosophy* 9:333–356.

Beaumont, P.B., H. de Villiers, and J.C. Vogel 1978 Modern man in sub-Saharan Africa prior to 49,000 years B.P.: a review and evaluation with particular reference to Border Cave. *South African Journal of Science* 74:409–419.

Becker, C.L. 1932 *The Heavenly City of the Eighteenth Century Philosophers.* Yale University Press, New Haven.

Bell, R. 1992 *Impure Science: Compromise and Political Influence in Scientific Research.* John Wiley & Sons, New York.

Bendyshe, T. 1865a The history of anthropology. *Memoirs of the Anthropological Society of London* 1:335–458.

Bendyshe, T. (ed) 1865b *The Anthropological Treatises of Johann Friedrich Blumenbach.* Longman, Green, Longman, Roberts, and Green, London.

Binford, L.R. 1981 *Bones: Ancient Men and Modern Myths.* Academic Press, New York.

Binford, L.R. 1984 *Faunal Remains from Klasies River Mouth.* Academic Press, New York.

Birdsell, J.B. 1957 Some population problems involving Pleistocene man. *Cold Spring Harbor Symposia on Quantitative Biology* 22:47–69.

Birdsell, J.B. 1972a *Human Evolution.* Rand McNally, Chicago.

Birdsell, J.B. 1972b The problem of the evolution of human races: classification or clines? *Social Biology* 19:136–162.

Black, D. 1930 On an adolescent skull of *Sinanthropus pekinensis* in comparison with an adult skull of the same species and with other hominid skulls, recent and fossil. *Palaeontologia Sinica,* Series D, Volume 7, Fascicle 2.

Boas, F. 1894 Human faculty as determined by race. *Proceedings of the American Association for the Advancement of Science* 43:301–327.

Boas, F. 1902 The foundation of a national anthropological society. *Science* 15:804–809.

Boas, F. 1911 *The Mind of Primitive Man.* Macmillan, New York.

Boas, F. 1912 Changes in the bodily form of descendants of immigrants. *American Anthropologist* 14:530–562.

Boas, F. 1924 The question of racial purity. *The American Mercury* 3:163–169.

Bock, W.J. 1994 Ernst Mayr, naturalist: his contributions to systematics and evolution. *Biology and Philosophy* 9:267–327.

Bolk, L. 1929 Origin of the racial characteristics in man. *American Journal of Physical Anthropology* 13(1):1–28.

Boorstin, D. 1948 *The Lost World of Thomas Jefferson.* Beacon Press, Boston.

Boule, M. 1913 *L'Homme Fossile de La Chapelle-aux-Saints.* Masson, Paris.

Boule, M. 1923 *Les Hommes Fossiles. Eléments de Paléontologie Humaine.* Masson, Paris.

Boule, M., and H.V. Vallois 1957 *Fossil Men.* Dryden, New York.

Bourdieu, P. 1975 The specificity of the scientific field and the social conditions of the progress of reason. *Social Science Information* 1:19–47.

Bowler, P.J. 1983 *The Eclipse of Darwinism: Anti-Darwinian Evolution Theories in the Decades Around 1900.* John Hopkins University Press, Baltimore.

Bowler, P.J. 1986 *Theories of Human Evolution: A Century of Debate, 1844–1944.* John Hopkins University Press, Baltimore.

Bowler, P.J. 1988 *The Non-Darwinian Revolution.* John Hopkins University Press, Baltimore.

Bowler, P. J. 1989 *The Mendelian Revolution.* Johns Hopkins University Press, Baltimore.

Bowler, P.J. 1992 From "savage" to "primitive": Victorian evolutionism and the interpretation of marginalized peoples. *Antiquity* 6:721–729.

Brace, C.L. 1964 The fate of the "Classic" Neanderthals: a consideration of hominid catastrophism. *Current Anthropology* 5:3–43 and 7:204–214.

Brace, C.L. 1981 Tales of the phylogenetic woods: the evolution and significance of evolutionary trees. *American Journal of Physical Anthropology* 56(4):411–429.

Brace, C.L. 1982 The roots of the race concept in American physical anthropology. In F. Spencer (ed): *A History of American Physical Anthropology 1930–1980.* Academic Press, New York. pp. 11–29.

Bräuer, G. 1984 The "Afro-European *sapiens* hypothesis" and hominid evolution in East Asia during the late middle and upper Pleistocene. In P. Andrews and J.L.

Franzen (eds): *The Early Evolution of Man, with Special Emphasis on Southeast Asia and Africa. Courier Forschungsinstitut Senckenberg* 69:145–165.

Bräuer, G. 1989 The evolution of modern humans: a comparison of the African and non-African evidence. In P. Mellars and C.B. Stringer (eds): *The Human Revolution: Behavioural and Biological Perspectives on the Origins of Modern Humans.* Edinburgh University Press, Edinburgh. pp. 123–154.

Bräuer, G., F. Zipfel, and H.J. Deacon 1992 Comment on the new maxillary finds from Klasies River, South Africa. *Journal of Human Evolution* 23(5):419–422.

Brockman, J. 1995 *The Third Culture.* Simon & Schuster, New York.

Brose, D.S., and M.H. Wolpoff 1971 Early upper Paleolithic man and late middle Paleolithic tools. *American Anthropologist* 73:1156–1194.

Brown, M.H. 1990 *The Search for Eve.* Harper and Row, New York.

Brown, P. 1990 Osteological definitions of "anatomically modern" *Homo sapiens:* a test using modern and terminal Pleistocene *Homo sapiens.* In L. Freedman (ed): *Is Our Future Limited by Our Past?* Proceedings of the Third Conference of the Australasian Society of Human Biology. Centre for Human Biology, University of Western Australia, Nedlands. pp. 51–74.

Brücher, H. 1935 Ernst Haeckel, ein Wegbereiter biologischen Staatdenkens. *Nationalsozialistische Monatshefte* 6:1087–1098.

Brues, A. 1972 Models of clines and races. *American Journal of Physical Anthropology* 37:389–399.

Burkhardt, R.W., Jr. 1977 *The Spirit of System: Lamarck and Evolutionary Biology.* Harvard University Press, Cambridge.

Burkhardt, R.W., Jr. 1994 Ernst Mayr: biologist-historian. *Biology and Philosophy* 9:359–371.

Butzer, K.W., F.H. Brown, and D.L. Thurber 1969 Horizontal sediments of the lower Omo valley: Kibish Formation. *Quaternaria* 11:15–29.

Cann, R.L. 1987 In search of Eve. *The Sciences* 27:30–37.

Cann, R.L., M. Stoneking, and A.C Wilson 1987 Mitochondrial DNA and human evolution. *Nature* 325:31–36.

Cartmill, M. 1992 New views on primate origins. *Evolutionary Anthropology* 1(3):105–111.

Cartmill, M., D. Pilbeam, and G. Ll. Isaac 1986 One hundred years of paleoanthropology. *American Scientist* 74(4):410–420.

Caspari, E. 1933 Über die Wirkung eines pleiotropen Gens bei der Mehlmotte Ephestia kühniella. *Zeitschrift für Entwicklungsmechanik der Organismen* 130:353–381.

Caspari, E. 1948 Cytoplasmic inheritance. *Advances in Genetics* 2:1–66.

Caspari, R., and M.H. Wolpoff 1990 The morphological affinities of the Klasies River Mouth skeletal remains (abstract). *American Journal of Physical Anthropology* 81(2):203.

Caspari, R., and M.H. Wolpoff 1995 The pattern of human evolution. In H. Ullrich (ed): *Man and the Environment in the Paleolithic. Études et Recherches Archéologiques de l'Université de Liège* 62:19–27.

Cavalli-Sforza, L.L., P. Menozzi, and A. Piazza 1993 Demic expansions and human evolution. *Science* 259:639–646.

Cavalli-Sforza, L.L., A. Piazza, P. Menozzi, and J. Mountain 1988 Reconstruction of human evolution: bringing together genetic, archaeological, and linguistic data. *Proceedings of the National Academy of Sciences USA* 85:6002–6006.

Chambers, R. 1844 *Vestiges of the Natural History of Creation.* Churchill, London. Published anonymously.

Charlesworth, B., R. Lande, and M. Slatkin 1982 A Neo-Darwinian commentary on macroevolution. *Evolution* 36:474–498.

Chen Tiemei, Yang Quan, and Wu en 1994 Antiquity of *Homo sapiens* in China. *Nature* 368:55–56.

Churchill, F.B. 1968 August Wesimann and a break from tradition. *Journal of the History of Biology* 1:91–112.

Churchill, S.E., O.M. Pearson, F.E. Grine, E. Trinkaus, and T.W. Holliday 1996 Morphological affinities of the proximal ulna from Klasies River Main Site: archaic or modern? *Journal of Human Evolution* 31(3): 213–237.

Clark, G.A. 1992 Continuity or replacement? Putting modern human origins in an evolutionary context. In H. Dibble and P. Mellars (eds): *The Middle Paleolithic: Adaptation, Behavior, and Variability*. University of Pennsylvania Museum, Philadelphia. pp. 183–205.

Clark, G.A. 1993 Paradigms in Science and Archaeology. *Journal of Archaeological Research* 1(3):203–234.

Clark, G.A., and C.M. Willermet 1995 *In search of the Neandertals: some conceptual issues with special reference to the Levant. Cambridge Archaeological Journal* 5(1):153–156.

Clark, J.D. 1992 African and Asian perspectives on the origins of modern humans. *Philosophical Transactions of the Royal Society*, Series B, 337:201–215.

Coleman, W. 1964 *Georges Cuvier, Zoologist*. Harvard University Press, Cambridge.

Coleman, W. 1980 Morphology in the evolutionary synthesis. In E. Mayr and W.B. Provine (eds): *The Evolutionary Synthesis: Perspectives on the Unification of Biology*. Harvard University Press, Cambridge. pp. 174–180.

Coon, C.S. 1939 *The Races of Europe*. Macmillan, New York.

Coon, C.S. 1950 Human races in relation to environment and culture, with special reference to the influence of culture upon genetic changes in human evolution. *Cold Spring Harbor Symposia on Quantitative Biology* 15:247–258.

Coon, C.S. 1962 *The Origin of Races*. Knopf, New York.

Coon, C.S. 1965 *The Living Races of Man*. Knopf, New York.

Coon, C.S. 1968 Comment on "bogus science." *Journal of Heredity* 59:275.

Coon, C.S. 1981 *Adventures and Discoveries. The Autobiography of Carleton S. Coon*. Prentice-Hall, Englewood Cliffs, NJ.

Coon, C.S. 1982 *Racial Adaptations. A Study of the Origins, Nature, and Significance of Racial Variations in Humans*. Nelson-Hall, Chicago.

Coon, C.S., S.M. Garn, and J.B. Birdsell 1950 *Races: A Study of the Problems of Race Formation in Man*. C.C. Thomas, Springfield, IL.

Corruccini, R.S. 1984 Interpretation of metrical variables in multivariate analysis. In G.N. van Vark and W.W. Howells (eds): *Multivariate Statistical Methods in Physical Anthropology*. D. Reidel, Dordrecht.

Corruccini, R.S. 1987 Shape in morphometrics: comparative analyses. *American Journal of Physical Anthropology* 73(3):289–303.

Corruccini, R.S. 1992 Metrical reconsideration of the Skhul IV and IX and Border Cave 1 crania in the context of modern human origins. *American Journal of Physical Anthropology* 87(4):433–445.

Cosans, C. 1994 Anatomy, metaphysics, and values: the ape brain debate reconsidered. *Biology and Philosophy* 9:129–165.

Crummett, T. 1994 The three dimensions of shovel-shaping. In J. Moggi-Cecchi and P. Luckett (eds): *Proceedings of the Ninth International Symposium on Dental Anthropology*. Angelo Pontecorboli Editore, Florence.

Dart, R.A., with D. Craig 1959 *Adventures with the Missing Link.* Viking, New York.

Davidson, I., and W. Noble 1992 Why the first colonization of the Australian region is the earliest evidence of modern human behavior. *Archaeology in Oceania* 27(3): 113–119.

Day, M.H., and C.B. Stringer 1982 A reconsideration of the Omo Kibish remains and the erectus-sapiens transition. In H. deLumley (ed): *L'Homo erectus et la Place de l'Homme de Tautavel parmi les Hominidés Fossiles.* Louis-Jean Scientific and Literary Publications, Nice. Volume 2, pp. 814–846.

Day, M.H., and C.B. Stringer 1991 Les restes crâniens d'Omo-Kibish et leur classification à l'intérieur de Genre *Homo. L'Anthropologie* 95(2/3):573–594.

Deacon, H. 1988 The origins of anatomically modern people and the South African evidence. *Paleoecologica Africana* 19:193–200.

Deacon, H.J. 1992 Southern Africa and modern human origins. *Philosophical Transactions of the Royal Society,* Series B, 337:177–183.

Deacon, H.J., and V.B. Geleijnse 1988 The stratigraphy and sedimentology of the main site sequence, Klasies River, South Africa. *South African Archaeological Bulletin* 43:5–14.

Deacon, H.J., and R. Shuurman 1993 The origins of modern people: the evidence from Klasies River. In G. Bräuer and F.H. Smith (eds): *Continuity or Replacement? Controversies in Homo sapiens Evolution.* Balkema, Rotterdam. pp. 121–129.

Dennett, D.C. 1995 *Darwin's Dangerous Idea: Evolution and the Meaning of Life.* Simon & Schuster, New York.

Desmond, A. 1982 *Archetypes and Ancestors.* University of Chicago Press, Chicago.

Desmond. A. 1989 *The Politics of Evolution.* University of Chicago Press, Chicago.

Desmond, A., and J.R. Moore 1991 *Darwin: the Life of a Tormented Evolutionist.* W.W. Norton, New York.

Diamond, J.M. 1990 A pox on our genes. *Natural History* 99(2):26–30.

Dixon, R.B. 1923 *The Racial History of Man.* Scribners, New York.

Dobzhansky, Th. 1937 *Genetics and the Origin of Species.* Columbia University Press, New York.

Dobzhansky, Th. 1944 On species and races of living and fossil man. *American Journal of Physical Anthropology* 2(3):251–265.

Dobzhansky, Th. 1950 Comment on "The problem of the earliest claimed representatives of *Homo sapiens,*" by T.D. Stewart. *Cold Spring Harbor Symposia on Quantitative Biology* 15:106–107.

Dobzhansky, Th. 1962 *Mankind Evolving: the Evolution of the Human Species.* Yale University Press, New Haven.

Dobzhansky, Th. 1963 The possibility that *Homo sapiens* evolved independently 5 times is vanishingly small. *Scientific American* 208(2):169–172.

Dobzhansky, Th. 1968 More bogus "science" of race prejudice. *Journal of Heredity* 59:102–104.

Dobzhansky, Th. 1980 The birth of the genetic theory of evolution in the Soviet Union in the 1920's. In E. Mayr and W.B. Provine (eds): *The Evolutionary Synthesis: Perspectives on the Unification of Biology.* Harvard University Press, Cambridge. pp. 229–242.

Dodds, J.W. 1973 *The Several Lives of Paul Fejos: A Hungarian-American Odyssey.* Wenner Gren Foundation, New York.

Dorit, R.L., H. Akashi, and W. Gilbert 1995 Absence of polymorphism at the ZFY locus on the human Y chromosome. *Science* 268:1183–1185.

Drapeau, M. 1992 *Franz Weidenreich: Quelques aspects de son œuvre.* Master's dissertation, University of Montreal.

Dubois, E. 1894 *Pithecanthropus erectus:* eine menschenähnliche Übergangsform von Java. Landesdruckerei, Batavia.

Dunn, L.C. 1965 *A Short History of Genetics.* McGraw-Hill, New York.

Eckhardt, R.B. 1987 Evolution east of Eden. *Nature* 326:749.

Eckhardt, R.B. 1989 Evolutionary morphology of human skeletal characteristics. *Anthropologischer Anzeiger* 47(3):193–228.

Eckhardt, R.B. 1989 Matching molecular and morphological evolution. *Human Evolution* 4(4):317–319.

Eicher, E.M. 1987 Ernst W. Caspari: geneticist, teacher, and mentor. In J.G. Scandalios (ed): *Molecular Genetics of Development. Advances in Genetics* 24:xv–xxix.

Eisley, L.C. 1949 Franz Weidenreich, 1873–1948. *American Journal of Physical Anthropology* 7(2):241–253.

Excoffier, L. 1990 Evolution of human mitochondrial DNA: evidence for a departure from a pure neutral model of populations at equilibrium. *Journal of Molecular Biology* 30:125–139.

Excoffier, L., and A. Langaney 1989 Origin and differentiation of human mitochondrial DNA. *American Journal of Human Genetics* 44:73–85.

Ferembach, D. 1986 History of human biology in France. Part 1: The early years. *International Association of Human Biologists Occasional Papers* 2(1).

Flower, W.H. 1888 Description of two skeletons of Akkas, a Pygmy race from central Africa. *Journal of the Royal Anthropological Institute* 18:3–18.

Foley, R. 1987 *Another Unique Species. Patterns in Human Evolutionary Ecology.* Wiley, New York.

Franciscus, R.G., and E. Trinkaus 1988 Nasal morphology and the emergence of *Homo erectus. American Journal of Physical Anthropology* 75(4):517–527.

Frayer, D.W. 1986 Cranial variation at Mladeč and the relationship between Mousterian and Upper Paleolithic hominids. In V.V. Novotný and A. Mizerová (eds): *Fossil Man. New Facts, New Ideas. Papers in Honor of Jan Jelínek's Life Anniversary. Anthropos* (Brno) 23:243–256.

Frayer, D.W. 1993 Evolution at the European edge: Neanderthal and Upper Paleolithic relationships. *Préhistoire Européenne* 2:9–69.

Frayer, D.W., M.H. Wolpoff, F.H. Smith, A.G. Thorne, and G.G. Pope 1993 The fossil evidence for modern human origins. *American Anthropologist* 95(1):14–50.

Frayer, D.W., M.H. Wolpoff, A.G. Thorne, F.H. Smith, and G.G. Pope 1994 Getting it straight. *American Anthropologist* 96(2):424–438.

Freedman, L., W.F.C. Blumer, and M. Lofgren 1991 Endocranial capacity of Western Australian Aboriginal crania: comparisons and association with stature and latitude. *American Journal of Physical Anthropology* 84(4):399–405.

Friedrichs, H.F. 1932 Schädel und Unterkiefer von Piltdown: neuer Untersuchen. *Zeitschrift für Anatomie und Entwicklungsgeschichte* 98:199–262.

Frisancho, A.R. 1979 *Human Adaptation.* C.V. Mosby, St. Louis.

Gamble, C.S. 1994 *Timewalkers. The Prehistory of Global Colonization.* Harvard University Press, Cambridge.

Garn, S.M. (ed) 1964 *Culture and the Direction of Human Evolution.* Wayne State University Press, Detroit.

Gasman, D. 1971 *The Scientific Origins of National Socialism: Social Darwinism in Ernst Haeckel and the German Monist League.* Elsevier, New York.

Gates, R.R. 1937 Genetics and race. *Man* 28–32.

Gates, R.R. 1944 Phylogeny and classification of hominids and anthropoids. *American Journal of Physical Anthropology* 2:279–292.

Gates, R.R. 1947 Specific and racial characters in human evolution. *American Journal of Physical Anthropology* 5:221–224.

Gates, R.R. 1948 *Human Ancestry from a Genetical Point of View.* Harvard University Press, Cambridge.

Gee, H. 1994 What is our line? *London Review of Books*, January 27:19.

Gee, H. 1996 Box of bones "clinches" identity of Piltdown palaeontology hoaxer. *Nature* 381:261–262.

Ghiselin, M.T. 1980 The failure of morphology to assimilate Darwinism. In E. Mayr and W.B. Provine (eds): *The Evolutionary Synthesis: Perspectives on the Unification of Biology.* Harvard University Press, Cambridge. pp. 180–193.

Gibbons, A. 1995 Old dates for modern behavior. *Science* 268:495–496.

Gieseler, W., and E. Breitinger 1956 Geleitwort. *Anthropologischer Anzeiger* 20.

Gillispie, C.C. 1959 *Genesis and Geology: A Study of the Relations of Scientific Thought, Natural Theology, and Social Opinion in Great Britain, 1790–1850.* Harper and Row, New York.

Giovanni, D.-B. 1993 Migrations, genetic variability, and DNA polymorphisms. *Current Anthropology* 34(5):765–775.

Glass, B., O. Temkin, and W.L. Strauss, Jr. (eds) 1959 *Forerunners of Darwin.* John Hopkins University Press, Baltimore.

Glausiusz, J. 1995 Hidden benefits. *Discover* 16(3):30–31.

Golding, W. 1955 *The Inheritors.* Faber and Faber, London.

Goldman, N., and N. Barton 1992 Human origins: genetics and geography. *Nature* 357:440–442.

Gorjanović-Kramberger, D. 1906 *Der diluviale Mensch von Krapina in Kroatia. Ein Beitrag zur Paläoanthropologie.* Kreidel, Wiesbaden.

Gorjanović-Kramberger, D. 1908 Zur Kinnbildung des *Homo primigenius.* Bericht über die Prähistoriker-Versammlung 23.–31. Juli zur Eröffnung des Anthropologischen Museums in Köln. pp. 109–113.

Gorjanović-Kramberger, D. 1909 Der vordere Unterkieferabschnitt des Altdiluvialen Menschen in seinen genetischen Beziehungen zum Unterkiefer des rezenten Menschen und jenem der Anthropoiden. *Zeitschrift für Inductive Abstammungs- und Vererbungslehre* 1:403–439.

Gorjanović-Kramberger, D. 1910 Über *Homo aurignacensis Hauseri. Verhandlungen der geologischen Reichstalt* 14:300–303, and *Homo aurignacensis Hauseri* in Krapina? *Verhandlungen der geologischen Reichstalt* 14:312–317.

Gould, S.J. 1976 Ladders, bushes, and human evolution. *Natural History* 85(4):24–31.

Gould, S.J. 1977 *Ontogeny and Phylogeny.* Harvard University Press, Cambridge.

Gould, S.J. 1978 Morton's ranking of races by cranial capacity. *Science* 200:503–509.

Gould, S.J. 1980 The Piltdown controversy. *Natural History* 89(8):8–28.

Gould, S.J. 1981 *The Mismeasure of Man.* Norton, New York.

Gould, S.J. 1982 The Hottentot Venus. *Natural History* 91(10):20–27.

Gould, S.J. 1987 Bushes all the way down. *Natural History* 96(6):12–19.

Gould, S.J. 1988 Honorable men and women. *Natural History* 97(3):16–20.

Gould, S.J. 1989 Grimm's greatest tale. *Natural History* 98(2):20–28.

Gould, S.J. 1990 Men of the thirty-third division. *Natural History* 99(4):12–24.

Gould, S.J. 1991a The smoking gun of eugenics. *Natural History* 100(12):8–17.

Gould, S.J. 1991b Petrus Camper's Angle. In *Bully for Brontosaurus.* Norton, New York pp. 229–240.
Gould, S.J. 1992 We are all monkey's uncles. *Natural History* 101(6):14–21.
Gould, S.J. 1994 The geometer of race. *Discover* 15(11):65–69.
Gould, S.J., and N. Eldredge 1977 Punctuated equilibria: the tempo and mode of evolution reconsidered. *Paleobiology* 3:115–151.
Gould, S.J., and N. Eldredge 1993 Punctuated equilibrium comes of age. *Nature* 366:223–227.
Gould, S.J., and R.C. Lewontin 1979 The spandrels of San Marco and the Panglossian paradigm: a critique of the adaptationalist programme. *Proceedings of the Royal Society, London,* Series B, 205:581–598.
Gould, S.L., and E. Vrba 1982 Exadaptation—a missing term in the science of form. *Paleobiology* 8:4–15.
Greene, J. C. 1954 Some early speculations on the origin of human races. *American Anthropologist* 56:31–41.
Greene, J.C. 1959 *The Death of Adam: Evolution and Its Impact on Western Thought.* Iowa State University Press, Ames.
Gregory, W.K. 1949 Franz Weidenreich, 1873–1948. *American Anthropologist* 51:85–90.
Gribbin, J., and J. Cherfas 1982 *The Monkey Puzzle: Reshaping the Evolutionary Tree.* Bodley Head, London.
Grimaud-Hervé, D., and T. Jacob 1983 Les pariétaux de Pithécanthropine Sangiran 10. *L'Anthropologie* 87:469–474.
Gutin, J.A. 1995 Do Kenya tools root birth of modern thought in Africa? *Science* 270:1118–1119.
Habgood, P.J. 1989 The origin of anatomically modern humans in Australasia. In P. Mellars and C.B. Stringer (eds): *The Human Revolution: Behavioural and Biological Perspectives on the Origins of Modern Humans.* Edinburgh University Press, Edinburgh. pp. 245–273.
Haeckel, E. 1866 *Generelle Morphologie der Organismen.* 2 volumes. Reimer, Berlin.
Haeckel, E. 1883 *The History of Creation, or the Development of the Earth and Its Inhabitants by Natural Causes. A Popular Exposition of the Doctrine of Evolution in General, and That of Darwin, Goethe, and Lamarck in Particular.* Appleton, New York.
Haeckel, E. 1896 *The Evolution of Man: A Popular Exposition of Human Ontogeny and Phylogeny.* 2 volumes. Kegan, Trench, Trubner and Company, New York.
Haeckel, E. 1898a On our present knowledge of the origin of man. *Annual Report of the Smithsonian Institution* pp. 461–480.
Haeckel, E. 1898b *The Last Link.* A. and C. Black, London.
Haeckel, E. 1905 *The Wonders of Life.* Harper, New York.
Halliday, T.R. 1983 The study of mate choice. In P. Bateson (ed): *Mate Choice.* Cambridge University Press, Cambridge. pp. 3–32.
Hammer, M.F. 1995 A recent common ancestry for human Y chromosome. *Nature* 378:376–378.
Hammond, M. 1979 A framework of plausibility for an anthropological forgery: the Piltdown case. *Anthropology* 3:47–58.
Hammond, M. 1982 The expulsion of the Neanderthals from human ancestry: Marcellin Boule and the social context of scientific research. *Social Studies of Science* 12:1–36.

Hammond, M. 1988 The shadow of man paradigm in paleoanthropology, 1911–1945. In G.W. Stocking Jr. (ed): *Bones, Bodies, Behavior. Essays on Biological Anthropology.* University of Wisconsin Press, Madison. pp. 117–137.

Haraway, D.J. 1988 Remodeling the human way of life. Sherwood Washburn and the new physical anthropology. In G.W. Stocking, Jr. (ed): *Bones, Bodies, Behavior. Essays on Biological Anthropology.* University of Wisconsin Press, Madison. pp. 206–259.

Harpending, H.C. 1994 Signature of ancient population growth in a low-resolution mitochondrial DNA mismatch distribution. *Human Biology* 66(4):591–600.

Harpending, H.C., S.T. Sherry, A.R. Rogers, and M. Stoneking 1993 The genetic structure of ancient human populations. *Current Anthropology* 34(4):483–496.

Harris, M. 1968 *The Rise of Anthropological Theory.* Thomas Y. Crowell, New York.

Harwood, J. 1987 National styles in science: genetics in Germany and the United States between the wars. *Isis* 78(1987):390–414.

Harwood, J. 1993 *Styles of Scientific Thought: The German Genetics Community 1900–1933.* University of Chicago Press, Chicago.

Hausman, A.J. 1982 The biocultural evolution of Khoisan populations of southern Africa. *American Journal of Physical Anthropology* 58(3):315–330.

Hays, H.R. 1958 *From Ape to Angel. An Informal History of Social Anthropology.* Knopf, New York.

Hedges, S.B., S. Kumar, K. Tamurs, and M. Stoneking 1992 Human origins and analysis of mitochondrial DNA sequences. *Science* 255:737–739.

Heilborn, A. 1923 Introduction to *The Evolution and Progress of Mankind,* by H. Klaatsch. Fisher Unwin, London. pp. 15–29.

Hemmer, H. 1967 *Allometrie-Untersuchungen zur Evolution des menschlichen Schädels und seiner Rassentypen.* Fischer, Stuttgart.

Hemmer, H. 1969 A new view of the evolution of man. *Current Anthropology* 10(2–3):179–180.

Herrnstein, R.J., and C. Murray 1994 *The Bell Curve. Intelligence and Class Structure in American Life.* The Free Press, New York.

Hill, W.C. Osman 1940 Anthropological nomenclature. *Nature* 145:260–261.

Hiorns, R.W., and G.A. Harrison 1977 The combined effects of selection and migration in human evolution. *Man* 12:438–445.

Holliday, T.W., and B. Falsetti 1995 Lower limb length of European early modern humans in relation to mobility and climate. *Journal of Human Evolution* 29(2):141–153.

Holloway, R.L. 1969 Culture: a human domain. *Current Anthropology* 10:395–412.

Holloway, R.L. 1972 Australopithecine endocasts, brain evolution in the Hominoidea, and a model of hominid evolution. In R. Tuttle (ed): *The Functional and Evolutionary Biology of Primates.* Aldine, Chicago. pp. 185–203.

Holloway, R.L. 1974 The casts of fossil hominid brains. *Scientific American* 231(1): 106–116.

Holloway, R.L. 1975 The role of human social behavior in the evolution of the brain. Forty-third James Arthur Lecture on the Evolution of the Human Brain. American Museum of Natural History, New York.

Holloway, R.L. 1988 Brain. In I. Tattersall, E. Delson, and J. Van Couvering (eds): *Encyclopedia of Human Evolution and Prehistory.* Garland, New York. pp. 98–105.

Holt, N. 1975 Monists & Nazis: a question of scientific responsibility. *Hastings Center Report* 5:37–43.

Home, H. 1774 *Sketches of the History of Man.* Edinburgh.

Hooton, E.A. 1925 The asymmetrical character of human evolution. *American Journal of Physical Anthropology* 8:125–141.

Hooton, E.A. 1926 Methods of racial analysis. *Science* 63:75–81.

Hooton, E.A. 1930a Doubts and suspicions concerning certain functional theories of primate evolution. *Human Biology* 2:223–249.

Hooton, E.A. 1930b An unnamed anthropologist among the wilder whites. *The Harvard Alumni Bulletin.* Harvard University Publication, Cambridge.

Hooton, E.A. 1931 *Up From the Ape.* Macmillan, New York.

Hooton, E.A. 1936 Plain statement about race. *Science* 83:511–513.

Hooton, E.A. 1937 *Apes, Men, and Morons.* G.P. Putnam, New York.

Hooton, E.A. 1939a *Twilight of Man.* G.P. Putnam, New York.

Hooton, E.A. 1939b Should we ignore racial differences? Transcript from an NBC broadcast of Town Hall, 11/16/39. *Town Hall* 5(6):9–15, 19–29.

Hooton, E.A. 1940 *Why Men Behave Like Apes, and Vice Versa; Or, Body and Behavior.* Princeton University Press, Princeton, NJ.

Hooton, E.A. 1942 *Man's Poor Relations.* Doubleday, New York. 412 pp.

Hooton, E.A. 1946 *Up From the Ape*, revised edition. Macmillan, New York.

Hooton, E.A. 1949 Human Evolution: A Review of a New Theory of Evolution by Sir Arthur Keith. *Antiquity* 23(91):126–128.

Hooton, E.A. 1954 Comments on the Piltdown Affair. *American Anthropologist* 56(2):287–289.

Howell, F.C. 1951 The place of Neanderthal man in human evolution. *American Journal of Physical Anthropology* 9:379–416.

Howell, F.C. 1952 Pleistocene glacial ecology and the evolution of "Classic Neandertal" man. *Southwest Journal of Anthropology* 8:377–410.

Howell, F.C. 1957 The evolutionary significance of variation and varieties of "Neanderthal" man. *Quarterly Review of Biology* 32:330–347.

Howell, F.C. 1994a A chronostratigraphic and taxonomic framework of the origins of modern humans. In M.H. Nitecki and D.V. Nitecki (eds): *Origins of Anatomically Modern Humans.* Plenum Press, New York. pp. 253–319.

Howell, F.C. 1994b Thoughts on Eugène Dubois and the "*Pithecanthropus*" saga. In J.L. Franzen (ed): *100 years of Pithecanthropus: The Homo erectus problem. Courier Forschungsinstitut Senckenberg* 171:11–20.

Howells, W.W. 1942 Fossil man and the origin of races. *American Anthropologist* 44:182–193.

Howells, W.W. 1944 *Mankind So Far.* American Museum of Natural History, Science Series number 5. Doubleday, Doran, and Company, Garden City, NY. 319 pp.

Howells, W.W. 1950 Origin of the human stock. Concluding remarks of the chairman: present outlook on human origins. *Cold Spring Harbor Symposia on Quantitative Biology* 15:79–86.

Howells, W.W. 1959. *Mankind in the Making*. Doubleday, Garden City, NY.

Howells, W.W. 1973 Cranial variation in man. A study by multivariate analysis of patterns of difference among recent human populations. *Papers of the Peabody Museum of Archaeology and Ethnology* 67:1–259.

Howells, W.W. 1976 Explaining modern man: evolutionists versus migrationists. *Journal of Human Evolution* 5:477–496.

Howells, W.W. 1981 Franz Weidenreich, 1873–1948. *American Journal of Physical Anthropology* 56(4):407–410.

Howells, W.W. 1989 Skull Shapes and the Map: Craniometric Analyses in the Dispersion of Modern *Homo. Papers of the Peabody Museum of Archaeology and Ethnology* 79:1–189.

Howells, W.W. 1993 *Getting Here. The Story of Human Evolution.* Compass Press, Washington.

Hrdlička, A. 1927 The Neanderthal phase of man. *Journal of the Royal Anthropological Institute of Great Britain and Ireland* 57:249–274.

Hrdlička, A. 1930 *The Skeletal Remains of Early Man.* Smithsonian Miscellaneous Collections, Volume 83. Smithsonian Institution, Washington, D.C.

Hsieh, C.L., and H.E. Sutton 1992 Mitochondrial lineages and nuclear variants in a U.S. Black population: origins of a hybrid population. *Annals of Human Genetics* 56:105–112.

Hublin, J-J. 1987 Qui fut l'Ancêtre de l'*Homo sapiens? Pour la Science* 113:26–35.

Hull, D.L. 1994 Ernst Mayr's influence on the history and philosophy of science: a personal memoir. *Biology and Philosophy* 9:375–386.

Hulse, F.S. 1962 Race as an evolutionary episode. *American Anthropologist* 64:929–945.

Hunt, E.E., Jr. 1981 The old physical anthropology. *American Journal of Physical Anthropology* 56(4):339–346.

Huxley, J.S. 1942 *Evolution, the Modern Synthesis.* Allen, London.

Huxley, J.S., and A.C. Haddon 1935 *We Europeans: a Survey of "Racial" Problems.* Jonathan Cape, London.

Isaac, G.Ll. 1983 Review of "Bones: Ancient Men and Modern Myths." *American Antiquity* 48:416–419.

Jablonka, E., and M.J. Lamb 1995 *Epigenetic Inheritance and Evolution.* Oxford University Press, New York.

Jackson, Fatimah L.C. 1994 Human biological races in the context of the three major models of the origins of modern humans. Abstract of talk presented to the 1994 American Anthropological Association meetings.

Jelínek, J. 1966 Jaw of an intermediate type of Neandertal man from Czechoslovakia. *Nature* 212:701–702.

Jelínek, J. 1969 Neanderthal man and Homo sapiens in central and eastern Europe. *Current Anthropology* 10:475–503.

Jelínek, J. 1976 The *Homo sapiens neanderthalensis* and *Homo sapiens sapiens* relationship in Central Europe. *Anthropologie* (Brno) 14:79–81.

Jelínek, J. 1978 *Homo erectus* or *Homo sapiens? Recent Advances in Primatology* 3:419–429.

Jelínek, J. 1980a European *Homo erectus* and the origin of *Homo sapiens.* In L.K. Königsson (ed): *Current Argument on Early Man.* Pergamon, Oxford. pp. 137–144.

Jelínek, J. 1980b Variability and geography. Contribution to our knowledge of European and north African Middle Pleistocene hominids. *Anthropologie* (Brno) 18: 109–114.

Jelínek, J. 1981 Was *Homo erectus* already *Homo sapiens? Les Processus de l'Hominisation.* CNRS International Colloquium, No. 599:85–89, Paris.

Jelínek, J. 1982 The east and southeast Asian way of regional evolution. *Anthropologie* (Brno) 20:195–212.

Jelínek, J. 1983 The Mladeč finds and their evolutionary importance. *Anthropologie* (Brno) 21:57–64.

Johannsen, W. 1922 Hundert Jahre Vererbungsforschung. *Verhandlungen der Gesellschaft Deutscher Naturforscher und Ärzte* 87:70–104.

Johnston, W.M. 1972 *The Austrian Mind.* University of California Press, Berkeley.

Kamminga, J., and R.V.S. Wright 1988 The Upper Cave at Zhoukoudian and the origins of the Mongoloids. *Journal of Human Evolution* 17(8):739–767.

Keith, A. 1910 A new theory of the descent of man. *Nature* 85(2146):206.

Keith, A. 1911 Klaatsch's theory of the descent of man. *Nature* 85(2155):509–510.

Keith, A. 1915 *The Antiquity of Man.* Williams and Norgate, London.

Keith, A. 1916 On certain factors concerned in the evolution of human races. *Journal of the Royal Anthropological Society* 46:10–34.

Keith, A. 1919 The differentiation of mankind into racial types. *Annual Reports of the Smithsonian Institution* 1919:443–453.

Keith, A. 1925 *The Antiquity of Man*, revised edition. 2 volumes. Lippincott, Philadelphia.

Keith, A. 1936 *History from Caves. A New Theory of the Origin of Modern Races of Mankind.* British Speleological Association Publication, London.

Keith, A. 1948 *A New Theory of Human Evolution.* Philosophical Library, New York.

Keith, A. 1950 *An Autobiography.* Philosophical Library, New York.

Kidder, J.H., R.L. Jantz, and F.H. Smith 1992 Defining modern humans: a multivariate approach. In G. Bräuer and F.H. Smith (eds): *Continuity or Replacement? Controversies in Homo sapiens Evolution.* Balkema, Rotterdam. pp. 157–177.

Kimbel, W.H., and L.B. Martin (eds) 1993 *Species Concepts and Primate Evolution.* Plenum, New York.

Kimura, M. 1979 The neutral theory of molecular evolution. *Scientific American* 241(5):98–126.

King, W. 1864 On the Neanderthal Skull, or reasons for believing it to belong to the Clydian Period and to a species different from that represented by man. *British Association for the Advancement of Science, Notices and Abstracts for 1863:* 81–82.

Klaatsch, H. 1910 Menschenrassen und Menschenaffen. *Correspondenzblatt der Deutschen Gesellschaft für Anthropologie, Ethnologie, und Urgeschichte* 41:91–99.

Klaatsch, H. 1923 *The Evolution and Progress of Mankind.* Stokes, New York.

Klein, R.G. 1989 *The Human Career. Human Biological and Cultural Origins.* University of Chicago Press, Chicago.

Klein, R.G. 1992 The archaeology of modern human origins. *Evolutionary Anthropology* 1(1):5–14.

Klein, R.G. 1995 Anatomy, behavior, and modern human origins. *Journal of World Prehistory* 9(2):167–198.

Knight, D. 1981 *Ordering the World: A History of Classifying Man.* Burnett, London.

Kollmann, J. 1903 Die Pygmäen und ihre systematische Stellung innerhalb des Menschengeschlechts. *Verhandlungen der Naturforschenden Gesellschaft in Basel* 16:85–117.

Kollmann, J. 1905 Neue Gedanken über das alte Problem von der Abstammung des Menschen. *Globus* 87(7):141–148.

Konigsberg, L.W., A. Kramer, S.M. Donnelly, J.H. Relethford, and J. Blangero 1994 Modern human origins. *Nature* 372:228–229.

Kotter, M.J. 1974 From 48 to 46: cytological technique, perception, and the counting of human chromosomes. *Bulletin of the History of Medicine* 48:465–502.

Kowalski, C.J. 1972 A commentary on the use of multivariate statistical methods in anthropometric research. *American Journal of Physical Anthropology* 36(1):119–132.

Kühl, S. 1994 *The Nazi Connection: Eugenics, American Racism, and German National Socialism.* Oxford University Press, New York.

Kuhn, T.S. 1962 *The Structure of Scientific Revolutions.* University of Chicago, Chicago.
Lahr, M.M. 1994 The Multiregional model of modern human origins: a reassessment of its morphological basis. *Journal of Human Evolution* 26(1):23–56.
Lahr, M.M., and R. Foley 1995 Multiple dispersals and modern human origins. *Evolutionary Anthropology* 3(2):48–60.
Lam, Y. M., O. M. Pearson, and C. M. Smith 1996 Chin Morphology and Sexual Dimorphism in the Fossil Hominid Mandible Sample from Klasies River Mouth. *American Journal of Physical Anthropology* 100(4): 545–557.
Lamarck, J-B. 1809 *Philosophie zoologique.* Dentu, Paris.
Lande, R. 1986 The dynamics of peak shifts and the pattern of morphological evolution. *Paleobiology* 12(4):343–354.
Larsen, C.S., R.M. Matter, and D.L. Gebo 1991 *Human Origins. The Fossil Record*, second edition. Waveland Press, Prospect Heights.
Lasker, G.W. (ed) 1960 *The Process of Ongoing Human Evolution.* Wayne State University Press, Detroit.
Laszlo, E., and I. Masulli (eds) 1993 *The Evolution of Cognitive Maps: New Paradigms for the Twenty-First Century.* The World Futures General Evolution Studies Series, Volume 5. Gordon and Breach Science Publishers, Yverdon (Switzerland).
Leakey, R.E. 1969 Early *Homo sapiens* remains from the Omo River region of Southwest Ethiopia. *Nature* 222:1137–1138.
Leakey, R.E. 1989 Recent fossil finds from East Africa. In J.R. Durant (ed): *Human Origins.* Clarendon Press, Oxford. pp. 53–62.
Lenz, F. 1931 *Menschliche Auslese und Rassenhygiene (Eugenik).* Lehmann, Munich.
Lewin, R. 1993 *The Origin of Modern Humans.* Scientific American Library, New York.
Lewontin, R.C. 1984 *Human Diversity.* W.H. Freeman, San Francisco.
Li Wenhsiung and L. Sadler 1991 Low nucleotide diversity in man. *Genetics* 129: 513–523.
Lieberman, L., and F.L.C. Jackson 1995 Race and three models of human origin. *American Anthropologist* 97(2):231–242.
Littlefield, A., L. Liebermann, and L.T. Reynolds 1982 Redefining race: the potential demise of a concept in physical anthropology. *Current Anthropology* 23:641–656.
Livingstone, F.B. 1964 On the nonexistence of human races. In M.F.A. Montagu (ed): *The Concept of Race.* Free Press, New York. pp. 46–60.
Lovejoy, C.O., R.P. Mensforth, and G.J. Armelagos 1982 Five decades of skeletal biology as reflected in the *American Journal of Physical Anthropology.* In F. Spencer (ed): *A History of American Physical Anthropology 1930–1980.* Academic Press, New York. pp. 329–336.
Lurie, E. 1960 *Louis Agassiz: A Life in Science.* University of Chicago Press, Chicago.
Maddison, D.R. 1991 African origin of human mitochondrial DNA reexamined. *Systematic Zoology* 40:355–363.
Mann, A.E. 1975 *Some Paleodemographic Aspects of the South African Australopithecines.* University of Pennsylvania Publications in Anthropology, Number 1, Philadelphia.
Mann, A.E. 1981 The significance of the Sinanthropus casts, and some paleodemographic notes. In B.A. Sigmon and J.S. Cybulski (eds): *Homo erectus. Papers in Honor of Davidson Black.* University of Toronto Press, Toronto. pp. 41–62 (combined bibliography at end of volume).
Marks, J. 1992 Review of "Piltdown, a Scientific Forgery" and "The Piltdown Papers" by F. Spencer. *American Journal of Physical Anthropology* 87(3):376–380.
Marks, J. 1995 *Human Biodiversity: Genes, Race, and History.* Aldine de Gruyter, New York.

Mayr, E. 1950 Taxonomic categories in fossil hominids. *Cold Spring Harbor Symposia on Quantitative Biology* 15:108–118.
Mayr, E. 1960 The emergence of evolutionary novelties. In S. Tax (ed): *The Evolution of Life.* University of Chicago Press, Chicago. pp. 349–380.
Mayr, E. 1963 *Animal Species and Evolution.* Belknap Press of Harvard University Press, Cambridge.
Mayr, E. 1972 Lamarck revisited. *Journal of the History of Biology* 5:55–94.
Mayr, E. 1980 Prologue: some thoughts on the history of the evolutionary synthesis. In E. Mayr and W.B. Provine (eds): *The Evolutionary Synthesis: Perspectives on the Unification of Biology.* Harvard University Press, Cambridge. pp. 1–48.
Mayr, E. 1982a *The Growth of Biological Thought: Diversity, Evolution, and Inheritance.* Belknap Press of Harvard University Press, Cambridge.
Mayr, E. 1982b Reflections on human paleontology. In F. Spencer (ed): *A History of American Physical Anthropology 1930–1980.* Academic Press, New York. pp. 231–237.
Mayr, E. 1983 How to carry out the adaptationist program? *American Naturalist* 121(3):324–334.
Mayr, E. 1988 *Toward a New Philosophy of Biology, Observations of an Evolutionist.* Belknap Press of Harvard University Press, Cambridge.
Mayr, E. 1991 *One Long Argument: Charles Darwin and the Genesis of Modern Evolutionary Thought.* Harvard University Press, Cambridge.
Mayr, E. (ed) 1976 *Evolution and the Diversity of Life.* Belknap Press of Harvard University Press, Cambridge.
Mayr, E., and Provine, W.B. (eds.) 1980 *The Evolutionary Synthesis: Perspectives on the Unification of Biology.* Harvard University Press, Cambridge.
McCown, T.D., and A. Keith 1939 *The Stone Age of Mount Carmel: The Fossil Human Remains from the Levallois-Mousterian.* Volume II. Clarendon Press, Oxford.
McCown, T.D., and K.A.R. Kennedy 1972 Introduction, introductory sections, and summary. In T.D. McCown and K.A.R. Kennedy (eds): *Climbing Man's Family Tree: A Collection of Major Writings on Human Phylogeny, 1699–1971.* Prentice-Hall, Englewood Cliffs, NJ. pp. 1–41, 93–100, 189–195, 241–249, 355–359, 461–464.
McCown, T.D., and K.A.R. Kennedy (eds) 1972 *Climbing Man's Family Tree: A Collection of Major Writings on Human Phylogeny, 1699–1971.* Prentice-Hall, Englewood Cliffs, NJ.
McEachron, D.L. 1984 Hypothesis and explanation in human evolution. *Journal of Social and Biological Structures* 7:9–15.
Medvedev, Z.A. 1971 *The Rise and Fall of T.D. Lysenko.* Anchor, New York.
Meindl, R.S. 1987 Hypothesis: a selective advantage for cystic fibrosis heterozygotes. *American Journal of Physical Anthropology* 74:39–45.
Mellars, P., and C. Stringer 1989 Introduction. In P. Mellars and C.B. Stringer (eds): *The Human Revolution: Behavioural and Biological Perspectives on the Origins of Modern Humans.* Edinburgh University Press, Edinburgh. pp. 1–14.
Michael, J. S. 1988 A new look at Morton's craniological research. *Current Anthropology* 29:348–354.
Mikić, Z. 1981 *Stanja i Problemi Fizičke Anthropologije u Jugoslaviji.* Akademija Nauka I Umjetnosti Bosne i Hercegovine, Sarajevo.
Molnar, S. 1983 *Human Variation, Races, Types and Ethnic Groups.* second edition. Prentice-Hall, Englewood Cliffs, NJ.
Montagu, M.F.A. 1942 *Man's Most Dangerous Myth: The Fallacy of Race.* Columbia University Press, New York.

Montagu, M.F.A. 1963 What is remarkable about varieties of man is likeness, not differences. *Current Anthropology* 4(4):361–363.

Montagu, M.F.A. (ed) 1962 *Culture and the Evolution of Man.* Oxford University Press, New York.

Montagu, M.F.A. (ed) 1964 *The Concept of Race.* The Free Press, New York.

Moore, J.H. 1994a Ethnogenetic theory. *Research and Exploration* 10(1):10–23.

Moore, J.H. 1994b Putting anthropology back together again: the ethnogenetic critique of cladistic theory. *American Anthropologist* 96(4):925–948.

Moore, J.H. 1995 The end of a paradigm. *Current Anthropology* 36(3):530–531.

Moore, J.R. 1979 *The Post-Darwinian Controversies.* Cambridge University Press, Cambridge.

Moorehead, A. 1969 The Beagle. *New Yorker* 45(28):31–70 and 45(29):41–95.

Morell, V. 1993 Anthropology: nature-culture battleground. *Science* 261:1798–1801.

Morgan, E. 1972 *The Descent of Woman.* Stein and Day, New York.

Morgan, L.H. 1877 *Ancient Society.* Reprinted in 1964 by Harvard University Press, Cambridge.

Morris, A.G. 1986 Khoi and San craniology: a re-evaluation of the osteological reference samples. In R. Singer and J.K. Lundy (eds): *Variation, Culture, and Evolution in African Populations: Papers in Honour of Dr. Hertha de Villiers.* Witwatersrand University Press, Johannesburg. pp. 1–12.

Morris, A.G. 1992a *A Master Catalogue: Holocene Human Skeletons from South Africa.* Witwatersrand University Press, Johannesburg.

Morris, A.G. 1992b Biological relationships between Upper Pleistocene and Holocene populations in southern Africa. In G. Bräuer and F.H. Smith (eds): *Continuity or Replacement? Controversies in Homo sapiens Evolution.* Balkema, Rotterdam. pp. 131–143.

Morton, N.E., and J. Lalouel 1973 Topology of kinship in Micronesia. *American Journal of Human Genetics* 25:422–432.

Mulkay, M., and G.N. Gilbert 1982 Accounting for error: how scientists construct their social world when they account for correct and incorrect belief. *Sociology* 16:165–183.

Neumann, G.K. 1956 The Upper cave skulls from Chou-Kou-Tien in the light of Paleo-Amerindian material (abstract). *American Journal of Physical Anthropology* 14:380.

Nordenskiöld, E. 1929 *The History of Biology.* Kegan Paul, Trench, Trubner & Co., London.

Nott, J.C. 1843 The Mulatto hybrid—probable extermination of the two races of whites and blacks are allowed to intermarry. *American Journal of Medical Science* VI:252–256.

Oschinsky, L. 1964 *The Most Ancient Eskimos.* Canadian Research Centre for Anthropology, University of Ottawa, Ottawa.

Owen, R.C. 1965 The patrilocal band: a linguistically and culturally hybrid social unit. *American Anthropologist* 67:675–690.

Parsons, P.A. 1983 *The Evolutionary Biology of a Colonizing Species.* Cambridge University, Cambridge.

Penniman, T.K. 1936 *A Hundred Years of Anthropology.* Macmillan, New York.

Pfiefer, E.J. 1965 The genesis of American neo-Lamarckism. *Isis* 56:156–167.

Pilbeam, D.R. 1972 *The Ascent of Man.* Macmillan, New York.

Piveteau, J. 1957 Traité de Paléonologie. Volume 7, Les Primates et l'Homme. Masson, Paris.

Pope, G.G. 1992a Craniofacial evidence for the origin of modern humans in China. *Yearbook of Physical Anthropology* 35:243–298.

Pope, G.G. 1992b Replacement versus regional continuity models: the paleobehavioral and fossil evidence from East Asia. In T. Akazawa, K. Aoki, and T. Kimura (eds): *The Evolution and Dispersal of Modern Humans in Asia.* Hokusen-sha, Tokyo. pp. 3–14.

Popper, K.L. 1959 *The Logic of Scientific Discovery.* Hutchinson & Company, London.

Popper, K.L. 1962 *Conjectures and Refutations: The Growth of Scientific Knowledge.* Basic Books, New York.

Pouchet, C.H.G. 1864 *De la Pluralité des Races Humaines: Essai Anthropologique.* Ballière, Paris.

Proctor, R. 1988 From *Anthropologie* to *Rassenkunde* in the German anthropological tradition. In G.W. Stocking, Jr. (ed): *Bones, Bodies, Behavior. Essays on Biological Anthropology.* University of Wisconsin Press, Madison. pp. 138–179.

Proctor, R. 1989 *Racial Hygiene: Medicine Under the Nazis.* Harvard University Press, Cambridge.

Protsch, R. 1975 The absolute dating of Upper Pleistocene sub-Saharan fossil hominids and their place in human evolution. *Journal of Human Evolution* 4:297–322.

Provine, W.B. 1980 Epilogue. In E. Mayr and W.B. Provine (eds): *The Evolutionary Synthesis: Perspectives on the Unification of Biology.* Harvard University Press, Cambridge. pp. 399–411.

Provine, W.B. 1986 Genetics and race. *American Zoologist* 26:857–887.

Quennell, M., and C.H.B. Quennell 1945 *Everyday Life in the Old Stone Age.* Batsford, London.

Radovčić, J. 1988 *Dragutin Gorjanović-Kramberger and Krapina Early Man: The Foundation of Modern Paleoanthropology.* Skolska knjiga and Hrvatski prirodoslovni muzej, Zagreb.

Radovčić, J., F.H. Smith, E. Trinkaus, and M.H. Wolpoff 1988 *The Krapina Hominids: An Illustrated Catalog of the Skeletal Collection.* Mladost Press and the Croatian Natural History Museum, Zagreb.

Ranke, J. 1896 Der fossile Mensch und die Menschenrassen. Correspondenzblatt der Deutschen Gesellschaft für Anthropologie, Ethnologie und Urgeschichte. *Archiv für Anthropologie* 27:151–156.

Raup, D.M. 1986 *The Nemesis Affair: A Story of the Death of Dinosaurs and the Ways of Science.* Nelson, New York.

Reif, W-E. 1983 Evolutionary theory in German paleontology. In M. Grene (ed): *Dimensions of Darwinism.* Cambridge University Press, London. pp. 173–203.

Reif, W-E. 1986 The search for a macroevolutionary theory in German paleontology. *Journal of the History of Biology* 19:79–130.

Relethford, J.H. 1995 Genetics and modern human origins. *Evolutionary Anthropology* 4(2):53–63.

Relethford, J.H., and H.C. Harpending 1994 Craniometric variation, genetic theory, and modern human origins. *American Journal of Physical Anthropology* 95(3):249–270.

Relethford, J.H., and H.C. Harpending 1995 Ancient differences in population size can mimic a recent African origin of modern humans. *Current Anthropology* 36(4):667–674.

Rensch, B. 1980 Historical development of the present synthetic Neo-Darwinism in Germany. In E. Mayr and W.B. Provine (eds): *The Evolutionary Synthesis: Perspectives on the Unification of Biology.* Harvard University Press, Cambridge. pp. 284–303.

Rich, N. 1973 *Hitler's War Aims.* Norton, New York.

Richards, O. 1995 Analysis of the region V mitochondrial marker in two Black communities of Equador, and their parental populations. *Human Evolution* 10(1):5–16.

Richards, R.J. 1987 *Darwin and the Emergence of Evolutionary Theories of Mind and Behavior.* University of Chicago, Chicago.

Ridley, M. 1993 *Evolution.* Blackwell Scientific Publications, Oxford.

Rightmire, G.P. 1976 Relationships of Middle and Upper Pleistocene hominids from sub-Saharan Africa. *Nature* 260:238–240.

Rightmire, G.P. 1978 Human skeletal remains from the southern Cape Province and their bearing on the Stone Age prehistory of South Africa. *Quaternary Research* 9:219–230.

Rightmire, G.P. 1979 Implications of the Border Cave skeletal remains for later Pleistocene human evolution. *Current Anthropology* 20(1):23–35.

Rightmire, G.P. 1984 *Homo sapiens* in sub–Saharan Africa. In F.H. Smith and F. Spencer (eds): *The Origins of Modern Humans: A World Survey of the Fossil Evidence.* Alan R. Liss, Inc., New York. pp. 295–325.

Rightmire, G.P., and H.J. Deacon 1991 Comparative studies of late Pleistocene human remains from Klasies River Mouth, South Africa. *Journal of Human Evolution* 20(2):131–156.

Ringer, F. 1969 *The Decline of the German Mandarins: The German Academic Community, 1890–1933.* Harvard University Press, Cambridge.

Rink, W.J., H.P. Schwarcz, F.H. Smith, and J. Radovčić, 1995 ESR dates for Krapina hominids. *Nature* 378:24.

Robinson, J.T. 1967 Variation and taxonomy of the early hominids. *Evolutionary Biology* 1:69–100.

Roesler, H. 1981 Some historical remarks on the theory of cancellous bone structure (Wolff's law). In S.C. Cowin (ed): *Mechanical Properties of Bone.* AMD volume 45:27–42. American Society of Mechanical Engineers, New York.

Rogers, A. 1995 How much can fossils tell us about regional continuity? *Current Anthropology* 36(4):674–676.

Rogers, A.R., and H. Harpending 1992 Population growth makes waves in the distribution of pairwise genetic differences. *Molecular Biology and Evolution* 9:552–569.

Rogers, A.R., and L.B. Jorde 1995 Genetic evidence on modern human origins. *Human Biology* 67(1):1–36.

Rousseau, D.L. 1992 Case studies in pathological science. *American Scientist* 80(1): 54–63.

Ruff, C.B. 1994 Morphological adaptation to climate in modern and fossil hominids. *Yearbook of Physical Anthropology* 37:65–107.

Santa Luca, A.P. 1978 A re-examination of presumed Neandertal-like fossils. *Journal of Human Evolution* 7:619–636.

Sapp. J. 1987 *Beyond the Gene: Cytoplasmic Inheritance and the Struggle for Authority in Genetics.* Oxford University Press, New York.

Sartono, S. 1980 Sagittal cresting in *Meganthropus palaeojavicus* (von Koenigswald). *Modern Quaternary Research in Southeast Asia* 7:201–210.

Sauer, N.J. 1992 Forensic anthropology and the concept of race: if races don't exist, why are forensic anthropologists so good at identifying them? *Social Sciences and Medicine* 34(2):107–111.

Schiller, F. 1979 *Paul Broca: Founder of French Anthropology, Explorer of the Brain.* University of California Press, Berkeley.

Schwalbe, G. 1899 Studien über *Pithecanthropus erectus* Dubois. *Zeitschrift für Morphologie und Anthropologie* 1:16–240.

Schwalbe, G. 1906 *Studien zur Vorgeschichte des Menschen. Zeitschrift für Morphologie und Anthropologie* 1:5–228.

Schwalbe, G. 1914 Kritische Besprechung von Boule's Werk: "L'homme Fossile de la Chapelle-aux-Saints" mit eigenen Untersuchungen. *Zeitschrift für Morphologie und Anthropologie* 16:527–610.

Schwarz, E., and R. Chambers 1947 Wilhelm Caspari 1872–1944. *Science* 105: 613.

Scientific American Special 1992 issue on "mind and brain." 267(3).

Shapiro, H.L. 1939 *Migration and Environment.* Oxford University Press, New York.

Shapiro, H.L. 1974 *Peking Man.* Simon and Schuster, New York.

Shapiro, H.L. 1981 Earnest A. Hooton, 1887–1954, *in Memoriam cum amore. American Journal of Physical Anthropology* 56(4):431–434.

Sharrock, S.R. 1974 Crees, Cree-Assiniboines, and Assiniboines: interethnic social organization on the far Northern Plains. *Ethnohistory* 21:95–122.

Shipman, P. 1994 *The Evolution of Racism. Human Differences and the Use and Abuse of Science.* Simon & Schuster, New York.

Shreeve, J. 1995 The Neanderthal Peace. *Discover* 16(9):71–81.

Shreeve, J. 1996 *The Neandertal Enigma: Solving the Mystery of Human Origins.* William Morrow, New York.

Singer, R., and J. Wymer 1982 *The Middle Stone Age at Klasies River Mouth in South Africa.* University of Chicago Press, Chicago.

Smith, F.H. 1992 Models and realities in modern human origins: the African fossil evidence. *Philosophical Transactions of the Royal Society,* Series B, 337:243–250.

Spencer, F. 1979 Aleš Hrdlička, M.D., 1869–1943: A Chronicle of the Life and Work of an American Physical Anthropologist. Ph.D. dissertation. University of Michigan, Ann Arbor.

Spencer, F. 1981 The rise of academic physical anthropology in the United States (1880–1980): A historical overview. *American Journal of Physical Anthropology* 56(4):353–364.

Spencer, F. 1983 Samuel George Morton's doctoral thesis on body pain: the probable source of Morton's polygenism. *Transactions and Studies of the College of Physicians of Philadelphia (Medicine and History)* V(4):321–338.

Spencer, F. 1986 *Ecce Homo: An Annotated Bibliographic History of Physical Anthropology. Bibliographies and Indexes in Anthropology* 2. Greenwood Press, New York.

Spencer, F. 1990 *Piltdown. A Scientific Forgery.* Oxford University Press, New York.

Spencer, F. (ed.) 1982 *A History of American Physical Anthropology, 1930–1980.* Academic Press, New York.

Spencer, F., and F.H. Smith 1981 The significance of Aleš Hrdlička's "Neanderthal phase of man": A historical and current assessment. *American Journal of Physical Anthropology* 56(4):435–459.

Spuhler, J.N. (ed) 1959 *The Evolution of Man's Capacity for Culture.* Wayne State University Press, Detroit.

Stanton, W. 1960 *The Leopard's Spots: Scientific Attitudes Toward Race in America 1815–59.* University of Chicago Press, Chicago,

Stein, G.J. 1988 Biological science and the roots of Nazism. *American Scientist* 76(1):50–58.

Stern, C. 1930 Entgegnung auf die Bemerkungen von Franz Weidenreich zu meinem Aufsatz "Erzeugung von Mutationen durch Röntgenstrahlen." *Natur und Museum* 60:133–134.

Stewart, T.D. 1950 The problem of the earliest claimed representatives of *Homo sapiens. Cold Spring Harbor Symposia on Quantitative Biology* 15:97–107.

Stewart, T.D. 1981 Aleš Hrdlička, 1869–1943. *American Journal of Physical Anthropology* 56(4):347–351.

Stiner, M.C. 1995 *Honor Among Thieves: A Zooarchaeological Study of Neandertal Ecology.* Princeton University Press, Princeton.

Stocking, G.W., Jr. 1968 *Race, Culture, and Evolution: Essays on the History of Anthropology.* Free Press, New York.

Stocking, G.W., Jr. 1971 What's in a name? The origins of the Royal Anthropological Institute. *Man* 6:369–390.

Stocking, G.W., Jr. 1976 Anthropology as *Kulturkampf:* science and politics in the career of Franz Boas. In W. Goldschmidt (ed): *The Uses of Anthropology.* Washington D.C.

Stocking, G.W., Jr. 1987 *Victorian Anthropology.* Free Press, New York.

Stocking, G.W., Jr. (ed) 1988 *Bones, Bodies, Behavior. Essays on Biological Anthropology.* University of Wisconsin Press, Madison.

Stoneking, M. 1993 DNA and recent human evolution. *Evolutionary Anthropology* 2(2):60–73.

Stoneking, M. 1994 In defense of "Eve": a response to Templeton's critique. *American Anthropologist* 96(1):131–141.

Stoneking, M., and R.L. Cann 1989 African origins of human mitochondrial DNA. In P. Mellars and C.B. Stringer (eds): *The Human Revolution: Behavioural and Biological Perspectives on the Origins of Modern Humans.* Edinburgh University Press, Edinburgh. pp. 17–30.

Stringer, C.B. 1982 Towards a solution to the Neanderthal problem. *Journal of Human Evolution* 11:431–438.

Stringer, C.B. 1989a Documenting the origin of modern humans. In E. Trinkaus (ed): *The Emergence of Modern Humans. Biocultural Adaptations in the Later Pleistocene.* Cambridge University Press, Cambridge. pp. 67–96, and combined references for the volume on pp. 232–276.

Stringer, C.B. 1989b *Homo sapiens:* single or multiple origin? In J.R. Durant (ed): *Human Origins.* Clarendon Press, Oxford. pp. 63–80.

Stringer, C.B. 1989c The origin of early modern humans: a comparison of the European and non-European evidence. In P. Mellars and C.B. Stringer (eds): *The Human Revolution: Behavioural and Biological Perspectives on the Origins of Modern Humans.* Edinburgh University Press, Edinburgh. pp. 232–244.

Stringer, C.B. 1990 The Asian connection. *New Scientist* (November 17):33–37.

Stringer, C.B. 1992a Reconstructing recent human evolution. *Philosophical Transactions of the Royal Society,* Series B, 337:217–224.

Stringer, C.B. 1992b Replacement, continuity, and the origin of *Homo sapiens.* In G. Bräuer and F.H. Smith (eds): *Continuity or Replacement? Controversies in Homo sapiens Evolution.* Balkema, Rotterdam. pp. 9–24.

Stringer, C.B. 1993 New views on modern human origins. In D.T. Rasmussen (ed): *The Origin and Evolution of Humans and Humanness.* Jones and Bartless, Boston. pp. 75–94.

Stringer, C.B. 1994 Out of Africa—a personal history. In M.H. Nitecki and D.V. Nitecki (eds): *Origins of Anatomically Modern Humans.* Plenum Press, New York. pp. 149–172.

Stringer, C.B., and P. Andrews 1988 Genetic and fossil evidence for the origin of modern humans. *Science* 239:1263–1268; 241:773–774.

Stringer, C.B., and G. Bräuer 1994 Methods, misreading, and bias. *American Anthropologist* 96(2):416–424.

Stringer, C.B., and C. Gamble 1993 *In Search of the Neanderthals.* Thames and Hudson, London.

Stringer, C.B., and R. McKie 1996 *African Exodus: the Origins of Modern Humanity.* Jonathan Cape, London.

Tattersall, I. 1992 Species concepts and species identification in human evolution. *Journal of Human Evolution* 22(4/5):341–350.

Tattersall, I. 1993 *The Human Odyssey: Four Million Years of Evolution.* Prentice-Hall, New York.

Tattersall, I. 1994a How does evolution work? *Evolutionary Anthropology* 3(1):2–3.

Tattersall, I. 1994b What does it mean to be human—and why does it matter? *Evolutionary Anthropology* 3(4):114–116.

Tattersall, I. 1995a *The Fossil Trail: How We Know What We Think We Know about Human Evolution.* Oxford University Press, New York.

Tattersall, I. 1995b *The Last Neanderthal: the Rise, Success, and Mysterious Extinction of our Closest Human Relatives.* Macmillan, New York.

Templeton, A.R. 1982 Genetic architectures of speciation. In C. Barigozzi (ed): *Mechanisms of Speciation.* Alan R. Liss, New York. pp. 105–121.

Templeton, A.R. 1993 The "Eve" hypotheses: A genetic critique and reanalysis. *American Anthropologist* 95(1):51–72.

Templeton, A.R. 1994 "Eve": hypothesis compatibility versus hypothesis testing. *American Anthropologist* 96(1):141–147.

Templeton, A.R. 1996 Gene lineages and human evolution. *Science* 272: 1363.

Thackeray, A.I. 1989 Changing fashions in the Middle Stone Age: the stone artifact sequence from Klasies River main site, South Africa. *South African Archaeological Review* 7:35–59.

Theunissen, B. 1988 Eugene Dubois and the Ape-Man from Java. Dordrecht, Reidel.

Thorne, A.G. 1981 The centre and the edge: the significance of Australasian hominids to African paleoanthropology. In R.E. Leakey and B.A. Ogot (eds): *Proceedings of the 8th Panafrican Congress of Prehistory and Quaternary Studies*, Nairobi, September 1977. TILLMIAP, Nairobi. pp. 180–181.

Thorne, A.G., and M.H. Wolpoff 1981 Regional continuity in Australasian Pleistocene hominid evolution. *American Journal of Physical Anthropology* 55:337–349.

Thorne, A.G., and M.H. Wolpoff 1992 The multiregional evolution of humans. *Scientific American* 266(4):76–83.

Thorne, A.G., M.H. Wolpoff, and R.G. Eckhardt 1993 Genetic variation in Africa. *Science* 261:1507–1508; 262:973–974.

Tobias, P.V. 1960 The Kanam jaw. *Nature* 185:946–947.

Tobias, P.V. 1978 The San: an evolutionary perspective. In P.V. Tobias (ed): *The Bushmen. San Hunters and Herders of Southern Africa.* Human and Rousseau., Cape Town. pp. 16–32.

Tobias, P.V. 1992 Piltdown: an appraisal of the case against Sir Arthur Keith. *Current Anthropology* 33(3):243–293.

Topinard, P. 1885 *Elements d'Anthropologie Générale.* Delahaye and Lecrosnier, Paris.

Toulmin, S., and J. Goodfield 1965 *The Discovery of Time.* Harper and Row, New York.

Trinkaus, E. 1993 A note on the KNM-ER 999 hominid femur. *Journal of Human Evolution* 24(6):493–504.

Trinkaus, E., and W.W. Howells 1979 The Neanderthals. *Scientific American* 241(6):118,122–133.

Trinkaus, E., and P. Shipman 1993 *The Neanderthals: Changing the Image of Mankind.* Knopf, New York.

Tuchman, B. 1966 *The Proud Tower: A Portrait of the World Before the War: 1890–1914.* Bantam, New York.

Tyler, D.E. 1995 The current picture of hominid evolution in Java. *Acta Anthropologica Sinica* 14(4):315–323.

Ullrich, H. 1982 Neuere paläanthropologische Untersuchen am Material aus der Höhle Vindija (Kroatien, Jugoslawien). *Palaeontologia Jugoslavica* 29:1–44.

Vallois, H.V. 1929 Les preuves anatomiques de l'origine monophylétique de l'homme. *L'Anthropologie* 39:77–101.

Vallois, H.V. 1952 Monophyletism and polyphyletism in man. *South African Journal of Science* 49:69–79.

Vallois, H.V. 1954 Neanderthals and presapiens. *Journal of the Royal Anthropological Institute* 84:111–130.

Van Valen, L.M. 1966 On discussing human races. *Perspectives in Biology and Medicine* 9:377–383.

Van Valen, L.M. 1986 Speciation and our own species. *Nature* 322:412.

Van Vark, G.N., and A. Bilsborough 1991 Shaking the family tree. *Science* 253: 834.

Van Vark, G.N., and A. Bilsborough 1994 Human cranial variability past and present. *American Journal of Physical Anthropology* 95(1):89–91.

Van Vark, G.N., and J. Dijkema 1988 Some notes on the origin of the Chinese people. *Zeitschrift für die Vergleichende Forschung am Menschen* 39(3–4):143–148.

Van Vark, G.N., and W.W. Howells (eds) 1984 *Multivariate Statistical Methods in Physical Anthropology.* D. Reidel, Dordrecht.

Vandermeersch, B. 1981 *Les hommes fossiles de Qafzeh (Israël).* Cahiers de Paléoanthropologie, Centre National de la Recherche Sciéntifique, Paris.

Vigilant, L., M. Stoneking, H. Harpending, K. Hawkes, and A.C. Wilson 1991 African populations and the evolution of human mitochondrial DNA. *Science* 233:1303–1307.

Vogt, C. 1864 *Lectures on Man: His Place in Creation and in the History of the Earth.* Longman, Green, Longman and Roberts, London.

von Baer, K.E. 1828 *Entwicklungsgeschichte der Tiere: Beobachtung und Reflexion.* Bornträger, Königsberg.

von Bonin. G. 1911 Klaatsch's theory of the descent of man. *Nature* 85:508–509.

von Koenigswald, G.H.R. 1956 *Meeting Prehistoric Man.* Thames and Hudson, London.

von Koenigswald, G.H.R. 1958 Der Solo-Mensch von Java: ein tropischer Neanderthaler. In G.H.R. von Koenigswald (ed): *Hundert Jahre Neanderthaler.* Böhlau, Köln. pp. 21–26.

von Koenigswald, G.H.R., and F. Weidenreich 1939 The relationship between *Pithecanthropus* and *Sinanthropus. Nature* 144:926–929.

Waddington, C.H. 1943 Canalization of the development and the inheritance of acquired characters. *Nature* 150:563–565.

Waddington, C.H. 1952 Genetic assimilation of an acquired character. *Evolution* 7:118–126.

Waddle, D.M. 1994 Matrix correlation tests support a single origin for modern humans. *Nature* 368:452–454.

Walker, A.C., M.R. Zimmerman, and R.E. Leakey 1982 A possible case of hypervitaminosis A in *Homo erectus. Nature* 296:248–250.

Wallace, A.R. 1864 The origin of human races and the antiquity of man. *Journal of the Anthropological Society of London* 2.

Wallace, R. 1992 Spacial mapping and the origin of language: a paleoneurological model. In A. Jonker (ed): *Studies in Language Origins,* Volume 4. John Benjamins, Philadelphia.

Washburn, S.L. 1950 The analysis of primate evolution with particular reference to the origin of man. *Cold Spring Harbor Symposia on Quantitative Biology* 15:67–78.

Washburn, S.L. 1951 The new physical anthropology. *Transactions of the New York Academy of Sciences,* Series 2, 13(7):298–304.

Washburn, S.L. 1963 The study of race. *American Anthropologist* 65:521–532.

Weidenreich, F. 1913 Über das Hüftbein und das Becken der Primaten und ihre Umformung durch den aufrechten Gang. *Anatomischer Anzeiger* 44:497–513.

Weidenreich, F. 1922a Der Menschenfuss. *Zeitschrift für Morphologie und Anthropologie* 22:51–282.

Weidenreich, F. 1922b Über die Beziehungen zwischen Muskelapparat und Knochen und den Character des Knochengewebes. *Verhandlungen der Anatomischen Gesellschaft Erlangen* 24:157–189.

Weidenreich, F. 1922c Die Typen- und Artenlehre der Vererbungswissenschaft und die Morphologie. *Paläontologische Zeitschrift* 5:276–289.

Weidenreich, F. 1923 Evolution of the human foot. *American Journal of Physical Anthropology* 6:1–10.

Weidenreich, F. 1924 Die Sonderform des Menschenschädels als Anpassung an den aufrechten Gang. *Zeitschrift für Morphologie und Anthropologie* 24:157–189.

Weidenreich, F. 1925a Wie kommen funktionelle Anpassungen der Aussenformen des Knochenskelettes zustande? *Paläontologische Zeitschrift* 7:34–44.

Weidenreich, F. 1925b Domestikation und Kultur in ihrer Wirkung auf Schädelform und Körpergestalt. *Zeitschrift für Konstitutionslehre* 11:1–51.

Weidenreich, F. 1927 *Rasse und Körperbau.* Springer, Berlin.

Weidenreich, F. 1928a Die Morphologie des Schädels. In F. Wiegers, F. Weidenreich, and F. Schuster (eds): *Der Schädelfund von Weimar-Ehringsdorf.* Fischer, Jena. pp. 41–135.

Weidenreich, F. 1928b Entwicklungs- und Rassentypen des *Homo primigenius. Natur und Museum* 58:1–13, 51–62.

Weidenreich, F. 1929 Vererbungsexperiment und vergleichende Morphologie. *Paläontologische Zeitschrift* 11:275–286.

Weidenreich, F. 1930 Bemerkungen zu dem Aufsatz von C. Stern. *Natur und Museum* 60:47.

Weidenreich, F. 1931 Der primäre Greifcharacter der menschlichen Hände und Füsse und seine Bedeutung für das Abstammungsproblem. *Verhandlungen der Gesellschaft für Physische Anthropologie* 5:97–110.

Weidenreich, F. 1932 Lamarckismus. *Natur und Museum* 62:298–300.

Weidenreich, F. 1935 The *Sinanthropus* population of Choukoutien (Locality 1) with a preliminary report on new discoveries. *Bulletin of the Geological Society of China* 14:427–468.

Weidenreich, F. 1936a The mandibles of *Sinanthropus pekinensis:* a comparative study. *Palaeontologia Sinica,* Series D, Volume 7, Fascicle 3.

Weidenreich, F. 1936b Observations on the form and the proportions of the endocranial casts of *Sinanthropus pekinensis,* other hominids and the great apes: a comparative study of brain size. *Palaeontologia Sinica,* Series D, Volume 7, Fascicle 4.

Weidenreich, F. 1937 The dentition of *Sinanthropus pekinensis:* a comparative odontography of the hominids. *Palaeontologia Sinica,* New Series D, Number 1 (Whole Series 101).

Weidenreich, F. 1938a *Pithecanthropus and Sinanthropus. Nature* 141:376–379.

Weidenreich, F. 1938b The face of the Peking woman. *Natural History* 41:338–360.

Weidenreich, F. 1939a On the earliest representatives of modern mankind recovered on the soil of East Asia. *Peking Natural History Bulletin* 13:161–174.

Weidenreich, F. 1939b Six lectures on *Sinanthropus pekinensis* and related problems. *Bulletin of the Geological Society of China* 19:1–110.

Weidenreich, F. 1940a Man or ape? *Natural History* 45:32–37.

Weidenreich, F. 1940b Some problems dealing with ancient man. *American Anthropologist* 42(3):375–383.

Weidenreich, F. 1940c The torus occipitalis and related structures and their transformation in the course of human evolution. *Bulletin of the Geological Society of China* 19:480–546.

Weidenreich, F. 1941a The brain and its role in the phylogenetic transformation of the human skull. *Transactions of the American Philosophical Society* 31:321–442.

Weidenreich, F. 1941b The extremity bones of *Sinanthropus pekinensis. Palaeontologia Sinica,* New Series D, Number 5 (Whole Series 116).

Weidenreich, F. 1943a The Neanderthal man and the ancestors of *Homo sapiens. American Anthropologist* 45:39–48.

Weidenreich, F. 1943b The skull of *Sinanthropus pekinensis:* A comparative study of a primitive hominid skull. *Palaeontologia Sinica,* New Series D, Number 10 (whole series No. 127).

Weidenreich, F. 1945a Giant early man from Java and south China. *Anthropological Papers of the American Museum of Natural History* 40:1–134.

Weidenreich, F. 1945b The brachycephalization of recent mankind. *Southwestern Journal of Anthropology* 1:1–54.

Weidenreich, F. 1946a *Apes, Giants, and Man.* University of Chicago Press, Chicago.

Weidenreich, F. 1946b Report on the latest discoveries of early man in the Far East. *Experienta* 2:265–272.

Weidenreich, F. 1946c Generic, specific, and subspecific characters in human evolution. *American Journal of Physical Anthropology* 4:413–431.

Weidenreich, F. 1947a Facts and speculations concerning the origin of *Homo sapiens. American Anthropologist* 49:187–203.

Weidenreich, F. 1947b Are human races in the taxonomic sense "races" or "species"? *American Journal of Physical Anthropology* 5(3):369–371.

Weidenreich, F. 1947c Some particulars of skull and brain of early hominids and their bearing on the problem of the relationship between man and anthropoids. *American Journal of Physical Anthropology* 5:387–418.

Weidenreich, F. 1947d The trend of human evolution. *Evolution* 1:221–236.

Weidenreich, F. 1948 The human brain in the light of its phylogenetic development. *Scientific Monthly* 67:103–109.

Weidenreich, F. 1949 Interpretations of the fossil material. In W.W. Howells (ed): *Early Man in the Far-East.* Wistar Press, Philadelphia. pp. 149–157.

Weidenreich, F. 1951 Morphology of Solo man. *Anthropological Papers of the American Museum of Natural History* 43(3):205–290.

Weiner, J.S. 1955 *The Piltdown Forgery.* Oxford University Press, London.

Weinert, H. 1928 *Pithecanthropus erectus. Zeitschrift für Anatomie und Entwicklungsgeschichte* 87:429–547.

Weinert, H. 1947 *Menschen der Vorzeit.* Enke Press, Stuttgart.

Weinreich, M. 1946 *Hitler's Professors, the Part of Scholarship in Germany's Crimes Against the Jewish People.* Yiddish Scientific Institute, New York.

Weismann, A. 1882 Studies in the theory of descent. Samson and Low, London.

Weismann, A. 1885 *Die Continuität des Keimplasmas als Grundlage einer Theorie der Vererbung.* Fischer, Jena.

Weiss, K.M., and T. Maruyama 1976 Archaeology, population genetics, and studies of human racial ancestry. *American Journal of Physical Anthropology* 44(1):31–50.

Wells, H.G. 1921 *The Grisly Folk.* (Reprinted). Penguin, New York.

Wendt, H. 1956 *In Search of Adam.* Houghton Mifflin, Boston.

White, C. 1799 *An Account of the Regular Gradation of Man, and in Different Animals and Vegetables; and from the Former to the Latter.* Dilly, London.

White, T.D. 1987 Cannibals at Klasies? *Sagittarius* 2(2):6–9.

Whitfield, L.S., J.E. Sulston, and P.N. Goodfellow 1995 Sequence variation of the human Y chromosome. *Nature* 378:379–380.

Wilder, H.H. 1926 *The Pedigree of the Human Race.* Henry Holt and Company, New York.

Wiley, E.O. 1978 The evolutionary species concept reconsidered. *Systematic Zoology* 27:17–26.

Wiley, E.O. 1981 *Phylogenetics. The Theory and Practice of Phylogenetic Systematics.* John Wiley & Sons, New York.

Wilford, J.N. 1995 Genetic signs of an evolutionary Adam. *The New York Times* (5/26/95):A9.

Williams, B.J. 1979 *Evolution and Human Origins.* Harper and Row, New York.

Wills, C. 1995 When did Eve live? An evolutionary detective story. *Evolution* 49(4):593–607.

Wilson, A.C., and R.L. Cann 1992 The recent African genesis of humans. *Scientific American* 266(4):68–73.

Winkler, H. 1924 Über die Rolle von Kern und Protoplasma bei der Vererbung. *Zeitschrift für induktive Abstammungs- und Vererbungslehre* 33:238–253.

Wobst, M.A. 1990 Minitime and megaspace in the Paleolithic at 18K and otherwise. In O. Soffer and C. Gamble (eds): *The World at 18,000 BP.* Unwin Hyman, London. pp. 331–343.

Wolf, E.R. 1994 Perilous ideas: race, culture, people. *Current Anthropology* 35(1):1–12.

Wolpoff, M.H. 1968a Climatic influence on the skeletal nasal aperture. *American Journal of Physical Anthropology* 29:405–424.

Wolpoff, M.H. 1968b "Telanthropus" and the single species hypothesis. *American Anthropologist* 70:447–493.

Wolpoff, M.H. 1976 Multivariate discrimination, tooth measurements, and early hominid taxonomy. *Journal of Human Evolution* 5(4):339–344.

Wolpoff, M.H. 1984 Evolution in *Homo erectus:* the question of stasis. *Paleobiology* 10(4):389–406.

Wolpoff, M.H. 1985 Human evolution at the peripheries: the pattern at the eastern edge. In P.V. Tobias (ed): *Hominid Evolution: Past, Present, and Future. Proceedings of the Taung Diamond Jubilee International Symposium.* Alan R. Liss, Inc., New York. pp. 355–365.

Wolpoff, M.H. 1986a Describing anatomically modern *Homo sapiens:* a distinction without a definable difference. In V.V. Novotný and A. Mizerová (eds): *Fossil Man. New Facts, New Ideas. Papers in Honor of Jan Jelínek's Life Anniversary, Anthropos* (Brno) 23:41–53.

Wolpoff, M.H. 1986b Stasis in the interpretation of evolution in *Homo erectus:* a reply to Rightmire. *Paleobiology* 12(3):325–328.

Wolpoff, M.H. 1989a Multiregional evolution: the fossil alternative to Eden. In P. Mellars and C.B. Stringer (eds): *The Human Revolution: Behavioural and Biological Perspectives on the Origins of Modern Humans.* Edinburgh University Press, Edinburgh. pp. 62–108.

Wolpoff, M.H. 1989b The place of the Neandertals in human evolution. In E. Trinkaus (ed): *The Emergence of Modern Humans. Biocultural Adaptations in the Later Pleistocene.* Cambridge University Press, Cambridge. pp. 97–141, and combined references for the volume on pp. 232–276.

Wolpoff, M.H. 1991 Comment on "Isolation and evolution in Tasmania" by C. Pardoe. *Current Anthropology* 32(1):17–18.

Wolpoff, M.H. 1992a Theories of modern human origins. In G. Bräuer and F.H. Smith (eds): *Continuity or Replacement? Controversies in Homo sapiens Evolution.* Balkema, Rotterdam. pp. 25–63.

Wolpoff, M.H. 1992b Levantines and Londoners. *Science* 255:142.

Wolpoff, M.H. 1994a How does evolution work? *Evolutionary Anthropology* 3(1): 4–5.

Wolpoff, M.H. 1994b Review of *The Origin of Modern Humans and the Impact of Chronometric Dating,* edited by M.J. Aitken, C.B. Stringer, and P.A. Mellars. American Journal of Physical Anthropology 93(1):131–137.

Wolpoff, M.H. 1994c What does it mean to be human—and why does it matter? *Evolutionary Anthropology* 3(4):116–117.

Wolpoff, M.H. 1994d Yes it is, . . . no it isn't: a reply to van Vark and Bilsborough. *American Journal of Physical Anthropology* 95(1):92–93.

Wolpoff, M.H. 1995 Wright for the wrong reasons. *Journal of Human Evolution* 29(2):185–188.

Wolpoff, M.H. 1996 Human Paleontology, *1996 edition.* McGraw-Hill, New York.

Wolpoff, M.H., and R. Caspari 1990 On Middle Paleolithic/Middle Stone Age hominid taxonomy. *Current Anthropology* 31(4):394–395.

Wolpoff, M.H., and R. Caspari 1996 The modernity mess. *Journal of Human Evolution* 30(2):167–172.

Wolpoff, M.H., F.H. Smith, M. Malez, J. Radovčić, and D. Rukavina 1981 Upper Pleistocene human remains from Vindija Cave, Croatia, Yugoslavia. *American Journal of Physical Anthropology* 54(4):499–545.

Wolpoff, M.H., J.N. Spuhler, F.H. Smith, J. Radovčić, G. Pope, D.W. Frayer, R. Eckhardt, and G. Clark 1988 Modern human origins. *Science* 241:772–773.

Wolpoff, M.H., A.G. Thorne, J. Jelínek, and Zhang Yinyun 1993 The case for sinking *Homo erectus:* 100 years of *Pithecanthropus* is enough! In J.L. Franzen (ed): *100 years of Pithecanthropus: The Homo erectus problem.* Courier Forschungsinstitut Senckenberg 171:341–361.

Wolpoff, M.H., A.G. Thorne, F.H. Smith, D.W. Frayer, and G.G. Pope 1994 Multiregional Evolution: a world-wide source for modern human populations. In M.H. Nitecki and D.V. Nitecki (eds): *Origins of Anatomically Modern Humans.* Plenum Press, New York. pp. 175–199.

Wolpoff, M.H., Wu Xinzhi, and A.G. Thorne 1984 Modern *Homo sapiens* origins: a general theory of hominid evolution involving the fossil evidence from east Asia. In F.H. Smith and F. Spencer (eds): *The Origins of Modern Humans: A World Survey of the Fossil Evidence.* Alan R. Liss, New York. pp. 411–483.

Wood, B.A. 1992 Evolution of australopithecines. In S. Jones, R. Martin, and D. Pilbeam (eds): *The Cambridge Encyclopedia of Human Evolution.* Cambridge University Press, Cambridge. pp. 231–240.

Wood, B.A. 1994 The problems of our origins. *Journal of Human Evolution* 27(6):519–529.

Wright, R.V.S. 1992 Correlation between cranial form and geography in *Homo sapiens:* CRANID—a computer program for forensic and other applications. *Archaeology in Oceania* 27:128–134.

Wright, R.V.S. 1995 The Zhoukoudian Upper Cave skull 101 and multiregionalism. *Journal of Human Evolution* 29(2):181–183.

Wright, S. 1943 Isolation by distance. *Genetics* 28:114–138.

Wright, S. 1945 Genes as physiological agents. *American Naturalist* 79:289–303.

Wright, S. 1969 *Evolution and the Genetics of Populations.* Volume 2. *The Theory of Gene Frequencies.* University of Chicago Press, Chicago.

Wright, S. 1986 The dynamics of peak shifts and the pattern of morphological evolution. *Paleobiology* 12(4):343–354.

Wu Xinzhi 1961 Study on the Upper Cave man of Choukoutien. *Vertebrata PalAsiatica* 3:202–211.

Wu Xinzhi and Wu Maolin 1985 Early *Homo sapiens* in China. In Wu Rukang and J.W. Olsen (eds): *Palaeoanthropology and Paleolithic Archaeology in the People's Republic of China.* Academic Press, New York. pp. 91–106.

Wub-e-ke-niew 1995 *We Have the Right to Exist.* Black Thistle Press, New York.

Xiong Weijun, Wenhsiung Li, I. Posner, T. Yamamura, A. Yamamoto, A.M. Gotto Jr., and L. Chan 1991 No severe bottleneck during human evolution: evidence from two apolipoprotein C–II deficiency alleles. *American Journal of Human Genetics* 48:383–389.

Yellen, J., A. Brooks, E. Cornelissen, M. Mehlman, and K. Stewart 1995 A Middle Stone Age worked bone industry from Katanda, Upper Semiliki Valley, Zaire. *Science* 268:553–556.

ACKNOWLEDGMENTS

We have been encouraged and helped by many of our friends and colleagues while writing this book, and gratefully acknowledge our special thanks to Marianne Bernstein-Weiner, Eugene Giles, Jan Jelínek, Jakov Radovčić, and the family of Franz Weidenreich for their many indispensable suggestions and corrections, and the insights they kindly gave us. We share the development of Multiregional evolution first and foremost with Alan Thorne and Wu Xinzhi, but also with David Frayer, Jan Jelínek, Geoff Pope, John Relethford, Alan Rogers, and Fred Smith. We are pleased to recognize our debt to them, and we have appreciatively incorporated their thinking in our own. We have had the fortunate opportunity to examine many fossil human remains and are deeply grateful for the institutions, and their curators and staffs, who so generously allowed us access and gave us space and their valuable time. We are particularly thankful to the Department of Archaeology at the University of Capetown (Rondebosch, South Africa), and especially to Andrew Sillen for inviting us to visit, and to Judy Sealy for kindly sharing data from her dissertation; to the University of Capetown Medical School, and especially Alan Morris; and to the curators and staff at the Croatian Natural History Museum (Zagreb, Croatia); the Moravian Museum (Brno, Czech Republic), the South African Museum (Capetown, South Africa), and the Natural History Museum, London, especially Chris Stringer. Finally we would like to thank Sarah Pinckney, our editor at Simon & Schuster, for being so very helpful while making our work much easier than it might have been.

INDEX

ABOUT THE AUTHORS

Milford Wolpoff received his Ph.D. from the University of Illinois and is professor of anthropology at the University of Michigan. He prides himself as having studied just about every human fossil there is. He is the recipient of major grants from the National Science Foundation, from the National Academy of Sciences exchange programs, and from the University of Michigan. Milford has written for *Scientific American*, and appeared on NPR and in paleoanthropology programs on the Discovery Channel, Learning Channel, A&E network, and PBS. His books include several paleoanthropology textbooks and a catalog of the Krapina Neandertals.

Rachel Caspari received her Ph.D. from the University of Michigan, where she now teaches and is a research scientist. She is also a consultant for the Cranbrook Institute of Science. Her research has been supported by the University of Michigan and the Care Foundation, and she has published papers in several encyclopedias, *Current Anthropology*, *Journal of Human Evolution*, and *American Journal of Physical Anthropology*. They are married and live with their children in Chelsea, Michigan.

9 781416 577966